Wildlife Research Monographs

The book series on "Wildlife Research Monographs" focuses on all aspects of wildlife biology to fill a vital niche at the intersection of several disciplines. The aim of the series is to summarize knowledge about interaction between wildlife species, their habitats, as well as human activities. Main areas are: wildlife ecology, diseases, toxicology, conservation, management, and the sustainable use of natural resources. The series addresses graduate students, university teachers, scientists and professionals.

More information about this series at http://www.springer.com/series/8826

Pedro P. Olea • Patricia Mateo-Tomás
José Antonio Sánchez-Zapata
Editors

Carrion Ecology and Management

Springer

Editors
Pedro P. Olea
Centro de Investigación en Biodiversidad y Cambio Global (CIBC-UAM)
Universidad Autónoma de Madrid
Madrid, Spain

Departamento de Ecología
Universidad Autónoma de Madrid
Madrid, Spain

Patricia Mateo-Tomás
Centre for Functional Ecology (CFE)
Department of Life Sciences
University of Coimbra
Coimbra, Portugal

Research Unit of Biodiversity
(UMIB, UO/CSIC/PA)
Mieres, Spain

José Antonio Sánchez-Zapata
Departamento de Biología Aplicada
Universidad Miguel Hernández
Elche, Spain

ISSN 2366-8733 ISSN 2366-8741 (electronic)
Wildlife Research Monographs
ISBN 978-3-030-16499-7 ISBN 978-3-030-16501-7 (eBook)
https://doi.org/10.1007/978-3-030-16501-7

This Springer imprint is published by the registered company Springer Nature Switzerland AG
The registered company address is: Gewerbestrasse 11, 6330 Cham, Switzerland

Foreword

The dog said to the bone: "If you are hard, I have time." Not in vain, one of my surnames is *Sancho*, as the manchego squire from Don Quixote, so I beg you to allow me to start with one anonymous sentence which I think is appropriate. Here, the "bone" is the ecology of scavenging, the central theme of the book you hold in your hands. The "dogs" (no offense intended) are all the researchers who have devoted much or all of their professional activity to the study of these processes. Seen from the current perspective of ecology, the topic is undoubtedly attractive and deserves to be at the forefront of cutting-edge science. This was not the case, however, just a few decades ago. Then, the "bone" was hard, and attempts to gnaw it sometimes involved risking the splintering of teeth. After all, scientists are not anomalies, but they are like anyone else, subject to the imponderables of the world in which they live. Among them, there are those that seek quick results and professional victories (lamentably more and more), which does not mesh well with the patience, observations, and setbacks that typically accompany investigation in the field. For this reason, working with carrion has not traditionally been on the radar of research teams in ecology. This is evident just reading the textbooks of three or four decades ago, when those of us who were already combing through gray hair were on the faculties of our venerable research institutions. Take that test and you will see.

But researchers arrived with a hard tooth, and after several decades of effort and tenacity, they are chipping away at the bone. The number of works on scavenging ecology has grown exponentially and continues to grow, thanks to the courage of researchers such as those that have edited and contributed to this book. I believe that I am not mistaken in affirming that a fundamental driver of the qualitative leap that has taken place has been the conservation of the great birds and mammals for whom scavenging is a way of life. The global crisis of biodiversity that we are facing has been ahead of the, until recently, flourishing populations of large vultures of the Old World as well as large carnivorous mammals. In parallel, the human obsession for achieving an "aseptic" world has nearly led to the end of traditional abandonment of carrion of domestic animals in the countryside. The impact of this change could affect the entire panoply of organisms that take advantage of discarded remains, including arthropods, bacteria, and many vertebrate facultative scavengers. This trend was born in the south of Europe, specifically in Spain, the refuge of the last healthy populations of carrion birds. Thus, it is not coincidental that the editors and many of the authors of these chapters carried out their work in the Mediterranean biome. I also think that, and this is a personal opinion of course, the appeal of "scavenging ecology" for many ecologists is anchored in something that is very much at our roots as human beings. There is no doubt that the use of carrion had a mark on our beginnings as a species: imagine the first humans, and our "cousins" the Neanderthals, harvesting the huge carcasses of mammoths and mastodons. What would it have meant for those groups of humans, always between life and death, to have access to a resource of this caliber? What challenges would they have been willing to face, such as carnivores and other beasts, to take this valuable booty? This, without a doubt, left its mark.

Prologues usually discuss the contents of subsequent chapters. I will not bore you more with descriptions and summaries of the corresponding texts. I also think that such a summary is unnecessary because the structure of the book is clear, and in a few pages, you will find a very detailed explanation of the contents. This is such a monumental work (it is difficult to find another word) that I am sure it will be a must for ecologists for many years to come. I invite you to immerse yourself in the texts and, for those who are less familiar with the topic, to enjoy learning. Let me finish with another reference, this time to a Cordovan philosopher born thousands of years ago in the empire that gave birth to the civilization in which we live. Seneca said that the effort calls the best. In the pages that follow, the best are represented, invigorating, with their hard work, a field of ecology for which they are passionate and which, beyond texts, chapters, and papers, transports us back, for a moment, to our beginnings: bipedal, erect, and trembling, making our way in a world of beasts guarding their precious treasures.

José Antonio Donázar Sancho
Department of Conservation Biology
Estación Biológica de Doñana (CSIC)
Sevilla, Spain

Acknowledgments

We thank the Series Editors on "Wildlife Research Monographs," who invited us to take part in this book series, and especially Jesús T. García, who offered us the opportunity to edit this volume on such a traditionally unpopular topic as carrion.

We wish to acknowledge all the contributors of this volume for their fine work and infinite patience with the accumulated delays during the process.

Thanks also to Eva López García for the brilliant pictures that illustrate this book and to José A. Donázar for writing the preface.

We greatly appreciate also the help of reviewers whose contributions have helped to improve the content of this text: Rafael Mateo (Institute of Game and Wildlife Research, Spain), Daniel Martín-Vega (University of Alcalá, Spain), Diego Méndez (Natural History Museum, Bolivia), and Marcos Moleón (University of Granada, Spain).

Finally, we would like to acknowledge the support of our research centers: Department of Ecology, Universidad Autónoma de Madrid, Spain; Centre for Functional Ecology, Department of Life Sciences, University of Coimbra, Portugal; Research Unit of Biodiversity (UO/CSIC/PA), Oviedo University, Spain; and Departamento de Biología Aplicada, Universidad Miguel Hernández, Spain.

Contents

About the Editors

Pedro P. Olea is professor and researcher at the Department of Ecology and at the Centro de Investigación en Biodiversidad y Cambio Global (CIBC-UAM), Universidad Autónoma de Madrid, Spain. His research focuses on understanding how worldwide human activities, such as hunting, livestock, and crop farming, affect patterns and processes in species, communities, and ecosystems at different spatial scales and how to apply this knowledge for their management and conservation. At present, one of his main research lines is focused on the emerging field of carrion ecology. The results from his research have been published in the most important scientific journals of ecology and conservation.

Patricia Mateo-Tomás is a postdoctoral researcher at the Centre of Functional Ecology at Coimbra University, Portugal, and Oviedo University, Spain. Her work focusses on the ecology and conservation of scavengers, especially vultures but also other facultative vertebrate scavengers, and their relationships with human activities such as hunting and livestock rearing (including transhumance) developed in natural ecosystems. She is also interested in exploring new ways for effective science communication.

José A. Sánchez Zapata is professor of ecology at the Department of Applied Biology, Universidad Miguel Hernández de Orihuela. His research initially focused in the ecology and conservation of raptors. During the last decade, he has broadened the topics including the role of scavenger guilds in ecosystem functioning and services under a socio-ecological framework. He has conducted research in Africa, America, Asia, Australia, and Europe reflected in scientific publications, most of them in ecology, biology, and biodiversity and conservation journals.

Introduction to the Topic of Carrion Ecology and Management

Pedro P. Olea, Patricia Mateo-Tomás, and José A. Sánchez-Zapata

Contents

Carrion Ecology: Key Concepts and State of the Art

Besides the "green world" composed of living organisms, another outstanding "brown world" exists which consists of dead organic matter, i.e. organic detritus (Swift et al. 1979; Moore et al. 2004). Detritus is ubiquitous in ecosystems as most of terrestrial plant biomass is not consumed by herbivores (up to 90%; Gessner et al. 2010) and a large number of animals die by causes other than predation (i.e. injuries, malnutrition, extreme weather conditions, parasites and disease; Young 1994, DeVault et al. 2003). For instance, >95% of reindeer in Svalbard (DeVault et al. 2003) and about 70% of all large ungulates in the African savannah die from causes other than predation (Houston 1979). In the Serengeti ungulate migrations,

P. P. Olea (✉)
Centro de Investigación en Biodiversidad y Cambio Global (CIBC-UAM), Universidad Autónoma de Madrid, Madrid, Spain

Departamento de Ecología, Universidad Autónoma de Madrid, Madrid, Spain
e-mail: pedrop.olea@uam.es

P. Mateo-Tomás
Centre for Functional Ecology (CFE), Department of Life Sciences, University of Coimbra, Coimbra, Portugal

Research Unit of Biodiversity, (UMIB, UO/CSIC/PA), Mieres, Spain

J. A. Sánchez-Zapata
Departamento de Biología Aplicada, Universidad Miguel Hernández, Elche, Spain
e-mail: toni@umh.es

P. P. Olea et al. (eds.), *Carrion Ecology and Management*, Wildlife Research Monographs 2, https://doi.org/10.1007/978-3-030-16501-7_1

6250 carcasses (~1100 tons of biomass) of drowned wildebeest (*Connochaetes taurinus*) enter the river each year when the mass herds cross the Mara River (Subalusky et al. 2017). In marine ecosystems, about 69,000 large whales die by natural causes each year (Smith and Baco 2003), providing between 6×10^5 and 2×10^6 tons of dead biomass (see chapter "Carrion Availability in Space and Time" for more details on how much carrion is produced in different terrestrial and marine ecosystems). Moreover, preys left uneaten by predators frequently enter the detritus pool too. Depending on the pack size, wolves (*Canis lupus*) can supply between 100 and 1200 kg of elk (*Cervus elaphus*) carrion from their captures during one winter in Yellowstone National Park (Wilmers et al. 2003a). This sizeable non-living organic component is subject to consumption, degradation or decomposition, i.e. a series of processes that ultimately recycle the energy and nutrients embodied into dead matter (see chapter "Carrion Decomposition"), thereby being critical to ecosystem functioning (Swift et al. 1979; Moore et al. 2004; Parmenter and MacMahon 2009; Barton et al. 2013; see chapter "Ecological Functions of Vertebrate Scavenging"). Despite the large amount of dead animal biomass present in the ecosystems, the study of decomposition of dead animals and its role in the functioning of ecosystems has received very little attention (but see Putman 1977, 1978a, b; Putman 1983; Carter et al. 2007; Parmenter and MacMahon 2009; Benbow et al. 2013; Subalusky et al. 2017) especially compared with that of dead plant matter (e.g. Swift et al. 1979; Putman 1983; Gessner et al. 2010; García-Palacios et al. 2013).

Detritus, *sensu lato*, is highly variable in origin and quality, from low-quality dead plant matter to high-quality dead animal tissue, i.e. carrion (Moore et al. 2004; Wilson and Wolkovich 2011). Carrion is yet a distinct component from the others that form the detritus pool in ecosystems (e.g. dead wood, dead plant biomass, animal dung), as its rate of decay is much faster, it is spatially patchy (Barton et al. 2013; chapter "Carrion Decomposition") and generally unpredictable and ephemeral (DeVault et al. 2003). Therefore, carrion is considered as a high-quality food pulsed resource (Yang et al. 2008) that usually appears in the form of carcasses in all aquatic and terrestrial ecosystems (Fig. 1). These carcasses are exploited by a plethora of invertebrate and vertebrate scavengers (see chapters "Invertebrate Scavenging Communities" and "Vertebrate Scavenging Communities") along with microorganisms (bacteria and fungi). It is therefore not surprising that competition for carrion occur among microorganisms, arthropods and vertebrates (Putman 1983; DeVault et al. 2003).

The unpredictable and ephemeral nature of carrion has constrained the evolution of true specialists on this resource (DeVault et al. 2003; Ruxton and Houston 2004; Kane et al. 2017). Species that rely entirely upon carrion must exhibit a high displacement speed at a low locomotion cost to search over large areas in order to attain a rate of encounter with the resource enough to meet their energetic demands (Carbone et al. 2011; Kane et al. 2017). Other adaptations required would be a high ability of sensory detection of carcasses and low metabolism in order to extend the inter-feeding time (Kane et al. 2017). Once localized the carcass, an efficient scavenger must possess a large body size in order to outcompete other competitors and

Fig. 1 Carrion, dead animal matter, appears as a pulsed food resource in the form of carcasses in all aquatic and terrestrial ecosystems. In human-modified ecosystems, fallen stock from livestock grazing is left in the open in many countries; in the pictures, carcasses of domestic ungulates in mountain ecosystems in northern Spain. Photo credits: José Vicente López-Bao

process the carrion rapidly to avoid its decay and the toxins from microorganisms decomposing the carrion (Kane et al. 2017). Consequently, a limited number of species have strictly specialized in the sole consumption of carrion, i.e. as obligate scavengers. Among vertebrates, no mammals or reptiles have evolved into obligate scavengers, and among extant birds only large soaring vultures are obligate carrion consumers (22 species, Houston 1979; DeVault et al. 2003; Ruxton and Houston 2004; Fig. 2). However, many other species use this resource optionally (facultative scavengers; Figs. 2 and 3) including early humans.

A wide diversity of species ranging from small primary consumers to top predators (e.g. lions, eagles and sharks) use carrion (Fallows et al. 2013; Mateo-Tomás et al. 2015). Terrestrial insects found in carcasses range from 30 to 522 species (Tabor et al. 2005), although it can include predators and parasites of scavengers and other species that use the carrion as habitat. In marine ecosystems, as many as 407 species of vertebrates and invertebrates can be found on whale falls (Smith and Baco 2003). Reptiles are less studied but at least twenty six species of snakes have been documented scavenging (DeVault and Krochmal 2002). Mammals and birds are most frequently registered consuming carrion. In Bialowieza Forest (Poland) and in Spanish ecosystems as many as 33 and 31 vertebrate species scavenged upon ungulate carcasses respectively (Mateo-Tomás et al. 2015). At a global scale, a review of the literature reported that at least 69 vertebrate species consumed carrion of wild ungulates with variable frequency (Mateo-Tomás et al. 2015). However, it is still unknown how many species show scavenger behavior globally. Depending on the biome considered, between 2 and 11% of the total terrestrial vertebrates present in an ecosystem consume carrion (Mateo-Tomás et al. 2015), suggesting that hundreds of vertebrate species may be facultative scavengers worldwide (i.e. 675–3700

Fig. 2 Vultures, with 22 species present in the world, are the only obligate scavengers among vertebrates as they are spezialized in the rapid location of carrion, e.g. the top picture shows griffon (*Gyps fulvus*) and cinereous (*Aegypius monachus*) vultures gathering at the guts left by hunters in the center of Spain. Besides vultures, carrion attracts many other facultative species (e.g. red fox *Vulpes vulpes* in the bottom image). Photo credits: Patricia Mateo-Tomás/Pedro P. Olea

Fig. 3 Carrion creates a local hotspot of biodiversity as many species from large vertebrates to invertebrates gather around the carcasses. Chimangos *Milvago chimango*, caracaras *Caracara plancus*, black *Coragyps atratus* and turkey *Cathartes aura* vultures feeding on a cow carcass in Argentina (above); tiny invertebrates such as the European bone-skipper fly *Thyreophora cynophila* on a horse carcass in Spain (bottom left image). Photo credits: María Eugenia Cabrera-García (above) and Jesús Fernández Carro (bottom left)

species from the 33,770 species of terrestrial vertebrates described worldwide; IUCN 2014). Altogether, these figures underscore the relevant role of carrion as a fundamental food resource for biodiversity worldwide.

The presence of carrion in the ecosystems attracts a wide diversity of species of invertebrates and vertebrates to specific locations creating a hotspot of biodiversity (Figs. 2 and 3). Not only there are species that visit the carrion to feed on the carcass, other ones come to search for prey, hosts or habitat (Tabor et al. 2005; Carter et al. 2007). Also, species visiting the carrion leave remains in the carcass (e.g. feathers, hair, urine, excrements, exuviae), attracting in turn more species and creating altogether a biological activity center in and around the carcass (Carter et al. 2007; Moleón and Sánchez-Zapata 2016). This intense animal activity along with the chemical and microbial processes contributes to decomposition of the carcass creating what is called a "cadaver decomposition island" (CDI; Carter et al. 2007). This CDI is usually visible as dead plant and bare soil appear due to trampling and the release of carcass fluids and maggot activity (Putman 1983; Carter et al. 2007). Carrion decomposition is a fundamental ecological process through which occurs the breakdown of the dead animal, and the recycling of the nutrients and energy embodied in the carcass. Therefore, "life concentrates biologically limiting resources", and death disperses them (chapter "Carrion Decomposition"). Dispersal of nutrients such as carbon, nitrogen or phosphorus away from carrion is conducted by arthropod and vertebrate scavengers (Barton et al. 2013; Subalusky et al. 2017). The movement of carrion nutrients can also occur across ecosystem or habitat boundaries (Payne and Moore 2006). In the Serengeti, for example, transfer of nutrients from terrestrial into aquatic ecosystem occurs via mass drowned wildebeests in the Mara River (see above; Subalusky et al. 2017). On the contrary, carcasses of Pacific salmon (*Oncorhynchus spp*) in Alaska are transported by biotic vectors such as wolves, bears and gulls from marine and fluvial ecosystem to terrestrial ecosystems (Payne and Moore 2006). Despite this dispersal of nutrients away from the carrion, other important part of the carcass nutrients are locally retained and led belowground by leakage of fluids and transport of carrion tissues by invertebrates. Thus, soil properties (pH and nutrient content) change. Fungi and bacteria from belowground participate in the degradation of complex organic molecules and conversion from organic to inorganic elements (mineralization), making it available to plants. Therefore, the CDI or pulse of fertility that form the carcass create a natural disturbance, which along with the remains of the invertebrate and vertebrate activity and the "passive" response of plant communities increase heterogeneity in the landscape (Carter et al. 2007).

The high diversity of species that gather around carrion point out to scavenging (i.e. carrion consumption) as a widespread feeding strategy in natural ecosystems (DeVault et al. 2003; see chapters "Invertebrate Scavenging Communities" and "Vertebrate Scavenging Communities") and underscores the importance of carrion for biodiversity conservation worldwide (see chapter "Ecological Functions of Vertebrate Scavenging"). Despite it, the role of scavengers in food webs has been traditionally underestimated by 16% (estimated by the number of trophic links involving scavenging; Wilson and Wolkovich 2011). Also, the energy transferred per scavenger link is 124-fold higher than per predation link; therefore, considering scavenging in food webs would lead to more accurate estimates of energy flows (Wilson and Wolkovich 2011). Provided that facultative scavengers consume multiple prey species, the inclusion of scavenging makes more complex reticulate webs, thereby stabilizing the food webs (Wilson and Wolkovich 2011).

Rapid carrion consumption by scavenger communities limits disease spreading and prevents soil and water contamination (Buechley and Şekercioğlu 2016; see chapter "Ecological Functions of Vertebrate Scavenging"). All these important ecological functions of carrion eaters provide additional benefits to humans, i.e. ecosystem services, such as waste disposal and disease regulation (O'Bryan et al. 2018), thus reducing the economic costs of maintaining public health or of carcass transport and disposal (Markandya et al. 2008; Morales-Reyes et al. 2015).

Increasing consumption by a growing human population is altering ecosystems worldwide at an unprecedented pace and extent (Corlett 2015). Besides directly threatening the conservation of scavenging species and their habitats (e.g. hunting, poaching, pollution, land degradation; Dirzo et al. 2014), the increasing consumption of natural resources by humans gives rise to an increasing production of waste (Hoornweg et al. 2013), including human-mediated carrion from, for example, livestock farming, hunting or fisheries (Figs. 1 and 3; Oro et al. 2013; Mateo-Tomás et al. 2015; see chapter "Human-Mediated Carrion: Effects on Ecological Processes"). Human-mediated carrion emerges therefore as a resource subsidizing scavengers across the world ecosystems (Wilmers et al. 2003b; Mateo-Tomás et al. 2015; Mateo-Tomás et al. 2016). Hunting activity produces ~1.0×10^8 tons of carrion per year in European ecosystems, and ~6.9×10^5 tons per year in USA (Vicente et al. 2011; Oro et al. 2013). Fisheries discards (i.e. fish catch which is thrown back, often dead, into the sea) account for ~7.3×10^6 tons per year across the world (Bicknell et al. 2013). Anthropogenic carrion does not only support scavengers but also is frequently related with conservation threats for biodiversity. Livestock carcasses treated with veterinary drugs (e.g. diclofenac) or hunting remains containing fragments of lead ammunition are known sources of toxicity for many scavenging species (Oaks et al. 2004; Finkelstein et al. 2012; Mateo-Tomá et al. 2016; see chapter "What Makes Carrion Unsafe for Scavengers? Considerations for Appropriate Regulatory Policies and Sound Management Practices"). Additionally, human-mediated carrion depends on sectorial policies such as sanitary or economic regulations that constrain its flow into ecosystems and thus impact also on scavenging dynamics and scavenger conservation. Sound examples of this are the sanitary regulations affecting the disposal of livestock carcasses after the outbreak of the Bovine Spongiform Encephalopathy (BSE) or "mad cow disease" in Europe in the 1990s (Tella 2001; Donázar et al. 2009) or the recent reform of the Common Fisheries Policy (CFP) regulating fisheries discards (Bicknell et al. 2013). To effectively deal with the increasing human impacts on the environment under the current scenario of global environmental change, it is necessary to know how the decomposer system works, as this is a crucial piece of ecosystem functioning (Swift et al. 1979).

What Is in This Book

This book focuses on carrion (i.e. dead animal matter; see Glossary in Box 1.1) and its role in ecosystems. The scope of the book is mainly ecological, but it provides also some insights into the conservation and management of scavengers and scavenging. The role of carrion in the environment is examined at different ecological levels, from individuals to populations and communities, as well as on the functioning of the ecosystem. Provided that ecosystems are increasingly altered by human activities and subsidized by anthropogenic carrion (e.g. from livestock, big game or fisheries), this book also deals with human-mediated carrion, examining its potential effects on the conservation of species and ecosystems and how to manage them. This book deals with carrion and their ecological processes involved in both terrestrial and aquatic (freshwater and marine) ecosystems, although a greater attention have been paid to terrestrial ecosystems according to the larger amount of evidence available.

In this book, carrion refers to dead (vertebrate) animal tissues ranging in size from small (e.g. mouse, shrew) to large vertebrates (e.g. elephant, whale; see Box 1.1). Although an abundant dead biomass of invertebrates exists in ecosystems (Nowlin et al. 2008), we have not considered it in this book, due in part to the scarce information still available on the topic (Barton et al. 2013). Yet, an important part of the book is devoted to invertebrates as key consumers of carrion along with vertebrates. Microorganisms that participate in the process of carrion decomposition are mentioned across the chapters, but a more detailed analysis is outside the scope of this book.

This introduction maintains a holistic conceptual view, considering carrion within the detritus pool as in, for example, Swift et al. (1979), Moore et al. (2004), Wilson and Wolkovich (2011) and Barton et al. (2013); thereby the carrion concept is rooted in the general ecological theory. Nonetheless, we consider carrion as a clearly distinct component from other ones (e.g. dead plant matter) within the detritus pool in ecosystems (see above; Wilson and Wolkovich 2011; Barton et al. 2013). The breakdown of dead organic matter into smaller fragments through biological processes that lead to its transformation and mass loss is usually referred to as "decomposition". Here we consider decomposition of carrion in a broad sense, which includes the consumption of carrion by invertebrate and vertebrate scavengers (i.e. scavenging) within the decomposition processes (Wilson and Wolkovich 2011), yet, recognizing the distinct role of scavengers and microorganisms in this process (Putman 1983; Barton et al. 2013). Following Putman (1983) there are two groups of organisms considered as decomposers: scavengers and true decomposers. Scavengers are those organisms (i.e. invertebrates and vertebrates) with holozoic nutrition, i.e. they feed by ingesting and digesting pieces of other organisms (e.g. removing fragments of the carrion); while saprophytes and saprozoic organisms (true decomposers, microorganisms) show saprozoic nutrition through which organisms release externally digestive enzymes, then absorbing from the environment the products resulting from the enzymatic action (reducing it *in situ*). Provided that both kinds of organisms and feeding strategies take part in carrion decay and the recycling of nutrients, it may be considered that both groups of scavengers form part altogether of the decomposition ecological processes of dead animal matter (Putman 1983). It should also be noted that both invertebrates and vertebrates share a similar feeding strategy (i.e. digestion after ingestion; see above), provided that only the *per-capita* removed material size of the carcass would make the difference (Putman 1983). However, as the reader can observe throughout this book, different authors use different terms to refer to carrion decomposition and the organisms that take part in it. Some authors consider scavenging by vertebrates within the decomposition process (e.g. chapter "Carrion Decomposition"), while other authors clearly separate scavenging, as driven by vertebrate scavengers, from decomposition undertaken by invertebrates and microorganisms (e.g. chapter "Ecological Functions of Vertebrate Scavenging").

In the current context of scientific knowledge on carrion decomposition above described, this book details on carrion ecology, paying special attention to the increasing dead animal biomass produced worldwide as a result of human activities. The effects that human-mediated food subsidies such as fisheries and hunting discards or dead livestock on ecosystems are widely recognized from individual to ecosystem level (e.g. Oro et al. 2013), but still poorly understood. For instance, it is largely unknown how ecosystems are assimilating these subsides or how to manage them to avoid conservation problems. This book tries to fill this gap, putting together the last scientific knowledge on carrion ecology with that on carrion management for biodiversity conservation. In this regard, the second part of the book includes three chapters devoted to human-mediated carrion and its management. This is not yet a comprehensive review on this topic since much knowledge is still lacking to accomplish a deep understanding of how human-mediated carrion is impacting subsidized ecosystems.

This book offers an up-to-date scientific review on carrion ecology and management for researchers of these and related disciplines. This volume gathers also information of interests for teachers and graduate students who can benefit from the materials here presented dealing with a fundamental process in ecology and their management applications. Finally, this work emerges as an useful guide for wildlife and conservation managers. The chapters have underwent a double peer review process by the editors and external reviewers (see Acknowledgements).

Literature Review

Although carrion and scavenging have been traditionally understudied in the scientific literature as compared with other topics in ecology, the number of investigations on the field seems to have increased during the last years (Moleón and Sánchez-Zapata 2015). In this section we review and analyze this temporal evolution of the scientific literature about carrion ecology. We examine whether carrion research has recently received a greater relative attention within the ecology discipline and relative to other traditionally more studied topics such as predation.

A Scopus search for "(carrion OR carcass) AND (scaveng*)" in "Title, Abstract and Keywords" of scientific works published until 31 December 2017, and then filtered by the presence of the terms "ecology OR conservation" throughout the manuscript, retrieved a total of 944 publications. There has been a sharp increase in the number of scientific works yearly dealing with carrion and scavenging since the first publication in 1965 until the 84 works published in 2017 (mean ± SD: 18 ± 23 works/year; Fig. 4a). However, when standardized by the total number of papers published in the Scopus categories of "Environment Sciences" and "Agricultural and Biological Sciences", the number of articles dealing with carrion and scavenging

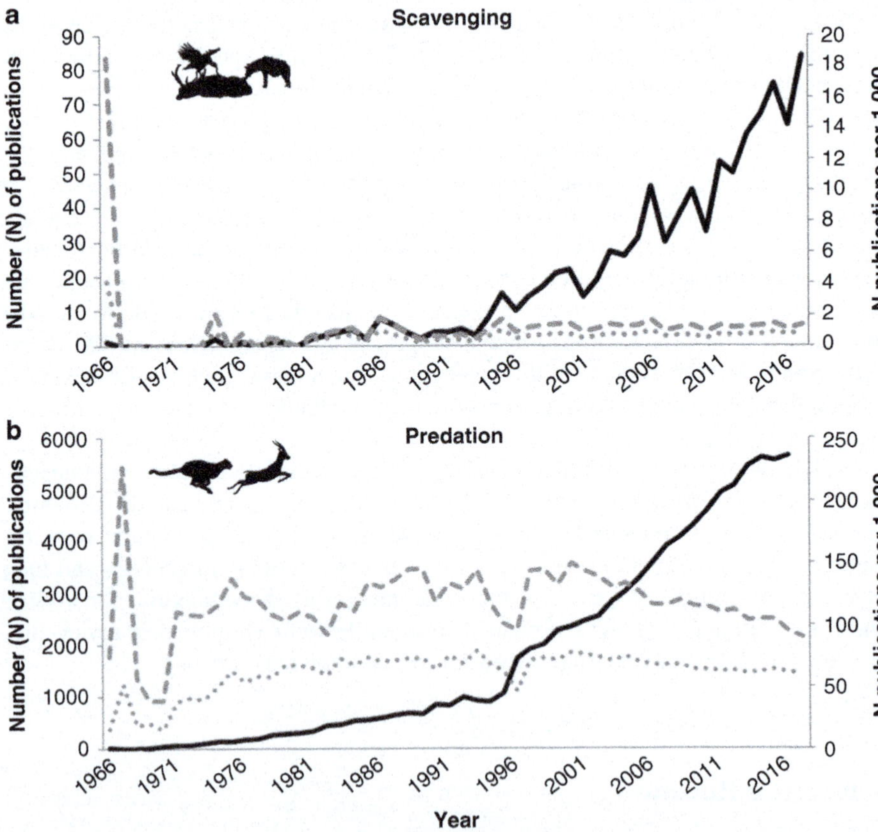

Fig. 4 Number of scientific publications (N) per year retrieved from a Scopus search for (**a**) "(carrion OR carcass) AND (scaveng*)" and (**b**) "predat*" in "Title, Abstract and Keywords" and filtered by the presence of the terms "ecology OR conservation" (solid black lines). The absolute number of publications per topic and year are standardized, i.e. dividing by each 1000 articles yearly published in the categories of "Environmental Sciences" (dashed grey line) and "Agricultural and Biological Sciences" (dotted grey line). Credits: the chapter authors

have been steady since the 1980s (i.e. 1–1.5 works per 1000 per year; Fig. 4a), indicating that the number of articles on this theme has grown at the same pace that the categories where they were included. Therefore despite its increase, there does not seem to be a higher relative intensity of research in the field of carrion and scavenging in the last years. Moreover, the same pattern of increase in the absolute number of publications is also observed in traditional research hot topics such as predation (i.e. as retrieved from a Scopus search for "predat*" in "Title, Abstract and Keywords" using the same parameters described above; Fig. 4b). Contrasting with the steady trend recorded for research in scavenging, the relative number of scientific publications on predation seems to have slightly decreased in the last years (i.e. from almost 150 works per 1000 per year in the 1990s to <100 in 2017; Fig. 4). However, the number of works yearly published on predation is 60–70 times higher than that of publications on carrion/scavenging, indicating the limited scientific attention still received by carrion ecology.

Book Organization

This section summarizes the book content to provide a synthetic picture of how it approaches the state of the art on carrion ecology and management. Additionally, through this summary, the reader can identify topics of interests within the two major sections of the book. The first section focusses on the main characteristics of carrion and the ecological processes triggered by its presence in ecosystems. The second part of the book deals with the role of humans as carrion providers through both the direct generation of carrion and the management of carrion in ecosystems. Finally, the book includes a methodological section intended to provide detailed descriptions and examples of some techniques frequently used in research in carrion ecology and management.

In chapter "Carrion Availability in Space and Time", Moleón et al. analyze the spatiotemporal availability of carrion in both terrestrial and marine ecosystems, including the terrestrial-aquatic interface. To disentangle how animal carcasses appear in ecosystems is of paramount importance for understanding the ecological processes and patterns observed at carrion. Here, the authors discuss the main natural causes for carrion appearance in ecosystems, and how different biotic and abiotic interactions may influence carrion availability for scavengers in space and time. From weather conditions, including extreme events such as El Niño–Southern Oscillation (ENSO; Greig et al. 2005) or hurricanes (Clua et al. 2014), to the key role of predation, accident or disease (see chapter "The Role of Scavenging in Disease Dynamics" for details on the role of disease in carrion ecology), several natural sources of mortality influence the generation of carrion in ecosystems. Similarly, habitat characteristics, weather conditions and inter- and intra-specific interactions (including competition among vertebrates, invertebrates and microorganisms; e.g. Houston 1979, DeVault et al. 2003) determine also carrion accessibility by different scavenging species. The authors highlight the role of species- and individual-specific characteristics such as life stage, body size and condition, sex or gregariousness that greatly determine how much carrion is available, where and when. Additionally, this chapter provides further insights in the different processes that determine carrion availability in terrestrial and marine ecosystems; highlighting, for example, the higher mobility of carrion in marine versus aquatic environments. The chapter focuses on naturally produced carrion (see chapter "Human-Mediated Carrion: Effects on Ecological Processes" for human-mediated carrion), including also extensive livestock because of their similar spatiotemporal availability to that of wild ungulates in many ecosystems around the world.

In chapters "Invertebrate Scavenging Communities" and "Vertebrate Scavenging Communities", Anderson et al. and Selva et al. describe the invertebrate and vertebrate communities, respectively, that consume carrion once it appears in ecosystems. According to the wide range of species that are known to scavenge (e.g. at least 2–11% of the vertebrate species in an ecosystem; Mateo-Tomás et al. 2015), to identify the species that consume carrion is key for biodiversity conservation, and thus for maintaining ecosystem functions and services (Wilson and Wolkovich 2011; see chapter "Carrion Decomposition").

In chapter "Invertebrate Scavenging Communities", Anderson et al. focus on the use of carrion by invertebrates in both terrestrial and aquatic (both freshwater and marine) ecosystems. The authors highlight the role of different groups of invertebrates (i.e. arthropods, nematodes, and molluscs), and the biotic and abiotic factors that influence their occurrence and succession at decaying carcasses. Insects, and especially flies, are the most important scavenging group among terrestrial invertebrates (Fig. 5), while, in marine ecosystems Crustacea have taken over the role of insects. Chapter "Invertebrate Scavenging Communities" shows that, contrasting with terrestrial ecosystems—where obligate and facultative invertebrate scavengers occur—no truly necrophagous aquatic insects exist that feed exclusively on carrion. Anderson et al. highlight the prominent role that temperature has in shaping invertebrate scavenging communities in terrestrial and aquatic ecosystems. A close two-way relationship exists among decomposition, as the process of chemical and material breakdown of carrion, and scavenging by invertebrates.

Fig. 5 The European bone-skipper fly *Thyreophora cynophila* (Panzer) is associated with animal carcasses. Once thought to be the first case of a dipteran species driven to extinction by humans, was rediscovered in 2007, after 160 years from its last observation in the 1840s (Martín-Vega et al. 2010 Syst. Entomol. 35(4): 607–613). Credit: Eva López García

Fig. 6 Among vertebrates, no mammals have evolved into obligate scavengers, yet the hyena is likely the most specialized mammal in the consumption of animal carcasses. Credit: Eva López García

In chapter "Vertebrate Scavenging Communities", Selva et al. describe the composition and structure of vertebrate scavenger communities from the main terrestrial biomes, from polar ecosystems to deserts and abyssal seabed. The authors show that birds and mammals dominate vertebrate scavengers worldwide, but reptiles, fishes and other taxa are also present within the large proportion of scavenging vertebrates that consume carrion as a high-quality food resource. The subchapter provides insights into the traits that characterize obligate (i.e. species exclusively feeding on carrion) versus facultative (i.e. species scavenging as part of a wider diet) vertebrate scavengers (Fig. 6); showing how obligate scavengers dominate carrion consumption in Africa, Asia and Mediterranean Europe while facultative scavengers replace them in higher latitudes. The authors highlight that, although scavenging has traditionally been considered a random and opportunistic process, vertebrate scavenger communities can show non-random complex patterns such as nestedness (i.e. the species feeding on carcasses visited by few consumers are subsets of those species feeding on carcasses visited by more consumers; Selva and Fortuna 2007).

In chapter "Carrion Decomposition", Barton and Bump focus on the decomposition of carrion, i.e. all biotic and abiotic processes that affect the breakdown of an animal carcass. Carrion decomposition is a central ecological process that releases energy and nutrients key to sustain biodiversity and ecosystem functioning (see chapter "Ecological Functions of Vertebrate Scavenging"). The authors detail how the previously described major groups of organisms (i.e. microorganisms, invertebrates and vertebrates; see chapters "Invertebrate Scavenging Communities" and "Vertebrate Scavenging Communities") gear together in carrion decomposition, which also implies intrinsic chemical and physical processes. The chapter provides a description of the temporal decay stages of carrion (i.e. fresh, bloat, active, advance, and dry decay; Michaud et al. 2015), which are pervasively used in entomological and forensic grounds. The contribution of different biotic and abiotic factors such as temperature, moisture or vegetation and habitat characteristics to carrion decomposition is also discussed. The authors further assess the role of scavengers in carrion decomposition in light of ecological theories such as, for example, succession, competition or coexistence and aggregation, as well as other processes operating at larger scales (e.g. movement of carrion across ecosystems, spatiotemporal patchiness of animal carcasses; see chapter "Carrion Availability in Space and Time"). The chapter finally discuss why and how new technologies (e.g. molecular analysis) and interdisciplinary collaboration should be incorporated in research in carrion decomposition.

In chapter "Ecological Functions of Vertebrate Scavenging", Beasley et al. discuss the main ecological functions and services (i.e. ecosystem functions that directly benefit humans) supported by scavengers in both terrestrial and aquatic ecosystems. The chapter highlights the important role of scavengers in food webs, summarizing how different advances in these conceptual frameworks have allowed new insights to better understand scavenging in ecosystems. Concretely, the authors provide detailed descriptions of key ecosystem functions supported by scavengers such as nutrient cycling, biodiversity maintenance, and disease control. These descriptions are illustrated with examples from well-studied systems, from the pristine habitats of the Białowieża Primeval Forest in Poland (Selva 2004) and the Yellowstone National Park in United States (Wilmers et al. 2003a) to farmlands in midwestern USA (DeVault et al. 2003). Although the chapter focuses on the biodiversity-ecosystem function (BEF) and -ecosystem services (BES) relationships of vertebrate scavengers, similar processes are described for invertebrates and microorganisms. Beasley et al. also discuss the impact that different anthropogenic activities influencing carrion availability exert on the ecosystem functions supported by scavengers, a topic of growing importance for scavenging dynamics under the current scenario of global change (see chapter "Human-Mediated Carrion: Effects on Ecological Processes"). The authors highlight the need for more integrative and manipulative research on scavenging ecology across many world ecosystems, especially island, arctic, tropical, and freshwater aquatic habitats.

In chapter "The Role of Scavenging in Disease Dynamics", Vicente and VerCauteren address the complex relationships that exist between wildlife disease dynamics and the epidemiological role of carrion and scavengers. Increasing the scarce knowledge on this topic is crucial to preserve biodiversity and ecosystem integrity but also to safeguard public health. The authors identify important aspects that are central to properly understand this interaction, i.e. the species that consume carrion and their exposure to disease, the role of scavenging for the spread and maintenance of pathogens in ecosystems, and the importance of human-related factors, e.g., for increasing the risks of disease transmission (Daszak et al. 2000). Vicente and VerCauteren highlight also the double dimension of the disease-scavenging interaction. A positive impact of disease on scavengers and scavenging exists; since diseases are important regulators of animal populations, death from disease can produce important pulses of carrion in ecosystems (see chapter "Carrion Availability in Space and Time"). On the other hand, Vicente and VerCauteren discuss also the potential role of carrion and scavengers for transmission and spread of infectious diseases or contaminants (Naranjo et al. 2008), noticing the key role of scavengers in limiting disease spreading (Ogada et al. 2012; see chapter

"Ecological Functions of Vertebrate Scavenging"). The authors review the state of the art in this challenging and often controversial topic, and update the current knowledge to identify major knowledge gaps that should guide future research. Furthermore, as stated in the *Human-mediated carrion* section of this book, this chapter points out the paramount role that humans play in the disease-scavenging dynamics through both directly managing carrion availability in ecosystems (Mateo-Tomás et al. 2015) and indirectly through managing populations of wild and domestic species (Gortázar et al. 2006). The authors call for collaboration between wildlife ecologists, veterinarians and public health professionals to properly tackling this topic.

In chapter "Human-Mediated Carrion: Effects on Ecological Processes" Moreno-Opo and Margalida provide a detailed description of how humans are becoming major providers of carrion worldwide (e.g. Mateo-Tomás et al. 2015) and how this may affect carrion ecology (Fig. 7). To assess how human-mediated carrion is subsidized into natural ecosystems provides further guidance for the effective management of scavengers and the ecosystem functions and services they support (see chapter "Ecological Functions of Vertebrate Scavenging"). The authors make a detailed review of how changes in carrion management influence from individuals to populations of scavenging species as well as the associated ecological processes at community and ecosystem levels (Oro et al. 2013). The chapter also discusses the consequences of these ecological alterations for human well-being (e.g. increasing human health issues and the associated economic costs, vultures' attacks to livestock; Markandya et al. 2008; Margalida et al. 2014).

Fig. 7 Griffon vultures *Gyps fulvus* are known to track the spatiotemporal distribution of potential food sources, such as transhumant livestock (Olea and Mateo-Tomás 2009). Credit: Eva López García

In chapter "What Makes Carrion Unsafe for Scavengers? Considerations for Appropriate Regulatory Policies and Sound Management Practices" Ogada et al. discuss how the quality of carrion have decreased worldwide due to human intervention, thus threatening scavenger conservation and the ecosystem functions and services associated (see chapter "Ecological Functions of Vertebrate Scavenging"). The authors outline a series factors that make carrion unsafe to scavengers, especially to obligate feeders such as vultures. These threats include accidental toxicity due to the consumption of carrions with residues of non-steroidal anti-inflammatory drugs (NSAIDs) and other veterinary drugs used to treat livestock (Oaks et al. 2004), lead ammunition used to hunt wild ungulates (Finkelstein et al. 2012), or insecticides and rodenticides used in pest control (Olea et al. 2009). Additionally, the chapter also delves into the role of human-wildlife conflicts for scavenger conservation, highlighting the deliberate poisoning of carcasses to kill predators of livestock or game species such as lions, wolves, eagles and hyenas (Fig. 8). The authors analyze the relevance of all these threats for scavenger's conservation across the globe, alerting also of the increase in deliberate poisoning of vultures by ivory poachers in Africa (Ogada et al. 2016). Based on the conservation and management experience of the authors, the chapter offers some recommendations and guidelines for developing appropriate regulatory policies, and vigilant management practices to properly safeguard scavengers from contaminated carrion.

At the end of this book the reader can find some *Methodological approaches* detailing some of the most common and/or novel methods used to research on carrion ecology. In chapter "Methods for Monitoring Carrion Decomposition in Aquatic Environments", Anderson and Wallace describe the techniques and facilities used to study scavenging underwater in freshwater and marine ecosystems. The authors provide further details on the Victoria Experimental Network under Sea (VENUS) observatory (Ocean Networks Canada), an elaborate cabled underwater laboratory in the west coast of Canada and the US. In chapter "Studying Movement of Avian Scavengers to Understand Carrion Ecology", Duriez et al. review the state of the art regarding the use remote technologies for tracking scavengers, focusing on the use of GPS telemetry for vulture monitoring. The authors provide also detailed descriptions of some techniques used to trap and tag these birds and of the usefulness of the information retrieved from these telemetry studies.

Fig. 8 The red kite *Milvus milvus* is a facultative scanvenger highly affected by poison due to this species feeding on small carcasses that are frequently used as poisoned baits. Credit: Eva López García

Book editors provide in chapter "Synthesis and Future Perspectives on Carrion Ecology and Management" a synthesis of the book content; a general picture of carrion ecology and how the existing knowledge on the topic may support carrion management for conservation. The editors highlight the main concepts and topics outlined by the authors of each chapter, including the existing knowledge as well as future research pathways and needs. A visual summary of the book content across the globe is also provided here.

Box 1.1 Glossary

Cadaver: Dead mammal, frequently used to refer to dead human bodies.

Cadaver decomposition island (CDI): As described by Carter et al. 2007, this term refers to the biological, chemical and microbial activity center created in and around a carcass during its decomposition process.

Carcass: Dead corpse of an animal. A usual form for carrion to naturally appear in ecosystems.

Carrion: Dead animal tissues characterized for its high-quality in terms of nutrient richness and assimilation efficiency by consumers. Generally unpredictable and ephemeral, due to a much faster rate of decay than other detritus such as, for example, plant litter.

Consumption: The direct consumption (i.e. feeding) of dead animal tissues by living organisms is part of the process of carrion decomposition.

Decomposer: Living organism contributing to the process of carrion decomposition. Putman (1983) classifies decomposers into two groups of organisms: scavengers (see below) and true decomposers. True decomposers are saprophytes and saprozoic organisms (i.e. microorganisms) with saprozoic nutrition, i.e. they release externally digestive enzymes, then absorbing from the environment the products resulting from the enzymatic action.

Decomposition: the breakdown of dead organic matter into smaller fragments through biological processes that lead to its transformation and mass loss. This is a fundamental ecological process through which occurs the recycling of the nutrients and energy embodied in the dead organic matter.

Degradation: Physical and chemical breakdown of detritus resulting from the action of abiotic factors such as temperature, oxygen concentration, mechanical action of water, wind…

Detritus: Dead organic matter of different origin (e.g. dead plants and animals but also animal dung and other teguments) and quality, i.e. in terms of chemical composition and assimilation efficiency by consumers. Carrion is considered as a high-quality detritus.

Scavenger: Organism (invertebrates and vertebrates) that consumes carrion uniquely (obligate scavenger) or optionally (facultative scavenger).

Scavenging: The process of feeding on carrion by invertebrates and vertebrates.

References

Barton PS, Cunningham SA, Lindenmayer DB, Manning AD (2013) The role of carrion in maintaining biodiversity and ecological processes in terrestrial ecosystems. Oecologia 171:761–772

Benbow ME, Lewis AJ, Tomberlin JK, Pechal JL (2013) Seasonal necrophagous insect community assembly during vertebrate carrion decomposition. J Med Entomol 50:440–450

Bicknell AWJ, Oro D, Camphuysen KCJ, Votier SC (2013) Potential consequences of discard reform for seabird communities. J Appl Ecol 50:649–658

Buechley ER, Şekercioğlu ÇH (2016) The avian scavenger crisis: looming extinctions, trophic cascades, and loss of critical ecosystem functions. Biol Conserv 198:220–228

Carbone C, Turvey ST, Bielby J (2011) Intra-guild competition and its implications for one of the biggest terrestrial predators, *Tyrannosaurus rex*. Proc Biol Sci 278:2682–2690

Carter DO, Yellowlees D, Tibbett M (2007) Cadaver decomposition in terrestrial ecosystems. Naturwissenschaften 94:12–24

Clua EE, Manire CA, Garrigue C (2014) Biological data of pygmy killer whale (Feresa attenuata) from a mass stranding in New Caledonia (South Pacific) associated with hurricane Jim in 2006. Aquat Mamm 40:162–172

Corlett RT (2015) The Anthropocene concept in ecology and conservation. Trends Ecol Evol 30:36–41

Daszak P, Cunningham AA, Hyatt AD (2000) Emerging infectious diseases of wildlife—Threats to biodiversity and human health. Science 287:443–449

DeVault TL, Krochmal AR (2002) Scavenging by snakes: an examination of the literature. Herpetologica 58:429–436

DeVault TL, Rhodes OE Jr, Shivik JA (2003) Scavenging by vertebrates: behavioral, ecological, and evolutionary perspectives on an important energy transfer pathway in terrestrial ecosystems. Oikos 102:225–234

Dirzo R, Young HS, Galetti M, Ceballos G, Isaac NJB, Collen B (2014) Defaunation in the Anthropocene. Science 345:401–406

Donázar JA, Margalida A, Carrete M, Sánchez-Zapata JA (2009) Too sanitary for vultures. Science 326:664–664

Fallows C, Gallagher AJ, Hammerschlag N (2013) White sharks (*Carcharodon carcharias*) scavenging on whales and its potential role in further shaping the ecology of an apex predator. PLoS ONE 8:e60797

Finkelstein ME, Doak DF, George D, Burnett J, Brandt J, Church M, Grantham J, Smith DR (2012) Lead poisoning and the deceptive recovery of the critically endangered California condor. Proc Natl Acad Sci U S A 109(28):11449–11454

García-Palacios P, Maestre FT, Kattge J, Wall DH (2013) Climate and litter quality differently modulate the effects of soil fauna on litter decomposition across biomes. Ecol Lett 16:1045–1053

Gessner MO, Swan CM, Dang CK, McKie BG, Bardgett RD, Wall DH, Hättenschwiler S (2010) Diversity meets decomposition. Trends Ecol Evol 25:372–380

Gortázar C, Acevedo P, Ruiz-Fons F, Vicente J (2006) Disease risks and overabundance of game species. Eur J Wildl Res 52:81–87

Greig DJ, Gulland FMD, Kreuder C (2005) A decade of live california sea lion (*Zalophus californianus*) strandings along the central California coast: causes and trends, 1991–2000. Aquat Mamm 3:11–22

Hoornweg H, Bhada-Tata P, Kennedy C (2013) Waste production must peak this century. Nature 502:615–617

Houston DC (1979) The adaptations of scavengers. In: Sinclair ARE, Griffiths MN (eds) Serengeti, dynamics of an ecosystem. The University of Chicago Press, Chicago, pp 263–286

IUCN (2014) The IUCN red list of threatened species. http://www.iucnredlist.org. Accessed 15 June 2018

Kane A, Healy K, Guillerme T, Ruxton GD, Jackson AL (2017) A recipe for scavenging in vertebrates - the natural history of a behaviour. Ecography 40:324–334

Margalida A, Campión D, Donázar JA (2014) Vultures vs livestock: conservation relationships in an emerging conflict between humans and wildlife. Oryx 48:172–176

Markandya A, Taylor T, Longo A, Murty MN, Murty S, Dhavala K (2008) Counting the cost of vulture decline—an appraisal of the human health and other benefits of vultures in India. Ecol Econ 67(2):194–204

Martín-Vega D, Baz A, Michelsen V (2010) Back from the dead: *Thyreophora cynophila* (Panzer, 1798) (Diptera: Piophilidae) 'globally extinct' fugitive in Spain. Syst Entomol 35(4):607–613

Mateo-Tomá P, Olea PP, Jiménez-Moreno M, Camarero PR, Sánchez-Barbudo IS, Rodríguez Martín-Doimeadios RC, Mateo R (2016) Mapping the spatio-temporal risk of lead exposure in apex species for more effective mitigation. Proc Biol Sci 283:20160662

Mateo-Tomás P, Olea PP, Moleón M, Vicente J, Botella F, Selva N, Viñuela J, Sánchez-Zapata JA (2015) From regional to global patterns in vertebrate scavenger communities subsidized by big game hunting. Divers Distrib 21:913–924

Michaud JP, Schoenly KG, Moreau G (2015) Rewriting ecological succession history: Did carrion ecologists get there first? Q Rev Biol 90:45–66

Moleón M, Sánchez-Zapata JA (2015) The living dead: time to integrate scavenging into ecological teaching. Bioscience 65:1003–1010

Moleón M, Sánchez-Zapata JA (2016) Non-trophic functions of carcasses: from death to the nest. Front Ecol Environ 14(6):340–341

Moore JC, Berlow EL, Coleman DC et al (2004) Detritus, trophic dynamics and biodiversity. Ecol Lett 7:584–600

Morales-Reyes Z, Pérez-García JM, Moleón M, Botella F, Carrete M, Lazcano C, Moreno-Opo R, Margalida A, Donázar JA, Sánchez-Zapata JA (2015) Supplanting ecosystem services provided by scavengers raises greenhouse gas emissions. Sci Rep 5:7811

Naranjo V, Gortazar C, Vicente J, de la Fuente J (2008) Evidence of the role of European wild boar as a reservoir of Mycobacterium tuberculosis complex. Vet Res 127:1–9

Nowlin WH, Vanni MJ, Yang LH (2008) Comparing resource pulses in aquatic and terrestrial ecosystems. Ecology 89:647–659

O'Bryan CJ, Braczkowski AR, Beyer HL, Carter NH, Watson JEM, McDonald-Madden E (2018) The contribution of predators and scavengers to human well-being. Nat Ecol Evol 2:229–236

Oaks JL, Gilbert M, Virani MZ et al (2004) Diclofenac residues as the cause of vulture population decline in Pakistan. Nature 427:630–633

Ogada DL, Torchin ME, Kinnaird MF, Ezenwa VO (2012) Effects of vulture declines on facultative scavengers and potential implications for mammalian disease transmission. Conserv Biol 26:453–460

Ogada D, Botha A, Shaw P (2016) Ivory poachers and poisons: drivers of Africa's declining vulture populations. Oryx 50:593–596

Olea PP, Mateo-Tomás P (2009) The role of traditional farming practices in ecosystem conservation: the case of trashumance and vultures. Biol Conserv 142:1844–1853

Olea PP, Sánchez-Barbudo IS, Viñuela J, Barja I, Mateo-Tomás P, Piñeiro A, Mateo R, Purroy FJ (2009) Lack of scientific evidence and precautionary principle in massive release of rodenticides threatens biodiversity: old lessons need new reflections. Environ Conserv 36:1–4

Oro D, Genovart M, Tavecchia G, Fowler MS, Martínez-Abraín A (2013) Ecological and evolutionary implications of food subsidies from humans. Ecol Lett 16:1501–1514

Parmenter RR, MacMahon JA (2009) Carrion decomposition and nutrient cycling in a semiarid shrub-steppe ecosystem. Ecol Monogr 79:637–661

Payne LX, Moore JW (2006) Mobile scavengers create hotspots of freshwater productivity. Oikos 115:69–80

Putman RJ (1977) Dynamics of the blowfly, *Calliphora erythrocephala*, within carrion. J Anim Ecol 46:853

Putman RJ (1978a) Patterns of carbon dioxide evolution from decaying carrion: decomposition of small mammal carrion in temperate systems, part I. Oikos 31:47–57

Putman RJ (1978b) Flow of energy and organic matter from a carcase during decomposition: decomposition of small mammal carrion in temperate systems, part II. Oikos 31:58–68

Putman RJ (1983) Carrion and dung: the decomposition of animal wastes. Edward Arnold (Publishers) Limited, London
Ruxton GD, Houston DC (2004) Obligate vertebrate scavengers must be large soaring fliers. J Theor Biol 228:431–436
Selva N (2004) The role of scavenging in the predator community of Białowieża Primeval Forest (E Poland). Ph.D. thesis, University of Sevilla
Selva N, Fortuna MA (2007) The nested structure of a scavenger community. Proc Royal Soc B 274:1101–1108
Smith CR, Baco AR (2003) Ecology of whale falls at the deep-sea floor. Oceanogr Mar Biol 41:311–354
Subalusky AL, Dutton CL, Rosi EJ, Post DM (2017) Annual mass drownings of the Serengeti wildebeest migration influence nutrient cycling and storage in the Mara River. Proc Natl Acad Sci U S A 114(29):7647–7652
Swift MJ, Heal OW, Anderson JM (1979) Decomposition in terrestrial ecosystems. University of California Press, Berkeley
Tabor KL, Fell RD, Brewster CC (2005) Insect fauna visiting carrion in Southwest Virginia. Forensic Sci Int 150:73–80
Tella JL (2001) Action is needed now, or BSE crisis could wipe out endangered birds of prey. Nature 410:408–408
Vicente J, Carrasco R, Acevedo P, Montoro V, Gortazar C (2011) Big game waste production: sanitary and ecological implications. In: Kumar S (ed) Integrated waste management, vol II. InTech, Rijeka, pp 97–128
Wilmers CC, Crabtree RL, Smith DW, Murphy KM, Getz WM (2003a) Trophic facilitation by introduced top predators: grey wolf subsidies to scavengers in Yellowstone National Park. J Anim Ecol 72:909–916
Wilmers CC, Stahler DR, Crabtree RL, Smith DW, Getz WM (2003b) Resource dispersion and consumer dominance: scavenging at wolf- and hunter-killed carcasses in Greater Yellowstone, USA: resource dispersion and consumer dominance. Ecol Lett 6:996–1003
Wilson EE, Wolkovich EM (2011) Scavenging: how carnivores and carrion structure communities. Trends Ecol Evol 26:129–135
Yang LH, Bastow JL, Spence KO, Wright AN (2008) What can we learn from resource pulses. Ecology 89:621–634
Young TP (1994) Natural die-offs of large mammals: implications for conservation. Conserv Biol 8:410–418

Part I
Carrion Ecology: Description of the Main Ecological Patterns and Processes Occurring Around a Carrion

Carrion Availability in Space and Time

Marcos Moleón, Nuria Selva, Maria Martina Quaggiotto, David M. Bailey, Ainara Cortés-Avizanda, and Travis L. DeVault

Contents

M. Moleón (✉)
Departamento de Biología Aplicada, Universidad Miguel Hernández, Alicante, Spain

Department of Conservation Biology, Doñana Biological Station (EBD-CSIC), Seville, Spain

Department of Zoology, University of Granada, Granada, Spain

N. Selva
Institute of Nature Conservation, Polish Academy of Sciences, Kraków, Poland

M. M. Quaggiotto · D. M. Bailey
Institute of Biodiversity, Animal Health and Comparative Medicine, College of Medical, Veterinary and Life Sciences, University of Glasgow, Glasgow, UK
e-mail: david.bailey@glasgow.ac.uk

A. Cortés-Avizanda
Department of Conservation Biology, Doñana Biological Station (EBD-CSIC), Seville, Spain

Animal Ecology and Demography Group, IMEDEA (CSIC-UIB), Palma de Mallorca, Spain

T. L. DeVault
US Department of Agriculture, Animal and Plant Health Inspection Service, Wildlife Services, National Wildlife Research Center, Sandusky, OH, USA
e-mail: travis.l.devault@aphis.usda.gov

P. P. Olea et al. (eds.), *Carrion Ecology and Management*, Wildlife Research Monographs 2, https://doi.org/10.1007/978-3-030-16501-7_2

Introduction

Availability of carrion to scavengers is a central issue in carrion ecology and management, and is crucial for understanding the evolution of scavenging behaviour. Compared to live animals, their carcasses are relatively unpredictable in space and time in natural conditions, with a few exceptions (see below, especially Sect. "Carrion Exchange at the Terrestrial-Aquatic Interface"). Carrion is also an ephemeral food resource due to the action of a plethora of consumers, from microorganisms to large vertebrates, as well as to desiccation (i.e., loss of water content; DeVault et al. 2003; Beasley et al. 2012; Barton et al. 2013; Moleón et al. 2014). With a focus on vertebrate carcasses, here we give an overview of (a) the causes that produce carrion, (b) the rate of carrion production, (c) the factors affecting carrion quality, and (d) the distribution of carrion in space and time, both in terrestrial and aquatic environments (including their interface). In this chapter, we will focus on naturally produced carrion, whereas non-natural causes of animal mortality are described in chapter "Human-Mediated Carrion: Effects on Ecological Processes". However, throughout this chapter we also refer to extensive livestock carrion, because in the absence of strong restrictions such as those imposed in the European Community after the bovine spongiform encephalopathy crisis (Donázar et al. 2009; Margalida et al. 2010), the spatiotemporal availability of carrion of extensive livestock and wild ungulates is similar.

Causes of Carrion Production

Animals commonly die from natural causes such as predation, starvation, parasites, disease, adverse climatic conditions and accidents. Many animals die naturally during the most susceptible stages of life: early when vulnerability reduces their survival, late because of senescence. Also, animals may die from casualties related to humans (see chapter "Human-Mediated Carrion: Effects on Ecological Processes"), which include intentional mortality sources such as game hunting, fisheries, deliberate poisoning and poaching, as well as unintentional mortality causes such as collisions with infrastructures, road kills, ship strikes, emerging infectious diseases and environmental toxins (Burkholder et al. 1992; Harvell et al. 1999; Laist et al. 2001; Fisher et al. 2006; Lambertucci et al. 2010; Collins and Kays 2011; Koch et al. 2013; Jepson et al. 2013; Wright et al. 2013; Kühn et al. 2015; Benbow et al. 2016). Further, these causes of death often work in concert (Newton 1998). For instance, cold weather conditions may increase the requirement of energy intake and thus increase the risk of starvation (Begon et al. 2006; Conover et al. 2013). Moreover, animals weakened by disease are more prone to predation (Schaller 1972) and, in some cases, collisions with automobiles (Møller et al. 2011).

In terrestrial biomes, most vertebrates usually die due to predation and diseases. Food and water shortage are major drivers of herbivore mortality, either directly or indirectly via enhancing vulnerability to predation, pathogens and others (Pereira et al. 2014). Moreover, exhaustion suffered during long-distance migrations may increase ungulate, bat, and bird mortality rates through different mechanisms (e.g., Geluso et al. 1976; Mduma et al. 1999). Apex carnivores, in contrast, mostly die as a consequence of competition (including intraguild—both inter- and intraspecific—killing; Palomares and Caro 1999; Thompson et al. 2015), parasites, and senescence.

In aquatic ecosystems, for some anadromous fish species (e.g. Pacific salmonids *Onchorhynchus* spp.), reproduction coincides with death. Necropsies of stranded marine mammals and turtles revealed these animals are susceptible to a wide range of parasites and viruses (Harvell et al. 1999; Arbelo et al. 2013; Work et al. 2015). Toxic diatom blooms can produce massive mortalities among sea lions (Scholin et al. 2000). Furthermore, mortality events can be also associated with large-scale climatic perturbations (Evans et al. 2005). In the South Pacific, for instance, mass mortality of Pygmy killer whales (*Feresa attenuata*) was linked with extreme meteorological conditions, such as the hurricanes Marylyn and Jim (Clua et al. 2014). Furthermore, fish kills can be attributed to natural conditions of oxygen deficiency (Stachowitsch 1984), extreme cold waters (Marsh et al. 1999; Hoag 2003) or river water acidification (Fjellheim and Raddum 1990). Similar to large animals in terrestrial ecosystems, whales and large sharks are probably subject to reduced predation risk and thus senescence and disease must be the most important natural causes of mortality, at least for adults.

Estimates of Carrion Production

Quantifying how much carrion is produced in ecosystems is one of the most important, although probably the most overlooked, aspect of carrion ecology (Oro et al. 2013). This is mostly due to methodological difficulties in obtaining empirical data in both terrestrial and aquatic systems. Estimating how much carrion is produced temporally and spatially is a complex issue that requires intensive fieldwork (e.g., Selva 2004), sometimes in combination with sophisticated modelling approaches (e.g., Wilmers and Getz 2004, 2005; Margalida et al. 2011).

Overall

In terrestrial ecosystems, as we show in Table 1, only a few studies have provided estimates of carrion biomass production per units of time and space in natural conditions (provision to dumps and vulture feeding stations, intensive livestock debris and game hunting remains are considered in chapter "Human-Mediated Carrion: Effects on Ecological Processes" because of their human origin). All these studies have focused on ungulates (wild, domestic or both). When all mortality causes are considered together, carrion is produced at an annual rate of up to ca. 700 kg/km^2. In all cases, carrion supply ranged between tens to hundreds of kg/km^2, although estimation methodologies, as well as the species evaluated, were highly heterogeneous among the reviewed studies (Table 1).

In well-preserved coastal systems, carcass supply can be substantially higher than inland. On the Isle of May (Scotland), an average of ca. 7000 kg of grey seal (*Halichoerus grypus*) placenta and dead pups and adults are produced annually, which means localised concentrations of as much as ca. 15,000 kg/km^2 (Quaggiotto et al. 2018). Up to 530 kg/km of stranded carrion accumulates annually along the shoreline of the Gulf of California (Polis and Hurd 1996b), whereas seabird colonies within the same gulf provided carrion with densities ranging between 5 and 100 kg/km^2, depending on the nesting island (Sánchez-Piñero and Polis 2000).

Table 1 Carrion biomass (kg/km^2) produced in different terrestrial systems by natural causes such as predation or disease

Continent	Country	Location	Carrion biomass	Species	Period	Reference
Africa	Zimbabwe	Rural Zimbabwe	696	Domestic and wild ungulates	Annual	Butler and Du Toit (2002)
Europe	Poland	Białowieża Primeval Forest	89	All wild ungulates	Annual	Selva (2004)
	Spain	Continental Spain	195	Extensive livestock	Annual	Morales-Reyes et al. (2017a)
		Cantabrian Mountains	2.1–40.3	Extensive livestock (transhumant)	Annual	Olea and Mateo-Tomás (2009)
			1.5–8.7	Extensive livestock (resident)	Annual	Olea and Mateo-Tomás (2009)
	Sweden	Sarek National Park	0.6–9.6	Reindeer (wolverine and Eurasian lynx kills)	Annual	Mattisson et al. (2011)
	UK	Scottish Highlands	264	Sheep and red deer	Annual	Brown and Watson (1964)
		Pastoral land (Wales)	190	Sheep	Annual	Newton et al. (1982)
North America	USA	Yellowstone	69–104	Red deer	Cold season	Houston (1979)
			13–63	All ungulates	Cold season	Gese et al. (1996)
			9	Red deer (wolf kills)	Cold season	Wilmers et al. (2003a, b)
South America	Chile	Central Patagonia	0.2	All ungulates (puma kills)	Annual	Elbroch and Wittmer (2012)

Studies evaluating human-mediated carrion (e.g. dumps, vulture feeding stations, intensive livestock debris or game hunting) are considered separately in Chap. 8

Covering over 70% of the planet's surface with a volume of 1332 billion km^3, the marine system is both the largest environment on earth and also one of the least accessible and least understood (Glover et al. 2010; Ramirez-Llodra et al. 2010). Natural food falls and fish carcasses are, in fact, relatively unusual observations (Soltwedel et al. 2003). This is probably because most of the seafloor is deep (>200 m); an environment which is hard to observe and where carrion is rapidly consumed (Isaacs and Schwartzlose 1975; Bailey and Priede 2002). Considering the nine largest whale species, it has been calculated that at any given time there is a fresh (i.e., with soft tissue) whale carcass on the sea floor every 16–36 km (Smith and Baco 2003).

In some rivers of the North Atlantic and North Pacific regions, salmon runs may signify a massive input of nutrients. For example, rivers of the Bristol Bay in Alaska (400 km long and 290 km wide) receive annually around 5.4×10^7 kg of adult salmon from the ocean. These salmon, whose biomass was gained mostly in the sea, will die shortly after spawning, thus subsidizing a complex array of both terrestrial and fresh water scavengers and decomposers (Gende et al. 2002). Terrestrial animals can also subsidized aquatic ecosystems through the provision of carrion (Pringle 2017). For instance, the mass drowning of wildebeests (*Connochaetes taurinus*) crossing the Mara River in the Serengeti (Africa) during the annual migration provides an annual input of 1,100 tonnes of biomass entering the river system (Subalusky et al. 2017).

Per Mortality Cause

Both in terrestrial and aquatic systems, carrion biomass provision depends chiefly on the mortality cause. Among natural sources of mortality, the main distinction is made between carrion supplied in the form of predator kills and other sources of carrion (see Sect. "Causes of Carrion Production"). While the latter result in complete carcasses being available to scavengers, only a portion of predator kills (i.e., the part not consumed by the predator) can be accessed by scavengers. In Białowieża Forest (Poland), Selva (2004) found that 35% of ungulate carrion was provided by disease and/or starvation, whereas predation by wolves (*Canis lupus*) and Eurasian lynxes (*Lynx lynx*) was responsible of 30% and 1% of carrion supply respectively (the remaining 34% of total ungulate carrion biomass was carrion of harvested animals left in the forest). In the Serengeti (Tanzania), 55–92% (depending on the month) of ungulate carcass supply to vultures comes from causes other than predation (Houston 1974). The abundance and nature of the predator is also important in determining how much of the predator kill is available to scavengers (Moleón et al. 2014; Table 1). For instance, solitary carnivores and those that do not return to their kills once satiated (e.g., cheetah *Acinonyx jubatus*) may leave more carrion per individual carcass in general than social ones (e.g., lions *Panthera leo*; Hunter et al. 2006; Elbroch and Wittmer 2012) and those that guard their kills for future consumption (e.g., leopards *P. pardus*; Kruuk 1967; Hunter et al. 2006).

Carrion Production in Relation to Species and Individuals

Carrion supply can also be strongly influenced by species identity, as it will determine carcass size and the rate at which carcasses are produced. Thus, large species would provide more carrion than smaller ones, and highly productive species would produce more carcasses than less productive ones (Fig. 1). Houston (1985) showed that Neotropical forests provide a greater food supply to vertebrate scavengers than Afrotropical forests due to the higher biomass and smaller average size (and thus higher average turnover) of herbivorous mammals in the former. In addition, obviously, population abundance must be considered when determining how much carrion biomass can be produced by a given species. In Białowieża Forest, red deer (*Cervus elaphus*) and wild boar (*Sus scrofa*) supply more carrion per year (3522 and 2417 kg, respectively) than larger (moose *Alces alces*, 404 kg; European bison *Bison bonasus*, 1600 kg) and smaller ungulates (roe deer *Capreolus capreolus*, 100 kg; Selva 2004).

Megaherbivores, i.e., herbivores weighing above 1000 kg (Owen-Smith 1988), are normally free from predation, especially as adults. Thus, megaherbivore carrion supply, mostly in the form of whole carcasses from sick, malnourished and senescent individuals, is an important contribution to the total carrion biomass in some African ecosystems, especially once depression of populations of elephants (*Loxodonta africana*) and other large herbivores by humans ceases and a stationary age distribution is reached (Pereira et al. 2014). The same basic pattern is expected to occur in the oceans, where whales, large sharks, marine mammals and large turtles are scarcely subject to predation (e.g., Heithaus et al. 2008).

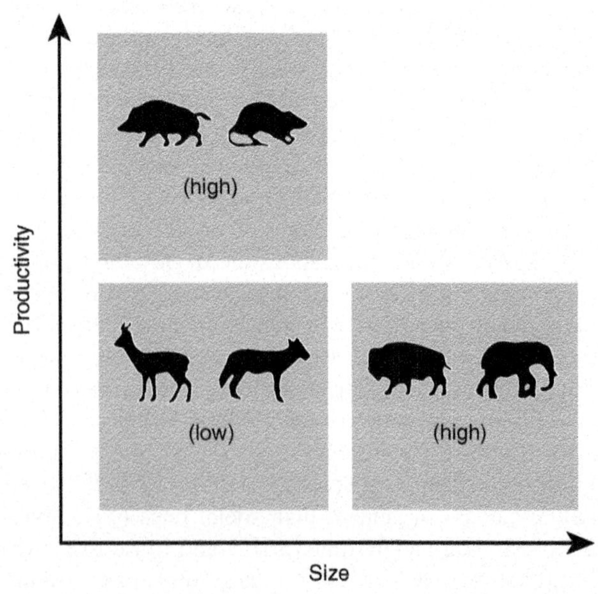

Fig. 1 Carrion supply magnitude (high *vs.* low) in relation to the size and productivity of animals under natural conditions. Note that there are no very large *and* very productive animals. Credits: the chapter authors

Another major factor influencing how much carrion is produced is individual identity. Sex, age and body condition strongly determines individuals' vulnerability to mortality (Pereira et al. 2014). For instance, newborns, calves and old animals (Husseman et al. 2003; Barber-Meyer et al. 2008), as well as pregnant females (Molinari-Jobin et al. 2004) and male ungulates competing for mating rights (FitzGibbon 1990), are especially vulnerable to predation.

Because of their high biomass, marine mammals are almost certainly one of the main contributors of vertebrate carrion to the marine system. However, the evidence of carcasses is rare. Marine mammal mortality is in fact estimated by monitoring their population through regular censuses, and losses from the population can be identified by simultaneous analysis of live resighting and dead recovery mark-recapture data. This method has been used to estimate an average first year survival in grey seals in Britain of ca. 20% for males and ca. 60% for females (Hall et al. 2001).

Factors Modulating Carrion Availability and Quality

The number of animals dying within and across ecosystems, the cause of death and the species and individual identity of carcasses are not the only factors influencing the amount of carrion biomass available to vertebrate scavengers. Carcass location, weather conditions and biotic interactions can greatly modulate carrion accessibility to scavengers in space and time. For scavengers that rely mainly on visual cues to detect food, vegetation cover and structure may crucially influence the rate and speed at which carrion is encountered (Ogada et al. 2012; Fig. 2). Some debilitated

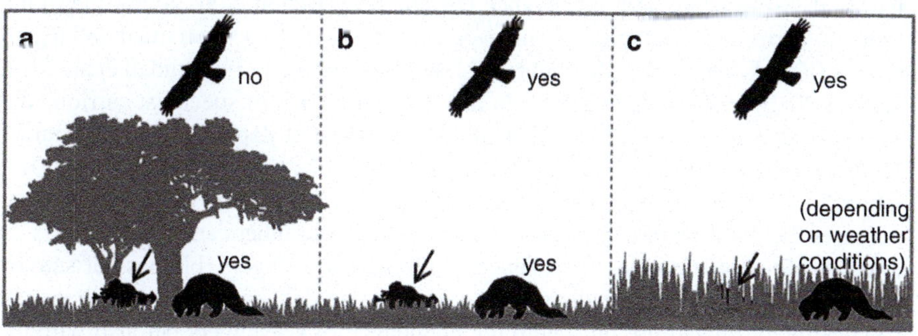

Fig. 2 Vegetation cover and structure may influence the ability of scavengers to locate carcasses (indicated by arrows). Soaring scavengers will rarely see a carcass located below dense tree canopy (**a**), whereas carcasses located in open habitat (**b**, **c**) may be easily found. Non-flying scavengers may readily see a carcass located within short grass, irrespective of the presence (**a**) or absence (**b**) of trees; however, they will find more difficulties in seeing a carcass located within tall grass (**c**). Difficulties in locating carcasses by sight can be overcome by acute sense of smell, provided that weather conditions do not dissipate odorants produced by carrion. Similar scenarios can be expected in aquatic environments. Credits: the chapter authors

animals such as malnourished megaherbivores often die near surface water (Conybeare and Haynes 1984). Due to vegetation encroachment along river and lake margins, such carcasses may be hardly accessible to certain scavengers such as large birds. Windy and rainy conditions will limit the capacity of soaring birds such as vultures and other large raptors to forage (Houston 1979; Selva et al. 2005). Rainfall can also reduce the olfactory capacity of mammalian carnivores (Savage 1977; Ruzicka and Conover 2012). In addition, carcass location by olfactory scavengers may be constrained by high temperatures and low humidity (Ruzicka and Conover 2012). In general, large carcasses are more accessible compared to carcasses of small animals that may appear in hollows inaccessible to most scavengers (Cox and Smith 1992, but see DeVault and Krochmal 2002).

Both inter- and intra-specific interactions, mediated by differences in behavioural, ecological, morphological and physiological traits among vertebrate obligate and facultative scavengers may influence the ability to find and consume carrion (Kendall 2013; Moleón et al. 2014; Sebastián-González et al. 2016; Moreno-Opo et al. 2016. First, mentioned above, predator kills may provide important quantities of food to scavengers (e.g., Kruuk 1967). Indeed, carrion provided by predators such as wolves is increasingly important in northern areas where milder winter weather conditions are reducing the number of carcasses produced by other causes during this critical period (Wilmers et al. 2003a, b; Selva et al. 2005; Wilmers and Getz 2005; Wilmers and Post 2006; Wikenros et al. 2013). Second, predators themselves and scavengers that rapidly arrive to the carcass may signal carcass location to other scavenger species and individuals (Attwell 1963; Kruuk 1967, 1972; Prior and Weatherhead 1991; Cortés-Avizanda et al. 2012; Kendall 2013; Cortés-Avizanda et al. 2014; Kane et al. 2014). Third, large mammalian carnivores and certain vultures may facilitate access to the interior of thick-skinned carcasses to less powerful scavengers by tearing open the tough hide (Attwell 1963; Stahler et al. 2002; Selva et al. 2003). Fourth, competition, both exploitative and interference (Birch 1957; Begon et al. 2006), may limit access to carrion by competitively inferior scavengers such as small mammalian carnivores (Kruuk 1967, 1972; Schaller 1972; Houston 1974, 1979; Kendall et al. 2012; Kendall 2013; Moleón et al. 2015). In this sense, the lack of specialist carrion consumers—such as *Gyps* vultures that quickly locate and deplete carrion resources (Houston 1979; Sebastián-González et al. 2013; Cortés-Avizanda et al. 2014; Sebastián-González et al. 2016; Morales-Reyes et al. 2017b)—in cold environments increases the duration of carcasses and thus could favour other carrion consumers. In British Columbia (Canada), in absence of obligate scavengers, biomass of salmon carrion—which is highly predictable and easy to handle—in estuarine environment was positively related not only with the abundance of scavengers but also with the diversity of the scavenging community (Field and Reynolds 2013).

The composition of invertebrate communities also can affect the temporal availability of carcass resources and, by extension, the composition of vertebrate scavenger communities. Houston (1985) showed how fly larvae completely consume carcasses of 2–10 kg within 3 days in Afrotropical forests, whereas in Neotropical forests a complex community of ants suppresses maggot infestations on carcasses, thereby extending the availability of carcasses to vertebrates to over 10 days

(Houston 1985, 1988). Houston (1985) speculated that this temporal regulation of the food supply by invertebrates contributed to the presence of vultures in forested regions in South America, as opposed to Africa, where vultures are confined to open savannahs and grasslands.

Microbial decomposers can quickly render carcasses inedible to most would-be scavengers by producing toxic compounds on carcasses (Janzen 1977; Burkepile et al. 2006; Shivik 2006). Microbes are more active as temperatures increase (Putman 1983; Carter et al. 2007), and several studies have demonstrated that vertebrates consume fewer available carcasses in warm weather compared to cold weather (DeVault et al. 2004; Selva et al. 2005; Parmenter and MacMahon 2009), likely due to increased activity of decomposers at higher temperatures. A factor closely related to microbial activity is carcass size: the larger the carcass, the longer the time needed to be consumed and, consequently, the greater the putrefaction (Pereira et al. 2014; Moleón et al. 2015). However, competition for food among vertebrate scavengers is reduced at very large carcasses (Moleón et al. 2015).

Similar to the terrestrial environment, the temporal availability of carrion to marine scavengers may be influenced by bacteria which affect carrion consumption. Unlike in terrestrial systems, internal decomposition may make the carcass buoyant so that it leaves the seafloor, removing carrion from the reach of some scavengers and giving access to others. Another difference in marine systems is that in large carrion items such as whale falls in deep water, the actions of bacteria result in overlapping stages of ecological succession (Bennett et al. 1994; Smith et al. 1998; Beasley et al. 2012). In these circumstances anaerobic bacterial decomposition of bone lipids (sulphophilic stage) occur after the removal of the soft tissue by necrophages (mobile-scavenger stage) and the colonisation of the bones and the nutrients-enriched sediments by assemblages of macro-benthic organisms (enrichment-opportunist stage). During decomposition, a chemosynthetic assemblage may develop where carbon is fixed from the water column using the energy released by the oxidation of sulphides in whale bones. In shallow waters, microbes were found to cover the skin of marine mammal carrion creating a bacterial mat that possibly prevents consumption by larger carrion consumers (Glover et al. 2010; Quaggiotto et al. 2016). Microbe-laden fish carrion, in fact, was colonised at lower extent than fresh carrion by scavengers such as crabs, demonstrating that bacteria can act as deterrent to scavenging activity (Burkepile et al. 2006).

Spatial Variation in Carrion Availability

The most likely place to find carrion is, of course, where live animals can be found, and more specifically, where they die (Olea and Mateo-Tomás 2009; Mateo-Tomás and Olea 2010). Thus, to determine spatial variation in carrion availability, one must consider where animals spend time during various activities. Also, because some individuals are consumed completely by predators immediately after their death and thus do not contribute to the pool of carrion available to scavengers (Houston 1974,

1979), spatial variation in carrion availability is also influenced by differences in cause-specific mortality across species and locations (see above). In this section, we briefly consider factors that affect carrion availability across space by examining where animals tend to die, and how the cause of death influences carrion availability across locations. First, we focus on terrestrial ecosystems, which differ from aquatic ecosystems in that animals that die on land typically do not move after death before consumption, whereas animal carcasses in water often sink, float or are moved by currents (Beasley et al. 2012; see Sect. "Aquatic Ecosystems").

Terrestrial Ecosystems

Some areas of the home range are more dangerous than others, and thus certain causes of death are associated with predictable locations. In an example highlighting particularly risky locations, male birds reduced nest visits to feed incubating females when perceived predation risk increased, suggesting that mortality is higher at the nest than elsewhere in the home range (Ghalambor and Martin 2002). Carrion also can be found predictably along migration routes and at grazing areas for wild and domestic ungulates (Houston 1974; Olea and Mateo-Tomás 2009; Kendall et al. 2014). Scavengers are known to take advantage of predictable carcass locations; for example, some scavenging birds preferentially forage for carrion along roads (Lambertucci et al. 2009).

Unfortunately, few data are available on locations of animal carcasses that die from exposure to toxins, parasites, disease, extreme elements, or starvation, as studies of cause-specific mortality rarely report this information (see Collins and Kays 2011). Even so, it seems likely that individuals dying from these causes might come to rest in less predictable locations than those that are killed by predators or human-related causes (e.g., collisions with vehicles and structures). For example, rodents exposed to anticoagulant poisons were found dead both in the open and in underground burrows (Cox and Smith 1992; Howald et al. 1999; Tuyttens and Stuyck 2002). We note that although underground carcasses might be unavailable to many scavengers, some species such as snakes could be well suited to detect, find, and consume such carcasses (Cowles and Phelan 1958; DeVault and Krochmal 2002). Sometimes, places of death are predictable at the microhabitat scale. For instance, weakened African elephants tend to die near water holes (Conybeare and Haynes 1984).

At a biogeographical scale, some patterns in carrion availability can be identified. Neotropical forests seem to provide a greater food supply to vertebrate scavengers than Afrotropical forests due to the higher biomass and average turnover of herbivorous mammals in the former (Houston 1985). In addition, as mentioned in Sect. "Factors Modulating Carrion Availability and Quality", carcasses remain available to vertebrates in Neotropical forests for longer periods than in Afrotropical forests likely due to interactions among the invertebrate community.

Aquatic Ecosystems

Fresh marine mammal carrion may not often be detectable because it is rapidly scavenged or decomposed at sea (Gulland 2006), making it challenging to determine its distribution in space. The fact that their skeleton or bones last longer could reveal some spatial patterns. However, the discrepancy occurring between where a marine mammal dies and where its body is transported, regardless it ends up with either a fall onto the seabed or a stranding on the shore, suggests that movements of carrion in the aquatic ecosystem involve a larger spatial and three-dimensional scale than carrion originated on land (Beasley et al. 2012). In water, mobility of carrion can be facilitated not only by currents and waves, but also by body density of the carcass and decomposition processes undergoing inside it. Body composition (blubber and lean) of the carcass determines its buoyancy (Peltier et al. 2012); putrefaction gases can cause its flotation (Reisdorf et al. 2012). Pressure and temperature also concur in facilitating movements of carrion: in particular, conditions of 15 atmosphere (atm) of hydrostatic pressure prevent the re-floating of large whales (>40 tons) in cold water (Smith and Baco 2003); whereas a carcass of a harbour porpoise (*Phoceona phocoena*, ca. 60–70 kg) at temperature above 4 °C is likely to surface in shallow waters (Moreno et al. 1992). Over the process of decomposition and consumption individual carrion items may be available to more than one type of scavenging assemblage depending on where the carcass is transported, with different consequences for the eventual fate of the energy and nutrients within it. Floating bodies may be exposed to scavenging birds when floating at the water surface or when stranded on the shore (Hewson 1995). Damage by scavenging birds may compromise the integrity of floating carcasses (Hewson 1995) and cause its sinking back to the seabed making it available again to the marine community. Overall, whale carcasses are expected to be most frequently found along migration routes and around feeding grounds, which use to be located near ocean margins (Smith and Baco 2003).

An environment where carrion is naturally available in high abundance is iceberg-scoured shallow coastal waters. Here benthic communities are periodically crushed and scraped from rocks, or ploughed into/from the sediment by the action of moving ice. Scavenger assemblages and activity are highly spatially variable. This variation can be between areas of different aspect (Smale et al. 2007) and depth (Dunlop et al. 2014). These differences presumably reflect the supply of carrion and the risk of death to the scavengers themselves. There is presumably an intermediate level of carrion impact, which is optimal for scavenging. At extremely high level of scouring there will never be much biomass to turn into carrion, with only biofilms having time to grow between impacts and the scavengers themselves could not survive. At very low levels of impact there will be little mortality to provide the carrion and so an intermediate level of iceberg impact is necessary (Dunlop et al. 2014).

In North Atlantic and North Pacific rivers, salmon carrion is expected to concentrate around spawning sites, where most adult mortality occurs (Gende et al.

2002). Naturally, currents, physical barriers and consumption rates by scavengers will largely determine to which extent this pulse of salmon carrion is expanded down the river.

Temporal Variation in Carrion Availability

Much like spatial variation in carrion availability, temporal variation in carrion availability is dependent on the cause of death. Starvation, predation, disease, exposure to extreme elements, and several human-related causes of death can fluctuate substantially in magnitude and thus provide carcasses to scavengers at differing rates across the year and between years (DeVault et al. 2003; Wilson and Wolkovich 2011; Pereira et al. 2014).

Terrestrial Ecosystems

Seasonality in carrion supply is a common feature of seasonal ecosystems, where ungulate mortality resulting from causes other than predation is related to the period of the year when food availability is limited and thermal stress is maximum. In northern latitudes, this takes place during the winter, while in African savannahs, it occurs toward the end of the dry season (see Pereira et al. 2014 and references therein). These limiting conditions can be especially harsh in some years, thus producing inter-annual oscillations in carcass production. For example, herbivore carcasses in Kruger National Park are very abundant during disease outbreaks and severe droughts, to the point that a number of them may remain uneaten (Owen-Smith and Mills 2008).

Although carrion resources are often produced in pulses, in some areas carrion is present consistently across the year and between years at the landscape scale. This consistency in available carrion is evident by the occurrence of resident vulture communities (e.g., across much of the Americas and parts of Africa, Asia and Europe), which subsist almost exclusively on carrion (Houston 1979; DeVault et al. 2003; Moleón et al. 2014).

Carrion availability through time can be strongly influenced by vertebrate predators. In one well documented example, Wilmers and colleagues (Wilmers et al. 2003a; Wilmers and Getz 2004, 2005; Wilmers and Post 2006) studied how carrion resources in Yellowstone National Park, USA, changed after the reintroduction of wolves. They found that predation on elk by wolves changed the nature of carrion available to scavengers from a predominantly pulsed resource resulting from a late-winter influx of whole carcasses (caused by starvation of elk and other herbivores in harsh environmental conditions) to a more temporally-stable resource resulting from partial consumption of elk killed by wolves. Wolves thus provided a temporal subsidy to their smaller congeners, coyotes (*Canis latrans*), and other vertebrate scavengers by reducing the variation in carrion availability within and across years.

Aquatic Ecosystems

Without human interference, carrion in the ocean is infrequent in time and space (Britton and Morton 1994). Carrion availability derived from marine mammals such as whales has changed over historical time (Smith and Baco 2003). Before whaling there would have been natural mortality, leading to the majority of cetacean carcasses being made available to scavengers. Nowadays, after implementation of conservative measures, whale falls may follow a seasonal pattern according to annual migrations (Smith and Baco 2003; Roman et al. 2014) and environmental conditions occurring at that time. For instance, the unusually high mortality of grey whales (*Eschrichtius robustus*) along the migratory route in 1999 was linked to the undernourished body conditions displayed by the stranded animals due to low amphipod biomass, their main food source. Higher sea temperatures, together with an increase in whale population density, were hypothesised to be the principal reason explaining the fatally low prey availability (e.g. Le Boeuf et al. 2000). Little is known about temporal availability of carrion in the ocean as most of the information available on this topic is the result of analysis on seasonal patterns of strandings (Peltier et al. 2013), whereas direct observations of carcasses at sea are rare (Smith et al. 2015).

Fish populations also change in predictable ways, probably providing temporal, but also spatial, variation in the carrion they supply. Deep sea fish populations appear to respond to changes in carrion availability due to migration. Inter-annual variation in abyssal fish numbers can be linked to variation of the Pacific hake (*Merluccius productus*) spawning aggregation in the waters overlying them (Drazen et al. 2012). Temporal trends in fish mass mortality can be predicted by determined environmental conditions more likely to occur seasonally (Stachowitsch 1984; Schoenebeck et al. 2012) or globally driven (Marsh et al. 1999). Salmon carcasses in rivers of the northern Hemisphere are available only in a short period of time after spawning (Gende et al. 2002).

Iceberg action leading to carcass production is also highly seasonal, dropping dramatically in winter when sheets of "fast ice" lock icebergs into position and occurring again in the spring. Cold winters with longer periods of fast ice presence result in lower levels of iceberg scouring (Smale et al. 2008).

Carrion Exchange at the Terrestrial-Aquatic Interface

Coastal regions are often massively affected by marine-derived inputs transferred from the ocean to the terrestrial ecosystem (Polis and Hurd 1996a, b; Polis et al. 1997; Rose and Polis 1998) by biotic factors such as marine currents, air circulation and weather conditions. Strandings of carrion, which are the evident results of mortality events happening at sea, occur when positively buoyant carcasses disperse from the original location of death and drift towards the coastline. Natural mortality of cetaceans, for instance, generates single unpredictable carcasses in the ocean, which can be possibly washed on the coast. Mass stranding can be classified as unusual mortality events (UMEs; Marine Mammal Protection Act) when for initially unknown reasons a large part of the population dies. These events, whose causes can be related to biotoxins, bacteria, parasites, human interactions and oceanographic conditions, are also spatially unpredictable phenomena, in which marine mammals strand in a delimited space and during a limited time (Wilkinson 1991). Stranding data can reveal supplementary information on the spatial distribution of fatalities and their post-mortem movements: from the position of the stranding it is possible, in fact, to estimate the original location where the animal died by reconstructing the animal's drift by modelling local sea currents and tides (Peltier et al. 2012; Santos et al. 2018).

Mass mortality caused by El Niño, instead, might be predicted as it periodically affects the east-central tropical Pacific Ocean. During the El Niño, California sea lion (*Zalophus californianus*) fatalities were dominated by malnourished juveniles (Greig et al. 2005), as the mothers spent more time at sea searching for food, which was drastically reduced by the environmental perturbation, instead of lactating the pups (Ono et al. 1987). The lack of food and the anomalous sea conditions created during El Niño can affect the survival of seabirds, which also die at sea and are washed ashore (Piatt and van Pelt 1997).

In addition to the abiotic transfer of resources, transport of marine nutrients can also occur when animals move from one system to another. Salmon, for instance, has already been shown to play a fundamental role in this process affecting different components of the ecosystem (Cederholm et al. 1999). Every year, in defined spawning areas up stream, salmon carcasses represent a source of carrion that enriches soils and plants, locally releasing nutrients (e.g. Quinn et al. 2009; Hocking and Reimchen 2009). This impact is long term as its signal is still detectable in historically salmon-bearing stream (Koyama et al. 2005). Moreover, predictability of salmon runs in riparian systems is responsible for the synchronisation between feeding strategy and food input in black bears (*Ursus americanus*) and more recently in wolves (Reimchen 2000; Darimont et al. 2003). Carcasses of moose, which widely forage on aquatic plants, mean an important nutrient transfer pathway from freshwater to terrestrial systems in the circumpolar region (Bump et al. 2009).

Likewise, marine mammals affect coastal systems and their impact is substantial when they assemble in colonies for breeding. Here, they provide sources of nutrients in terms of carrion which are used either directly by above ground secondary consumers, or indirectly by increasing the input of nutrients from excreta (Anderson and Polis 1998). In Scotland, for instance, placentae and dead seals produced at a grey seal (*H. grypus*) colony represent a valuable food resource for scavenging gulls during the cold season (Quaggiotto et al. 2018). Due to the proximity to the sea and the inherent tidal and wave action, carcasses may be dislodged and potentially affect also the intertidal and sublittoral marine scavengers.

Conclusions and Future Perspectives

Death is a fundamental component of ecosystems. Every year, many kilograms of carrion biomass (tens to hundreds according to the reviewed studies) are produced in terrestrial systems per km^2 (see chapter "Human-Mediated Carrion: Effects on Ecological Processes" for human-mediated carrion and the possible impact of global environmental change on the spatiotemporal availability of carrion resources). This figure can be substantially higher in well-preserved coastal environments with large colonies of marine mammals. Also, salmon and other fish spawning provide massive carrion pulses. The magnitude of carcass supply, as well as its quality and availability to scavengers, is the result of many interacting factors such as animal density and turnover rates, the cause of death, carcass location, weather conditions and biotic interactions (both among and within vertebrates, invertebrates and microorganisms). In addition, carrion biomass available to scavengers may vary among species and individuals. However, despite its paramount importance for understanding carrion ecology, carrion supply remains largely understudied compared to carrion consumption. For instance, the relationship between interspecific interactions and carrion provision and nutritive value deserves further scientific attention.

The spatial predictability of carrion varies greatly, depending on carcass type, cause of death, and ecosystem. Future research on cause-specific mortality (Collins and Kays 2011) should consider reporting the exact location of death (i.e., for radio-marked animals) so that spatial variation in carrion availability can be more accurately defined. The availability of carrion to scavengers can also vary according to the temporal scale considered. In some situations, temporal variation of carrion supply is linked to spatial variation. For example, shortly after spawning (once per year), semelparous salmon die in large numbers in rivers, and many scavengers take advantage of this resource in the limited area where it occurs (Bennetts and McClelland 1991; Hewson 1995). In contrast, animals in temperate and arctic regions commonly die from starvation during extended periods of cold weather (Wilmers et al. 2003a); these carcasses are produced in a temporal pulse, but can be found scattered across extensive areas.

The spatiotemporal availability and quality of carrion can substantially influence the density and diversity of scavengers occupying an area (DeVault et al. 2003; Wilson and Wolkovich 2011; Barton et al. 2013). The growing field of scavenging ecology would benefit from further investigations quantifying the spatiotemporal availability of carrion for vertebrate scavengers across regions. More studies are needed in some continents and on islands. Quantitative studies in aquatic systems are especially encouraged, as well as those evaluating carrion exchange between terrestrial and aquatic environments. Multi-species approaches would be especially welcome. For instance, quantifying carrion biomass produced by small and highly productive animals versus large and less productive animals may be particularly interesting. The influence of microhabitat characteristics on carcass availability and accessibility to scavengers that rely on vision versus those with high olfactory skills should be further investigated. In summary, the fields of carrion ecology and management would largely benefit from the study of the many mysteries related to the spatial and temporal availability of carrion that remain unresolved.

References

Anderson WB, Polis GA (1998) Marine subsidies of island communities in the Gulf of California: evidence from stable carbon and nitrogen isotopes. Oikos 81:75–80

Arbelo M, de los Monteros AE, Herráez P et al (2013) Pathology and causes of death of stranded cetaceans in the Canary Islands (1999-2005). Dis Aquat Org 103:87–99

Attwell RIG (1963) Some observations on feeding habits, behaviour and inter-relationships of Northern Rhodesian vultures. Ostrich 34:235–247

Bailey DM, Priede IG (2002) Predicting fish behaviour in response to abyssal food-falls. Mar Biol 141:831–840

Barber-Meyer SM, Mech LD, White PJ (2008) Elk calf survival and mortality following wolf restoration to Yellowstone National Park. Wildl Monogr 169:1–30

Barton PS, Cunningham SA, Lindenmayer DB et al (2013) The role of carrion in maintaining biodiversity and ecological processes in terrestrial ecosystems. Oecologia 171:761–772

Beasley JC, Olson ZH, DeVault TL (2012) Carrion cycling in food webs: comparisons among terrestrial and marine ecosystems. Oikos 121:1021–1026

Begon M, Townsend CR, Harper JL (2006) Ecology: from Individuals to ecosystems, 4th edn. Blackwell Publishing, Malden

Benbow ME, Tomberlin JK, Tarone AM (2016) Carrion ecology, evolution, and their applications. CRC Press, New York

Bennett BA, Smith CR, Glaser B et al (1994) Faunal community structure of a chemoautotrophic assemblage on whale bones in the deep northeast Pacific Ocean. Mar Ecol Prog Ser 108:205–223

Bennetts RE, McClelland BR (1991) Differences in the distribution of adult and immature bald eagles at an autumn concentration in Montana. Northwest Sci 65:223–230

Birch LC (1957) The meanings of competition. Am Nat 91:5–18

Britton JC, Morton B (1994) Marine carrion and scavengers. Oceanogr Mar Biol 32:369–434

Brown LH, Watson A (1964) The golden eagle in relation to its food supply. Ibis 106:78–100

Bump JK, Tischler KB, Schrank AJ et al (2009) Large herbivores and aquatic–terrestrial links in southern boreal forests. J Anim Ecol 78:338–345

Burkepile DE, Parker JD, Woodson CB et al (2006) Chemically mediated competition between microbes and animals: microbes as consumers in food webs. Ecology 87:2821–2831

Burkholder JM, Noga EJ, Hobbs CH et al (1992) New "phantom" dinoflagellate is the causative agent of major estuarine fish kills. Nature 358:407–410

Butler JRA, du Toit JT (2002) Diet of free-ranging domestic dogs (Canis familiaris) in rural Zimbabwe: implications for wild scavengers on the periphery of wildlife reserves. Anim Conserv 5:29–37

Carter DO, Yellowlees D, Tibbett M (2007) Cadaver decomposition in terrestrial ecosystems. Naturwissenschaften 94:12–24

Cederholm C, Kunze MD, Murota T et al (1999) Pacific salmon carcasses: essential contributions of nutrients and energy for aquatic and terrestrial ecosystems. Fisheries 24:6–15

Clua EE, Manire CA, Garrigue C (2014) Biological data of pygmy killer whale (*Feresa attenuata*) from a mass stranding in New Caledonia (South Pacific) associated with hurricane Jim in 2006. Aquat Mamm 40:162–172

Collins C, Kays R (2011) Causes of mortality in North American populations of large and medium-sized mammals. Anim Conserv 14:474–483

Conover MR, Dinkins JB, Haney MJ (2013) Impacts of weather and accidents on wildlife. In: Wildlife management and conservation: contemporary principles and practices. Johns Hopkins University Press, Baltimore, pp 144–155

Conybeare A, Haynes G (1984) Observations on elephant mortality and bones in water holes. Quat Res 22:189–200

Cortés-Avizanda A, Jovani R, Carrete M et al (2012) Resource unpredictability promotes species diversity and coexistence in an avian scavenger guild: a field experiment. Ecology 93:2570–2579

Cortés-Avizanda A, Jovani R, Donázar JA et al (2014) Birds sky networks: how do avian scavengers search for carrion resource. Ecology 95:1799–1808

Cowles RB, Phelan RL (1958) Olfaction in rattlesnakes. Copeia 1958:77–83

Cox P, Smith RH (1992) Rodenticide ecotoxicology: pre-lethal effects of anticoagulants on rat behavior. Proc Vertebr Pest Conf 15:165–170

Darimont CT, Reimchen TE, Paquet PC (2003) Foraging behaviour by gray wolves on salmon streams in coastal British Columbia. Can J Zool 81:349–353

DeVault TL, Krochmal AR (2002) Scavenging by snakes: an examination of the literature. Herpetologica 58:429–436

DeVault TL, Rhodes OE Jr, Shivik JA (2003) Scavenging by vertebrates: behavioral, ecological, and evolutionary perspectives on an important energy transfer pathway in terrestrial ecosystems. Oikos 102:225–234

DeVault TL, Brisbin IL Jr, Rhodes OE Jr (2004) Factors influencing the acquisition of rodent carrion by vertebrate scavengers and decomposers. Can J Zool 82:502–509

Donázar JA, Margalida A, Carrete M et al (2009) Too sanitary for vultures. Science 326:664

Drazen JC, Bailey DM, Ruhl HA et al (2012) The role of carrion supply in the abundance of deep-water fish off California. PLoS One 7:e49332

Dunlop KM, Barnes DKA, Bailey DM (2014) Variation of scavenger richness and abundance between sites of high and low iceberg scour frequency in Ryder Bay, West Antarctic Peninsula. Polar Biol 37:1741–1754

Elbroch LM, Wittmer HU (2012) Table scraps: inter-trophic food provisioning by pumas. Biol Lett 8:776–779

Evans K, Thresher R, Warneke RM et al (2005) Periodic variability in cetacean strandings: links to large-scale climate events. Biol Lett 1:147–150

Field RD, Reynolds JD (2013) Ecological links between salmon, large carnivore predation, and scavenging birds. J Avian Biol 44:9–16

Fisher I, Pain DJ, Thomas VA (2006) Review of lead poisoning from ammunition sources in terrestrial birds. Biol Conserv 131:421–432

FitzGibbon CD (1990) Why do hunting cheetahs prefer male gazelles? Anim Behav 40:837–845

Fjellheim A, Raddum GG (1990) Acid precipitation: biological monitoring of streams and lakes. Sci Total Environ 96:57–66

Geluso KN, Altenbach JS, Wilson DE (1976) Bat mortality: pesticide poisoning and migratory stress. Science 194:184–186

Gende SM, Edwards RT, Willson MF et al (2002) Pacific salmon in aquatic and terrestrial ecosystems. Bioscience 52:917–928

Gese EM, Ruff RL, Crabtree RL (1996) Foraging ecology of coyotes Canis latrans: the influence of extrinsic factors and a dominance hierarchy. Can J Zool 74:769–783

Ghalambor CK, Martin TE (2002) Comparative manipulation of predation risk in incubating birds reveals variability in the plasticity of responses. Behav Ecol 13:101–108

Glover AG, Higgs ND, Bagley PM et al (2010) A live video observatory reveals temporal processes at a shelf-depth whale-fall. Cah Biol Mar 51:375–381

Greig DJ, Gulland FMD, Kreuder C (2005) A decade of live California sea lion (*Zalophus californianus*) strandings along the central California coast: causes and trends, 1991-2000. Aquat Mamm 31:11–22

Gulland FMD (2006) Review of the marine mammal unusual mortality event response program of the National Marine Fisheries Service. U.S. Dept. of Commerce, NOAA Tech. Memo, NMFS-OPR-33

Hall AJ, Connell BJM, Barker RJ (2001) Factors affecting first-year survival in grey seals and their implications for life history strategy. J Anim Ecol 70:138–149

Harvell CD, Kim K, Burkholder JM et al (1999) Emerging marine diseases-climate links and anthropogenic factors. Science 285:1505–1510

Heithaus MR, Wirsing AJ, Thomson JA et al (2008) A review of lethal and non-lethal effects of predators on adult marine turtles. J Exp Mar Biol Ecol 356:43–51

Hewson R (1995) Use of salmonid carcasses by vertebrate scavengers. J Zool 235:53–65

Hoag H (2003) Atlantic cod meet icy death. Nature 422:792

Hocking MD, Reimchen TE (2009) Salmon species, density and watershed size predict magnitude of marine enrichment in riparian food webs. Oikos 118:1307–1318

Houston DC (1974) The role of griffon vultures *Gyps africanus* and *Gyps ruppellii* as scavengers. J Zool 172:35–46

Houston DC (1979) The adaptation of scavengers. In: Sinclair ARE, Griffiths N (eds) Serengeti, dynamics of an ecosystem. University of Chicago Press, Chicago, pp 263–286

Houston DC (1985) Evolutionary ecology of afrotropical and neotropical vultures in forests. In: Buckley PA, Foster MS, Morton ES et al (eds) Neotropical ornithology. The American Ornithologists Union, Washington, DC, pp 856–864

Houston DC (1988) Competition for food between neotropical vultures in forest. Ibis 130: 402–417

Howald GR, Mineau P, Elliott JE et al (1999) Brodifacoum poisoning of avian scavengers during rat control on a seabird colony. Ecotoxicology 8:431–447

Hunter JS, Durant SM, Caro TM (2006) Patterns of scavenger arrival at cheetah kills in Serengeti National Park, Tanzania. Afr J Ecol 45:275–281

Husseman JS, Murray DL, Power G et al (2003) Assessing differential prey selection between two sympatric large carnivores. Oikos 101:591–601

Isaacs JD, Schwartzlose RA (1975) Active animals of the deep-sea floor. Sci Am 233:85–91

Janzen D (1977) Why fruits rot, seeds mold, and meat spoils. Am Nat 111:691–713

Jepson PD, Deaville R, Acevedo-Whitehouse K et al (2013) What caused the UK's largest common dolphin (*Delphinus delphis*) mass stranding event? PLoS One 8:e60953

Kane A, Jackson AL, Ogada DL et al (2014) Vultures acquire information on carcass location from scavenging eagles. Proc R Soc B 281:20141072

Kendall C (2013) Alternative strategies in avian scavengers: how subordinate species foil de despotic distribution. Behav Ecol Sociobiol 67:383–393

Kendall C, Virani MZ, Kirui P et al (2012) Mechanisms of coexistence in vultures: understanding de patterns of vulture abundance at carcasses in Masai Mara National Reserve, Kenya. Condor 114:523–531

Kendall CJ, Virani MZ, Hopcraft JGC et al (2014) African vultures don't follow migratory herds: scavenger habitat use is not mediated by prey abundance. PLoS One 9:e83470

Koch V, Peckham H, Mancini A et al (2013) Estimating at-sea mortality of marine turtles from stranding frequencies and drifter experiments. PLoS One 8:e56776

Koyama A, Kavanagh K, Robinson A (2005) Marine nitrogen in central Idaho riparian forests: evidence from stable isotopes. Can J Fish Aquat Sci 62:518–526

Kruuk H (1967) Competition for food between vultures in East Africa. Ardea 55:171–193

Kruuk H (1972) The spotted Hyena. A study of predation and social behavior. University of Chicago Press, Chicago

Kühn S, Bravo-Rebolledo EL, Van Franeker JA (2015) Deleterious effects of litter on marine life. In: Bergmann M, Gutow L, Klages M (eds) Marine anthropogenic litter. Springer, Berlin, pp 75–116

Laist DW, Knowlton AR, Mead JG et al (2001) Collisions between ships and whales. Mar Mamm Sci 17:35–75

Lambertucci SA, Speziale KL, Rogers TE et al (2009) How do roads affect the habitat use of an assemblage of scavenging raptors? Biodivers Conserv 18:2063–2074
Lambertucci SA, Donazar JA, Hiraldo F (2010) Poisoning people and wildlife with lead ammunition: time to stop. Environ Sci Technol 44:7759–7760
Le Boeuf BJ, Pérez-Cortés M, Urban J et al (2000) High gray whale mortality and low recruitment in 1999: potential causes and implications. J Cetacean Res Manag 2:85–99
Margalida A, Donázar JA, Carrete M et al (2010) Sanitary versus environmental policies: fitting together two pieces of the puzzle of European vulture conservation. J Appl Ecol 47:931–935
Margalida A, Colomer MA, Sanuy D (2011) Can wild ungulate carcasses provide enough biomass to maintain avian scavenger populations? An empirical assessment using a bio-inspired computational model. PLoS One 6:e20248
Marsh R, Petrie B, Weidman CR et al (1999) The 1882 tilefish kill - a cold event in shelf waters off the North-Eastern United States? Fish Oceanogr 8:39–49
Mateo-Tomás P, Olea PP (2010) When hunting benefits raptors: a case study of game species and vultures. Eur J Wildl Res 56:519–528
Mattisson J, Andrén H, Persson J, Segerström P (2011) Influence of intraguild interactions on resource use by wolverines and Eurasian lynx. J Mammal 92:1321–1330
Mduma SAR, Sinclair ARE, Hilborn R (1999) Food regulates the Serengeti wildebeest: a 40-year record. J Anim Ecol 68:1101–1122
Moleón M, Sánchez-Zapata JA, Selva N et al (2014) Inter-specific interactions linking predation and scavenging in terrestrial vertebrate assemblages. Biol Rev 89:1042–1054
Moleón M, Sánchez-Zapata JA, Sebastián-González E et al (2015) Carcass size shapes the structure and functioning of an African scavenging assemblage. Oikos 124:1391–1403
Molinari-Jobin A, Molinari P, Loison A et al (2004) Life cycle period and activity of prey influence their susceptibility to predators. Ecography 27:323–329
Møller AP, Erritzøe H, Erritzøe J (2011) A behavioral ecology approach to traffic accidents: interspecific variation in causes of traffic casualties among birds. Zool Res 32:115–127
Morales-Reyes Z, Pérez-García JM, Moleón M et al (2017a) Evaluation of the network of protection areas for the feeding of scavengers in Spain: from biodiversity conservation to greenhouse gas emission savings. J Appl Ecol 54:1120–1129
Morales-Reyes Z, Sánchez-Zapata JA, Sebastián-González E, Botella F, Carrete M, Moleón M (2017b) Scavenging efficiency and red fox abundance in Mediterranean mountains with and without vultures. Aca Oecol 79:81–88
Moreno P, Benke H, Lutter S (1992) Behaviour of Harbour (Phocoena phocoena) carcasses in the German Bight: surfacing rate, decomposition and drift routes. Untersuchungen über Bestand, Gesundheitszustand und Wanderungen der Kleinwalpopulationen (Cetacea) in deutschen Gewässern.–Interim Report, WWF Fachbereich Wattenmeer and Nordseeschutz und Forschungs-und Technologiezentrum Westküste, Aussenstelle der Universität Kiel
Moreno-Opo R, Trujillano A, Margalida A (2016) Behavioral coexistence and feeding efficiency drive niche partitioning in European avian scavengers. Behav Ecol 27:1041–1052
Newton I (1998) Population limitation in birds. Academic, London
Newton I, Davis PE, Davis JE (1982) Ravens and buzzards in relation to sheep-farming and forestry in Wales. J Appl Ecol 19:681–706
Ogada DL, Torchin ME, Kinnaird MF et al (2012) Effects of vulture declines on facultative scavengers and potential implications for mammalian disease transmission. Conserv Biol 26:453–460
Olea PP, Mateo-Tomás P (2009) The role of traditional farming practices in ecosystem conservation: The case of trashumance and vultures. Biol Conserv 142:1844–1853
Ono KA, Boness DJ, Oftedal OT (1987) The effect of a natural environmental disturbance on maternal investment and pup behavior in the California sea lion. Behav Ecol Sociobiol 21: 109–118
Oro D, Genovart M, Tavecchia G et al (2013) Ecological and evolutionary implications of food subsidies from humans. Ecol Lett 16:1501–1514
Owen-Smith N (1988) Megaherbivores. The influence of very large body size on ecology. Cambridge University Press, Cambridge

Owen-Smith N, Mills MGL (2008) Predator-prey size survival rates in an African large-mammal food web. J Anim Ecol 77:173–183

Palomares F, Caro TM (1999) Interspecific killing among mammalian carnivores. Am Nat 153:492–508

Parmenter RR, MacMahon JA (2009) Carrion decomposition and nutrient cycling in a semiarid shrub- steppe ecosystem. Ecol Monogr 79:637–661

Peltier H, Dabin W, Daniel P et al (2012) The significance of stranding data as indicators of cetacean populations at sea: modelling the drift of cetacean carcasses. Ecol Indic 18:278–290

Peltier H, Baagøe HJ, Camphuysen KCJ et al (2013) The stranding anomaly as population indicator: the case of harbour porpoise *Phocoena phocoena* in North-Western Europe. PLoS One 8:e62180

Pereira LM, Owen-Smith N, Moleón M (2014) Facultative predation and scavenging by mammalian carnivores. Mammal Rev 44:44–55

Piatt JF, Van Pelt TI (1997) Mass-mortality of Guillemots (*Uria aalge*) in the Gulf of Alaska in 1993. Mar Pollut Bull 34:656–662

Polis GA, Hurd SD (1996a) Allochthonous input across habitats, subsidized consumers, and apparent trophic cascades: examples from the ocean-land interface. In: Polis GA, Winemiller KO (eds) Food webs: integration of patterns and dynamics. Chapman and Hall, Inc., New York, pp 275–285

Polis GA, Hurd SD (1996b) Linking marine and terrestrial food webs: allochthonous input from the ocean supports high secondary productivity on small islands and coastal land communities. Am Nat 147:396

Polis GA, Anderson WB, Holt RD (1997) Toward an integration of landscape and food ecology: the dynamics of spatially subsidized food webs. Annu Rev Ecol Syst 28:289–316

Pringle RM (2017) How large herbivores subsidize aquatic food webs in African savannas. Proc Natl Acad Sci U S A 114:7489–7491

Prior KA, Weatherhead PJ (1991) Competition at the carcass: opportunities for social foraging by turkey vultures in southern Ontario. Can J Zool 69:1550–1556

Putman RJ (1983) Carrion and dung: the decomposition of animal wastes. Edward Arnold, London

Quaggiotto MM, Burke LR, McCafferty DJ et al (2016) First investigation of the consumption of seal carcasses by terrestrial and marine scavengers. Glasgow Nat 26:33–52

Quaggiotto MM, Barton PS, Morris CD et al (2018) Seal carrion is a predictable resource for coastal ecosystems. Acta Oecol 88:41–51

Quinn TP, Carlson SM, Gende SM et al (2009) Transportation of Pacific salmon carcasses from streams to riparian forests by bears. Can J Zool 87:195–203

Ramirez-llodra E, Brandt A, Danovaro R et al (2010) Deep, diverse and definitely different: unique attributes of the world's largest ecosystem. Biogeosciences 7:2851–2899

Reimchen T (2000) Some ecological and evolutionary aspects of bear-salmon interactions in coastal British Columbia. Can J Zool 78:448–457

Reisdorf AG, Bux R, Wyler D et al (2012) Float, explode or sink: postmortem fate of lung-breathing marine vertebrates. Palaeobiol Palaeoenviron 92:67–81

Roman J, Estes JA, Morissette L et al (2014) Whales as marine ecosystem engineers. Front Ecol Environ 12:377–385

Rose MD, Polis GA (1998) The distribution and abundance of coyotes: the effects of allochthonous food subsidies from the sea. Ecology 79:998–1007

Ruzicka RE, Conover MR (2012) Does weather or site characteristics influence the ability of scavengers to locate food? Ethology 118:187–196

Sánchez-Piñero F, Polis G (2000) Bottom-up dynamics of allochthonous input: direct and indirect effects of seabirds on islands. Ecology 81:3117–3132

Santos BS, Kaplan DM, Friedrichs MA et al (2018) Consequences of drift and carcass decomposition for estimating sea turtle mortality hotspots. Ecol Indic 84:319–336

Savage RJG (1977) Evolution in carnivorous mammals. Palaeontology 20:237–271

Schaller GB (1972) The serengeti lion: a study of predator prey relations. University of Chicago Press, Chicago

Schoenebeck CW, Brown ML, Chipps SR et al (2012) Nutrient and algal responses to winterkilled fish-derived nutrient subsidies in eutrophic lakes. Lake Reserv Manage 28:189–199

Scholin CA, Gulland F, Doucette GJ et al (2000) Mortality of sea lions along the central California coastlinked to a toxic diatom bloom. Nature 403:80–84

Sebastián-González E, Sánchez-Zapata JA, Donázar JA et al (2013) Interactive effects of obligate scavengers and scavenger community richness on lagomorph carcass consumption patterns. Ibis 155:881–885

Sebastián-González E, Moleón M, Gibert JP et al (2016) Nested species-rich networks of scavenging vertebrates support high levels of interspecific competition. Ecology 97:95–105

Selva N (2004) The role of scavenging in the predator community of Białowieża Primeval Forest (Poland). PhD thesis, Univ. de Sevilla, Sevilla

Selva N, Jedrzejewska B, Jedrzejewski W et al (2003) Scavenging of European bison carcasses in Białowieża Primeval Forest (eastern Poland). Ecoscience 10:303–311

Selva N, Jedrzejewska B, Jedrzejewski W et al (2005) Factors affecting carcass use by a guild of scavengers in European temperate woodland. Can J Zool 83:1590–1601

Shivik JA (2006) Are vultures birds, and do snakes have venom, because of macro-and microscavenger conflict? Bioscience 56:819–823

Smale DA, Barnes DKA, Fraser KPP et al (2007) Scavenging in Antarctica: intense variation between sites and seasons in shallow benthic necrophagy. J Exp Mar Biol Ecol 349: 405–417

Smale DA, Brown KM, Barnes DKA et al (2008) Ice scour disturbance in Antarctic shallow waters. Science 321:371

Smith CR, Baco AR (2003) Ecology of whale-falls at the deep sea floor. Oceanogr Mar Biol 41:311–354

Smith CR, Maybaum HL, Baco AR et al (1998) Sediment community structure around a whale skeleton in the deep Northeast Pacific: macrofaunal, microbial and bioturbation effects. Deep-Sea Res 45:335–364

Smith CR, Glover AG, Treude T et al (2015) Whale-fall ecosystems: recent insights into ecology, paleoecology, and evolution. Annu Rev Mar Sci 7:571–596

Soltwedel T, von Juterzenka K, Premke K et al (2003) What a lucky shot! Photographic evidence for a medium-sized natural food-fall at the deep seafloor. Oceanol Acta 26:623–628

Stachowitsch M (1984) Mass mortality in the Gulf of Trieste: the course of community destruction. PSZNI Mar Ecol 5:243–264

Stahler D, Heinrich B, Smith D (2002) Common ravens, *Corvus corax*, preferentially associate with grey wolves, *Canis lupus*, as a foraging strategy in winter. Anim Behav 64: 283–290

Subalusky AL, Dutton CL, Rosi EJ et al (2017) Annual mass drownings of the Serengeti wildebeest migration influence nutrient cycling and storage in the Mara River. Proc Natl Acad Sci U S A 114:7647–7652

Thompson D, Onoufriou J, Brownlow A et al (2015) Preliminary report on predation by adult grey seals on grey seal pups as a possible explanation for corkscrew injury patterns seen in the unexplained seal deaths. Sea Mammal Research Unit, University of St Andrews, Report to Scottish Government, no. US1 and 6 addendum, St Andrews

Tuyttens FAA, Stuyck JJJM (2002) Effectiveness and efficiency of chlorophacinone poisoning for the control of muskrat (*Ondatra zibethicus*) populations. N Z J Zool 29:33–40

Wikenros C, Sand H, Ahlqvist P et al (2013) Biomass flow and scavengers use of carcasses after re-colonization of an apex predator. PLoS One 8:e77373

Wilkinson DM (1991) Program review of the marine mammal stranding networks. Report to Assistant Administrator for Fisheries. NOAA

Wilmers CC, Getz WM (2004) Simulating the effects of wolf-elk population dynamics on resource flow to scavengers. Ecol Model 177:193–208

Wilmers CC, Getz WM (2005) Gray wolves as climate change buffers in Yellowstone. PLoS Biol 3:571–576

Wilmers CC, Post E (2006) Predicting the influence of wolf-provided carrion on scavenger community dynamics under climate change scenarios. Glob Chang Biol 12:403–409
Wilmers CC, Crabtree RL, Smith DW et al (2003a) Trophic facilitation by introduced top predators: grey wolf subsidies to scavengers in Yellowstone National Park. J Anim Ecol 72:909–916
Wilmers CC, Stahler DR, Crabtree RL et al (2003b) Resource dispersion and consumer dominance: scavenging at wolf- and hunter-killed carcasses in Greater Yellowstone, USA. Ecol Lett 6:996–1003
Wilson EE, Wolkovich EM (2011) Scavenging: how carnivores and carrion structure communities. Trends Ecol Evol 26:129–135
Work TM, Balazs GH, Summers TM et al (2015) Causes of mortality in green turtles from Hawaii and the insular Pacific exclusive of fibropapillomatosis. Dis Aquat Org 115:103–110
Wright AJ, Maar M, Mohn C et al (2013) Possible Causes of a harbour porpoise mass stranding in Danish Waters in 2005. PLoS One 8(1–14):e55553

Invertebrate Scavenging Communities

Gail S. Anderson, Philip S. Barton, Melanie Archer, and John R. Wallace

Contents

G. S. Anderson (✉)
School of Criminology, Centre for Forensic Research, Simon Fraser University, Burnaby, BC, Canada
e-mail: ganderso@sfu.ca

P. S. Barton
Fenner School of Environment and Society, The Australian National University, Canberra, ACT, Australia
e-mail: philip.barton@anu.edu.au

M. Archer
Department of Forensic Medicine, Monash University, Southbank, VIC, Australia
e-mail: Melanie.Archer@vifm.org

J. R. Wallace
Department of Biology, Millersville University, Millersville, PA, USA
e-mail: John.Wallace@millersville.edu

P. P. Olea et al. (eds.), *Carrion Ecology and Management*, Wildlife Research Monographs 2, https://doi.org/10.1007/978-3-030-16501-7_3

Introduction

Carrion is a rich and ephemeral resource that supports a large number of sequentially colonizing microbial, invertebrate and vertebrate fauna that utilize the resource over time. As carrion degrades, it progresses through rapid biological, chemical and physical changes, while supporting an equally changing ecosystem. Many abiotic and biotic factors impact this process.

In this section of the chapter we focus exclusively on the invertebrate use of carrion in both terrestrial and aquatic ecosystems, and consider the role of different groups of invertebrates, along with the biotic and abiotic parameters that influence their succession. The most important invertebrate group on terrestrial carrion is the insects, and this will be a focus for the chapter. However, we will cover the role at carrion of other groups, such as the crustaceans, arachnids, nematodes and molluscs. The terrestrial and aquatic ecosystems (both freshwater and marine) will be considered separately in order to compare and contrast the roles of invertebrate groups in these different environments, and to help highlight how carrion invertebrate ecology is influenced by a mixture of biotic and environmental factors.

We will also briefly discuss the intimately linked concept of decomposition, which is the chemical and material breakdown of a carcass. Decomposition has been divided into discrete stages by a number of authors (e.g. Early and Goff 1986; Payne 1965; Smith 1986), although in practical terms, stages occur on a continuum, and their separation can be somewhat subjective (Early and Goff 1986; Michaud et al. 2015). Stages of decomposition can also differ between parts of the same carcass in some cases, which increases the complexity of assigning a definitive stage (Anton et al. 2011; Matuszewski et al. 2010). Decomposition stage change is a powerful driver of invertebrate succession because the species that attend carrion are often adapted to utilize it at certain times in its breakdown. Conversely, invertebrate activity is a major driver of decomposition (Payne 1965). This is especially true of the larvae of blow flies (Diptera: Calliphoridae) in the opening decay stages, because their feeding and mechanical activity has the potential to liquefy the tissues (Hobson 1932), as well as remove a significant amount of biomass (Putman 1977). Other invertebrate families, however, such as the hide beetles (Dermestidae), may also be responsible for significant biomass removal in the mid and later decomposition stages.

Terrestrial Colonization

The phenomenon of carrion colonization by invertebrates is imperfectly understood due to the complexity of the chemical and visual cues utilized (see below). However, there are both obligate species, which must use carrion to complete their lifecycle, and facultative species, which attend carrion opportunistically. Facultative species can often utilize a range of alternative decaying organic materials such as dung or rotting plant matter, and sometimes resources such as the flesh of living animals. There are also incidental species that do not visit carrion specifically, but may use it as a hunting ground, or carry out other activities on the carcass, such as mating.

It is becoming clear that carrion colonizing taxa use a cocktail of chemical attractants to track carrion from long range (Kalinová et al. 2009; Zhu et al. 2013). Short range cues, however are used where visual and tactile contact is made with carrion, and these cues may include red/brown colour, hair bearing texture, and heat. The value of these cues is certainly suggested by mimicry of carrion by flowers that use carrion insects as pollinators (Angioy et al. 2004). Although there is a diverse invertebrate fauna that attends carrion, some taxa are predictably associated with particular decomposition stages, whereas others may be more flexible in their visitation times. The latter is especially true of the facultative and incidental taxa (Archer 2003). The following section is a brief introduction to the main obligate and facultative taxa that are found on carrion.

Diptera

The Order Diptera (true flies) contains the species that are usually the first to colonize carrion, and their offspring also tend to remove the most biomass, as well as dominate the carcass numerically (Putman 1983; Smith 1986). Both adults and larvae may also be the main prey species of various beetle and wasp species. Dipteran species also have a long tenure on decomposing carcasses, with various species from many families adapted to the early, middle and late stages of decomposition. Dipteran species may also colonize carrion of various sizes, ranging from dead invertebrates (Seastedt et al. 1981), to small and large vertebrates (Blackith and Blackith 1990; Davies 1999).

Calliphoridae (Blow Flies) Calliphorid species are generally the first to arrive at carrion, although there is often a succession of species within the family (Norris 1965). Some species, such as *Chrysomya rufifacies* (Macquart), can also act as initial or later colonizers depending on the geographic area (O'Flynn and Moorehouse 1979). Carrion breeding calliphorids are primarily oviparous and have a wide global distribution; their larvae are usually responsible for removing most biomass from carrion (Putman 1983; Smith 1986). The most common ecological niche for the family is parasitizing earthworms and snails, although some species can also act as agents of myiasis (infestation of living vertebrates) by fly (Diptera) larvae, or breed in carrion and other organic matter (Colless and McAlpine 1991; Norris 1965). Obligate and facultative carrion breeding species may use carcasses not only as breeding pabulum, but also as a protein source for maturing egg batches and sperm, and as a gathering place for mating. The major carrion breeding species are found in the genera *Calliphora, Chrysomya, Lucilia, Protophormia, Phormia* and *Cochliomya* (Byrd and Castner 2001; Smith 1986).

Muscidae (House Flies and Allies) Muscids commonly breed in decaying organic matter, such as faeces and compost (Byrd and Castner 2001; Pont 1973). A number of species are facultative carrion breeders, while a few are carrion specialists, such as some species within *Australophyra* and *Hydrotaea*. Their arrival at carrion tends to be delayed until at least several days after death (Fuller 1934a; Smith 1986).

Fanniidae (Latrine and Lesser House Flies) Larvae of this small, primitive family mature in decaying organic matter, including carrion (Pont 1977; Smith 1986). The family has a widespread distribution, although it is predominantly temperate. Species such as the facultative *Fannia canicularis* (Linnaeus) (lesser house fly) are cosmopolitan, and often associated with human dwellings, urine and faeces (Pont 1977).

Sarcophagidae (Flesh Flies) Sarcophagids are viviparous or ovoviviparous, and some species breed in carrion (Colless and McAlpine 1991; Szpila et al. 2015). Recent Australian taxonomic research has highlighted the paucity of knowledge about the family, which is partially due to difficulties in identification at species level (Meiklejohn 2012). Sarcophagid maggot numbers on carrion are generally fairly low, and it has been suggested that this is caused by strong competition from calliphorids (Denno and Cothran 1976). Sarcophagids are widely distributed, but are found predominantly in the tropical and warm temperate regions (Byrd and Castner 2001). Those in cooler parts of the family's range will often over-winter as pupae (Chen et al. 1991; Fuller 1934a).

Miscellaneous Families A number of additional dipteran families have a less prominent role in succession, but will also utilize carrion for breeding, protein feeding and mating. In many cases where species of these families breed in carrion, their use is facultative due to their ability to breed in a wide range of other decaying organic material. Several families can produce high numbers of offspring on medium to large carcasses. The Piophilidae (cheese skippers) is a small family with members that breed mainly in decomposed carrion and fungus, and some species also produce maggots capable of jumping when disturbed (Pitkin 1989; Smith 1986). The larvae of one common carrion species *Piophila casei* (Linnaeus) may be on a carcass within days of death, but are most characteristic of the later decay stages (Archer 2014; Smith 1986). Several other species can also be important components of the carrion fauna, both as adults and larvae (Rochefort et al. 2015).

Phoridae (scuttle flies) and Psychodidae (moth flies) adults and larvae may also be found on vertebrate carrion (Manlove and Disney 2008; Smith 1986). Phorids can also breed in invertebrate carrion (Seastedt et al. 1981), and are well known for the association of several species, such as *Conicera tibialis* Schmitz (the coffin fly), with buried remains (Smith 1986). Some families will occasionally produce small numbers of maggots on carcasses, and/or use the carcass to aggregate for mating. These include the Sepsidae (ant flies), Sphaeroceridae (small dung flies), Stratiomyiidae (soldier flies), Sciadoceridae, Anisopodidae, Drosophilidae (vinegar flies), Ephydridae (shore flies), Heleomyzidae (sun flies), Platystomatidae (boatman flies), Syrphidae (bee flies), Ulidiidae (picture-wing flies), Trichoceridae and Scenopinidae (window flies) (Byrd and Castner 2001; Fuller 1934b, 1935; Smith 1986).

Coleoptera

The Order Coleoptera (beetles) forms an important part of the carrion fauna. Most are predators on other carrion inhabitants, although some feed on the carcass itself (Smith 1986). The first beetle species usually arrive within a few days of death, and adult and larval beetles of various taxa can remain associated with a large carcass for weeks to months after death (e.g. Anderson and VanLaerhoven 1996; Archer 2014).

Silphidae (Carrion and Burying Beetles)
The beetles that arrive earliest at carrion are the unique burying beetles of the family Silphidae. They specialize in the rapid burial of small mammal carcasses, which are formed into brood balls for their offspring (Eggert and Mulller 1997; Putman 1983). The silphids are distributed predominantly in the northern hemisphere (Matthews 1982), and many species that do not bury carrion also occur worldwide (e.g. Payne 1965; Peck 2001). Adults of non-burying species usually feed on flesh, dipteran maggots and dead insects, while their larvae can consume large amounts of putrefying flesh and skin (Peck 2001). Larvae of non-burying silphids, such as *Ptomaphila lacrymosa* Schreibers may be the main agents of tissue removal in late decay (Archer 2000).

Histeridae (Clown and Burying Beetles) and Staphylinidae (Rove Beetles)
Also often arriving during early putrefaction are the Histeridae and the Staphylinidae. Species from these predatory families arrive early to take advantage of the abundant dipteran eggs, larvae and pupae. But these families may have tenure times on large carcasses of over three months, with their larvae appearing during late decay (Archer 2014). Some species of the staphylinid genus *Aleochara* also parasitize the pupae of carrion breeding Diptera (Seevers and Herman 1978; Smith 1986).

Dermestidae (Hide Beetles)
Families, such as Dermestidae, that commonly occur during mid to late decomposition are often able to feed on dried flesh. They can also be found in the nests of both vertebrates and invertebrates, where they feed largely on remnants of prey left by the nesting species, or on nest occupants that have died (Hinton 1963).

Cleridae (Checkered Beetles)
Cleridae (checkered beetles) of the genus *Necrobia* will also feed on flesh, as well as being predaceous (Simmons 1925; Smith 1986). They are primarily present during late decay.

Fig. 1 Trogidae on a Kangaroo (*Macropus giganteus* Shaw) Carcass (Photo by P. Barton)

Trogidae (Skin Beetles)
Trogidae (skin beetles) occur in mid to late decay, and feed on skin, feathers and connective tissues. Their larvae develop in burrows under the carcass (Matthews 1984; Smith 1986). Trogids can sometimes occur in large numbers on the dried remains of carcasses (Fig. 1).

Miscellaneous Families
There are a number of other coleopteran families that also visit carrion occasionally on a facultative basis. The mainly predatory Carabidae (ground beetles) can associate with carcasses in early, mid or late decay (Anderson and VanLaerhoven 1996; Chapman and Sankey 1955; Smith 1986). Ptiliidae (featherwing beetles) feed on fungal spores and can be found on carrion (Easton 1966; Matthews 1982; Payne and King 1970). Both adult (Payne and King 1970) and larval Scarabaeidae (dung beetles) may be associated with carrion (Byrd and Castner 2001), with dung beetles often recorded at the carcasses of large herbivores where they may feed on the gut contents. The mainly aquatic family Hydrophilidae (water scavenger beetles) has terrestrial representatives, and *Cercyon* spp. adults are often collected in succession studies (Chapman and Sankey 1955; Easton 1966; Moore 1955). They are thought to prey on dipteran larvae (Smith 1986). Some species of Nitidulidae (sap beetles) are attracted to carrion (Byrd and Castner 2001; Moore 1955; Payne 1965), and will breed in it (Anderson and VanLaerhoven 1996). Adults of Leiodidae (round fungus beetles) are also commonly captured at carrion (Zwick 1979). Adults have been observed feeding on nematodes, and leiodid larvae may occur in the soil under largely decomposed carcasses (Archer 2002).

Other Invertebrates

Despite insects being the dominant taxon at carcasses, there are several other invertebrate taxa frequently encountered at carrion in terrestrial environments, including many different groups within the Phylum Arthropoda (such as mites and spiders) and other phyla (such as Nematoda and Mollusca). Some of these taxa are facultative and do not specialize on carrion, but will opportunistically scavenge on carrion or other insect-generated detritus. Other invertebrate taxa are predators that will take advantage of the concentrated and abundant invertebrates present at the remains of an animal. Importantly, some of the groups are considered incidental, do not occur in high numbers, and are not reliably observed at carcasses. Knowledge of their colonization and succession patterns at carrion is often poor, and they have not received as much research attention as the other main groups (see above).

Mites (**Acari**) can be the most numerous of all arthropods at carrion, and can reach densities of many tens of thousands of individuals at a medium sized mammal carcass (Perotti and Braig 2009). Many mites display phoresy, and depend on larger flies or beetles for their dispersal between carcasses. Mites will feed on nematodes, fly eggs, and probably also scavenge on small fragments of carcass and insect detritus. The most abundant are typically the Astigmata (O'Connor 2009). Several studies of Mesostigmata have shown that their arrival coincides with their insect hosts (Barton et al. 2014; Perotti and Braig 2009), but their abundance may peak later following their rapid reproduction and population growth at the carcass (Barton et al. 2014).

Spiders (**Arachnida**) are obligate predators and can be found sporadically at carcasses preying on other arthropod taxa (Chin et al. 2007). Spiders can be found more frequently at carrion where it has a disproportionally large importance as a resource, such as on small islands or in caves (Barton 2015).

Some minor insect groups often present at carcasses include springtails (**Collembola**), earwigs (**Dermaptera**) and cockroaches (**Blattodea**). These taxa are generalist soil or surface-dwelling detritus feeders (Bornemissza 1957). Their succession patterns are only weakly associated with particular decay stages due to their generalist diets.

Ants, bees and wasps (**Hymenoptera**) can all be found at carcasses, and often in large numbers. Many species of ant will opportunistically prey on the abundant arthropods at carrion, including fly larvae, but may also scavenge on the carrion itself (e.g. the Australian meat ant *Iridomyrmex purpureus* Smith). Compared with flies or beetles, ants cannot disperse rapidly to a carcass unless it is close to the nest (Barton et al. 2013a), and so their arrival patterns are typically delayed for several hours or days, and may be unpredictable. Bees have been observed at carcasses feeding on the liquids present, possibly to target water, sodium or even proteins (Santos et al. 2014; Silveira et al. 2005). Several groups of parasitoid wasp (e.g. Pteromalidae, Diapriidae, Braconidae) are known to target calliphorid flies attendant at carrion (Voss et al. 2009). *Nasonia vitripennis* Walker (Pteromalidae) is a common parasitoid of fly larvae and much work has been devoted to its ecology and development (Frederickx et al. 2013; Grassberger and Frank 2003). The host is important to parasitoid development times, and therefore emergence and carrion departure times (Voss et al. 2010).

Butterflies and moths (**Lepidoptera**) will opportunistically feed on the liquids present at carcasses (Boggs and Dau 2004), but this is limited to early stages of decay when fluids are present. Tineid moths, in contrast, are commonly reported at carcasses where they appear to specialize on keratin-based animal components, such as hair, hooves, and horns (Bornemissza 1957; Braack 1987). Tineid moths, therefore, are more common in the later, dry stages of decay when the resistant keratinous components are left (Bucheli et al. 2010). Some other typically phytophagous insects, such as grasshoppers (**Orthoptera**) or bugs (**Hemiptera**), have been observed feeding at carcasses. It is thought that this is may be an opportunistic behavior related to the acquisition of fluids, salts or protein (Baz et al. 2010).

Wood lice (**Isopoda**), also known as pill bugs or slaters, are generalist detritus feeders that may be encountered at animal remains in terrestrial environments (Braig and Perotti 2009). Other kinds of crustacean (e.g. **Amphipoda**) occur at carrion in either freshwater (Barrios and Wolff 2011) or marine and inter-tidal environments (Colombini and Chelazzi 2003).

Snails (**Gastropoda**) vary in their diet from generalist herbivores to detritivores, and may be found at a carcass feeding opportunistically on the detritus. This taxon is more important in marine environments where they are active and abundant detritivores (Britton and Morton 1994).

Soil-dwelling nematodes (**Nematoda**) are a hyper-abundant taxon with a very broad dietary range that depends on the species. Nematodes will migrate to the soil surface and opportunistically feed on the bacteria and fungi that proliferate beneath a carcass, although this follows no strong succession pattern (Carter et al. 2007). Some species, however, are more specialized. The nematode, *Rhabditis stammeri* Völk (Rhabtidae), for example, has been found to live exclusively on carrion and is dispersed among carcasses by the burying beetle, *Nicrophorus vespilloides* Herbst (Silphidae) (Richter 1993).

Aquatic Colonization

Freshwater Invertebrate Colonization of Carrion

The specialized necrophagous or scavenger terrestrial invertebrates that feed on carcasses have evolved restricted dietary regimes through a long evolutionary process that allowed for the specialization not only in the consumption of carrion but also its detection (Fenoglio et al. 2014). While aquatic insects have exhibited high diversity, abundance and distribution in both lotic (running water) and lentic (standing water) ecosystems, they have managed to fill all trophic niches save one: there are no truly necrophagous aquatic insects that feed exclusively on carrion (Fenoglio et al. 2014; Giller and Malmqvist 1998; Merritt and Wallace 2009). Several hypotheses have been proposed to explain the absence of such trophic specialization in aquatic systems including:

(1) due to the small proportion of all terrestrial insects being necrophagous, the probability of necrophagy would be low among aquatic insects;
(2) aquatic insects are more plastic in their feeding roles without being specialists;
(3) the difficulty of locating carrion as a food source limiting such specialization;

(4) the paucity both in carrion availability and rapidity of its breakdown in lotic aquatic systems compared with terrestrial would limit trophic specialization in aquatic systems;
(5) assuming this feeding strategy evolved from animal waste consumption, the diversity of animal waste in aquatic systems is not as available as it is in terrestrial systems and;
(6) crustacean exploitation of aquatic systems as non-specialized scavengers may have limited or excluded insects from diversifying to feed on carrion (Fenoglio et al. 2014).

Despite the lack of ecological and evolutionary drivers to support a necrophagous lifestyle in freshwaters, aquatic insects have been little studied except for their association with certain types of carrion e.g. rats (Keiper et al. 1997; Tomberlin and Adler 1998), pigs (Barrios and Wolff 2011; Haefner et al. 2004; Vance et al. 1995) and humans (Merritt and Wallace 2009). There has also been forensic research on how they may be used to establish postmortem submersion intervals (PMSI) (Haskell et al. 1989; Hobischak and Anderson 2002; Keiper et al. 1997; Wallace et al. 2008).The PMSI estimation requires an understanding of the stages of decomposition in aquatic systems and how faunal succession takes place in aqueous environments as carrion decomposes.

The stages of decomposition for carrion in terrestrial systems have been well established in the literature (Anderson and VanLaerhoven 1996; Payne 1965). However, since earlier work by Payne and King (Payne and King 1972), the stages of decomposition in aquatic (primarily lotic) systems have been reduced from six to five stages due to the influence of terrestrial insects on floating aquatic carcasses (Wallace 2015). This revised collection of stages includes: (1) submerged fresh, (2) early floating, (3) early floating decay, (4) advanced floating decay and (5) sunken remains (Merritt and Wallace 2009). Among these stages, the aquatic faunal players exhibit a succession of a microbial community consisting of bacteria and fungi to a suite of macroinvertebrate processors on a decomposing carcass (Wallace 2015).

As hypothesized by Fenoglio et al. (2014), aquatic insects exhibit a considerable amount of plasticity in their diet which has allowed aquatic ecologists to classify them into functional feeding groups (FFG). These groups are based on their morphological and behavioral adaptations associated with food acquisition including: shredders, scrapers, collector-filterers, collector-gatherers and predators (Cummins 1974; Cummins and Klug 1979). The primary issue associating aquatic invertebrates with a particular decomposition stage has been attributed to the fact that there are no truly necrophagous aquatic insects (Fenoglio et al. 2014). However, the functional feeding groups approach may be used to develop the necessary associations between aquatic insect succession and the stage of decomposition for the development of statistical models (Wallace 2015).

The faunal succession on carrion in freshwater ecosystems varies not only among stages of decomposition but also as a function of the microhabitat within e.g. lotic or streams systems a carcass is found or season (Wallace 2015). For example, shredder taxa such as case-building caddisflies (**Trichoptera**: Limnephilidae), stoneflies

(**Plecoptera**: Nemouridae) or crayfish—(**Decapoda**: Cambaridae) may burrow in and remove small pieces of tissue from a carcass e.g. salmon (Fenoglio et al. 2014; Minshall et al. 1991) or pig (Merritt and Wallace 2009). Depending on the season, the shredder functional feeding groups may be found more often on carcasses in pool habitats than riffles (Merritt and Wallace 2009). As the carcass begins to fragment, this fragmentation can occur more rapidly in lotic systems than lentic (Merritt and Wallace 2009), at which time, the fine particulate organic matter (FPOM) may attract collector-filterers such as black flies (**Diptera**: Simulidae) and net-spinning caddisflies (**Trichoptera**: Hydropsychidae) in riffle habitats (Wallace 2015). Collector gatherers may also be attracted in both riffle and pool habitats e.g. midges (**Diptera**: Chironomidae), mayflies (**Ephemeroptera**: Baetidae, Caenidae and Ephemerellidae) and some beetles (**Coleoptera**: Hydrophilidae) (Barrios and Wolff 2011; Chaloner and Wipfi 2002; Fenoglio et al. 2005; Merritt and Wallace 2009; Wipfi et al. 1998).

Scraper mayflies (**Ephemeroptera:** Heptageniidae) have been reported on pigs during the submerged fresh and early floating stages in riffle habitats, presumably grazing on the algal accumulation on the carcass during those stages (Haefner et al. 2004). Several predators in lotic systems such as stoneflies (**Plecoptera**: Chloroperlidae, Perlidae and Perlolidae), diving beetles (**Coleoptera**: Dytiscidae) and some dragonfly larvae (**Odonata**: Gomphidae and Calopterygidae) have been observed on dead fish. They were probably attracted by the abundance of prey (taxa listed above) associated with carrion (Ellis 1970; Fenoglio et al. 2005; Nicola 1968).

Marine Invertebrate Colonization of Carrion

The marine environment is the only environment which has not been exploited by insects. Other arthropods, primarily the Crustacea, have taken over this role. The taphonomy of large carcasses such as whales has been well studied for over 160 years. Unusually high species richness has been observed on whale falls, comparable to that seen at hot vents, and complete skeletonization may take years (Smith and Baco 2003). The scavenging of such large carcasses has been broken into four stages (Smith and Baco 2003). In the first stage, both vertebrate and invertebrate scavengers remove most of the soft tissue with the primary invertebrates being crustaceans, including **Amphipoda** (Lysianassidae) and **Decapoda**. By the second stage, the sediment surrounding the skeletonized carcasses becomes extremely nutrient rich, and this, together with the remaining bones, provide a resource for invertebrates including extremely high numbers of Polychaeta (**Annelida**) and juvenile bivalves (**Mollusca**). The remaining two stages are dominated by anaerobic bacteria and suspension feeders such as Serpulidae, Chaetopteridae and Sabelidae polychaetes (Smith and Baco 2003).

In 'smaller' carcasses, such as pigs or human bodies, the question of whether the carcass will go through common aquatic decomposition stages (e.g. bloat, active decay, advanced decay and remains) (Anderson 2010) or will be rapidly consumed,

Fig. 2 Pig carcass (*Sus scrofa* L.) submerged at a depth of 300 m in the Strait of Georgia in the Salish Sea being scavenged by Lyssianassidae amphipods (*Orchomene* complex). Image taken 35 h after submergence. Total skeletonization occurred by Day 4 (Ocean Network's Canada, VENUS observatory). Photo by G.S. Anderson, using VENUS observatory

primarily by arthropods, may depend on whether the body is free floating or in contact with the seabed. Carcasses of humans (Boyle et al. 1997; Haglund 1993; Magni et al. 2015) and pigs allowed to float into the water column (Anderson and Hobischak 2004) have been shown to go through the usual decomposition stages, whereas pig carcasses in direct contact with the seabed were rapidly consumed (Anderson and Bell 2014), probably due to the easier access for arthropods to carrion on the seafloor.

In these smaller carcasses, decapod and amphipod crustaceans are the major scavengers with numbers and species varying with depth, season and most notably, with oxygen levels (Anderson and Bell 2014). In shallow waters off the coast of Vancouver (7–15 m) a diverse array of arthropods, seastars (Class **Asteroidea**), and molluscs colonized the remains and complete skeletonization took several weeks (Anderson and Hobischak 2002). In deeper coastal waters (100–300 m) the scavengers were almost entirely malacostracan crustaceans (i.e. **Decapoda**, **Amphipoda**) and were limited to only three major species: *Metacarcinus magister* (Dana), *Munida quadrispina* Benedict, *Pandalus platyceros* Brandt and the Lyssianassidae amphipods (*Orchomene* complex) (Fig. 2). However, much greater numbers of organisms were recorded at greater depths (Anderson and Bell 2014, 2016).

Once the soft tissue of a carcass has been removed, the bones themselves support a rich assemblage of invertebrates, primarily Annelida. A new genus of marine worm, *Osedax*, (meaning bone-eating) was first described in 2004 in whale bones from the Monterey Bay area of California (Rouse et al. 2004) and since this time has been found in many oceans of the world (Fujikura et al. 2006; Glover et al. 2005, 2013). These worms burrow into the bones to feed on lipids. Some researchers believe these worms are exclusive to whale bones (Dahlgren et al. 2006; Fujikura et al. 2006; Glover et al. 2005) while others have shown that they can colonize other species' bones when experimentally placed (Jones et al. 2008). Defleshed whale, pig and cow bones deployed at 20 m in the Mediterranean Sea over a one year period have been shown to support a diverse successional assemblage of polychaete worms (Taboada et al. 2016).

Factors that Impact Invertebrate Colonization of Carrion

Carrion in Terrestrial Environs

There are many abiotic factors which impact colonization of carrion including decomposition stage, geographic region, season, habitat, carcass type as well as biotic factors.

As carrion decomposes, it changes in several important biochemical and physical characteristics. This underpins, in part, the succession of invertebrate taxa that use carrion as a resource or as habitat. Decomposition occurs along a continuum from the freshly dead animal to its eventual complete disintegration (Schoenly and Reid 1987), but it has been historically convenient to classify this process into different broad stages (Bornemissza 1957; Payne 1965). Progression through these stages is largely attributable to temperature, but has often been described with reference to change in resident insects at the carcass (Anderson and VanLaerhoven 1996; Matuszewski et al. 2008). A number of decomposition classification schemes have been proposed, but that described by Payne (1965), and summarized by Carter et al. (2007), covers six main stages (fresh, bloated, active decay, advanced decay, dry, remains). The most obvious changes in carrion through these stages is the loss of moisture and biomass, as well as the distinct biochemical change in the tissues through microbial activity (Carter et al. 2007). Each of these stages has an associated community of invertebrate fauna that is 'characteristic' of that stage, although the set of species present changes from region to region, as well as between seasons (Barton et al. 2013b; Braack 1987; Schoenly and Reid 1987). Further detailed information about the decomposition process is given in the chapter "Carrion Decomposition".

Geography and climate are key drivers of the spatial differences observed in invertebrate scavenger communities (Barton et al. 2013b; Pinero 1997). As for all ecological communities, spatial differences in environmental conditions among landscapes, regions or continents, leads to changes in the communities of organisms present in terms of the occurrence of species and their relative abundance (Barton et al. 2013c; Gaston 2000). This phenomenon is perhaps best known for patterns of insect biodiversity along gradients of latitude or aridity (Gaston and Lawton 1988). This is because temperature (latitude) and moisture (aridity) are two key climatic factors strongly affecting insect communities, and selecting for different tolerances among species.

An example of adaptation among carrion-associated insects to changes in aridity is the co-occurrence of blow fly species (Calliphoridae) in southern Australia. Molecular work has identified co-occurrence of sister-species within the Calliphoridae along an aridity gradient from west (inland, dry) to east (coastal, wet) (Wallman et al. 2005). Another important geographic gradient is the level of urbanization. Intact ecosystems consist of naturally occurring vegetation and soil that are characteristic of that region. However, urbanization leads to dramatic changes in vegetation, often including the domination of exotic species, and changed surfaces to include impervious roads and pavements that act as a barrier between carrion and soil. One study (Kavazos and Wallman 2012), for example, found that some calliphorid flies were present at higher abundances in urbanized sites during winter. This may be because urbanized areas provide more constant temperatures and act as thermal refuges.

The exact roles of temperature and moisture in carrion insect distribution are still under investigation, and the picture is likely to be complicated at least with respect to calliphorids. Some evidence exists that moisture related parameters, such as evaporation and humidity, have more of a role than maximum and minimum temperature in determining the distribution of certain South African blow flies. Risk of frost occurring is also an important influence on the distributions of these species (Richards et al. 2009a).

Even within the same geographic area, and within the same season, there may still be intra and inter day variation in temperature and moisture levels. This can produce a marked effect on adult carrion insect activity and therefore colonization, which can be delayed if temperatures are low (George et al. 2013; Norris 1966). Fluctuating temperature can also potentially cause variation in larval growth rates (Anderson 2000; Warren and Anderson 2013), although it is not clear whether this is of a magnitude great enough to affect succession rate.

Temperature in particular may also partially govern the competitive interactions within a carrion community associated with an individual carcass. Masses of calliphorid and sarcophagid maggots generate heat, which appears related largely to mass size (Slone and Gruner 2007), but also at least partially to larval age and perhaps food availability (Charabidze et al. 2011). There is also speculation that the ability to withstand high temperatures within maggot masses could confer a competitive advantage on the individuals of one species over those of another (Richards et al. 2009b). Additionally, the ability to generate mass heat has been suggested to act as a buffer against low temperature, which may extend seasonal activity times for larval stages (Deonier 1940). The mechanisms controlling mass temperature generation and maintenance are poorly understood (Rivers et al. 2011). Evidence is emerging, however, that calliphorid and sarcophagid larvae of several species have optimal temperatures, which they will self-select to an extent by movement within the maggot mass (Johnson et al. 2014).

Both temperature and rainfall are positively correlated with carrion mass loss rate, and speed of progression through decomposition stages (Archer 2004). In the case of higher temperature, this is likely to be due mainly to the acceleration of maggot activity, which will promote higher rates of biomass loss. In the case of rainfall, not only will decomposition products be washed away, but the carrion will stay moist, and therefore potentially prolong the tenure of maggots (Archer 2004). Also, re-colonization of dried remains by maggots has been known to occur after they become rehydrated (Anderson and VanLaerhoven 1996; Smith 1986).

The time of year or season in which the animal died will impact colonization. Calliphorids are the first colonizers in most parts of the world but their relative abundance changes depending on season as well as geography, with different species dominating over the year in different regions (Anderson 2009). In Maryland, USA, for example, *Calliphora vicina* Robineau-Desvoidy, *C. livida* Hall and *Lucilia illustris* (Meigen) were found only in spring, *L. coeruleiviridis* Macquart and *Phormia regina* (Meigen) were found in both spring and summer and *L. sericata* (Meigen) only in summer (Introna et al. 1991). Whereas in Mississippi, *P. regina* was found throughout the year (Goddard and Lago 1985). Seasonality of some calliphorid species is linked to geography in that some species are considered to be cold weather species in the southern US but are found throughout the insect season further north (Anderson and VanLaerhoven 1996; Dillon 1997). Later colonizing species are also impacted by season with uni-voltine species (those having one generation per year) showing very specific seasonality.

Adults of certain carrion species will diapause over the cold season, or aestivate over summer. This alters the patterns of taxon succession (Archer 2003), and may allow some less competitive groups that are better adapted to cold conditions in particular to gain a larger share of the carrion resource in winter (e.g. Erzinclioglu 1980). However, the rate of carrion mass loss is also slowed during the cool periods where calliphorid and sarcophagid larval activity is inhibited. In a reciprocal fashion, the speed and nature of decomposition itself may also be affected by insect succession activity, which is likely to be one factor underpinning the differences that occur in seasonal succession patterns between years (Archer 2003).

Many carrion insects have very specific and localized habitat preferences. Many species are synanthropic (i.e. live near and benefit from an association with humans) and hence found more commonly in urban than rural habitats. Some species are considered more common in rural areas, such as *C. vomitoria* (L.) and *L. illustris* while others are considered more common in urban areas, such as *C. vicina* and *L. sericata* (Anderson 2009). In an interesting study in Britain, baited trap collection from central London into the rural surroundings identified three distinct habitats based on fly species (Hwang and Turner 2005). Some species prefer sun exposed open areas, with others selectively choosing shaded or forested regions. This can again vary with geography. Although applying more specifically to human cadavers, whether a body is found inside a residence (Anderson 2011) or a vehicle (Anderson 2009; Voss et al. 2008) or is buried (VanLaerhoven and Anderson 1999) hanged (Shalaby et al. 2000) burned (Anderson 2005) or clothed (Kelly et al. 2009) will impact not only timing of arrival but also which species will colonize.

Carcass type and size can also impact colonization and succession. Most carrion-associated invertebrates have no preference for a species or type of carrion, yet some factors can affect colonization. Carcass size can impact insect succession and decomposition with some species not colonizing carcasses below a certain size because they cannot support the larval life cycle (Denno and Cothran 1976). A cheese fly (Piophilidae), *Thyreophora cynophila* (Panzer), prefers large carcasses and was considered to have been extinct in central Europe since the 1840s, allegedly due to the historical loss of wild megafauna and changes in livestock practices and increasingly sanitary carcass disposal, which result in a consequent dearth of large animal carcasses (Fontaine et al. 2007). Six specimens were recently re-discovered in Spain (Martin-Vega et al. 2010) and their renewed presence is thought to be due to the recent, somewhat controversial (Fielding et al. 2014), trend of maintaining supplementary feeding stations in an effort to protect and conserve endangered scavenger bird populations, which also depended previously on large wild or domestic animal carcasses (Martín-Vega and Baz 2011). Research from Norway suggests that the presence of large carcasses is important for habitat restoration and impacts species diversity and richness in European boreal forest ecosystems (Melis et al. 2004). At the other end of the spectrum, there are species which actively prefer smaller carcasses as they are easier to bury, for example, *Nicrophorus* sp. (burying beetles), which prefer small rodent carcasses (Meierhofer et al. 1999). In addition, the presence, absence and type of fur or pelt may also affect initial colonization, with evidence suggesting that gravid flies are attracted to darker colors which may be indicative of a common carcass pelt (Brodie et al. 2014). Vertebrate scavenging can facilitate initial colonization by insects as it exposes more tissue (Dillon and Anderson 1995).

A number of biotic factors can strongly affect the species composition and succession pattern of invertebrates at carrion, with the most notable being the various kinds of interactions among species, such as competition, predation and parasitism. A detailed discussion of some of the theory underpinning these biotic factors is given in the chapter "Carrion Decomposition". Competition among invertebrate scavengers at a carcass can be intense, and the timing of arrival of an individual and its competitive ability can determine if it will dominate its carcass resource. Aggregation and coexistence among carrion flies has been used as a model system for testing theory about the division of finite resources among organisms (Hanski 1987a; Ives 1991; Woodcock et al. 2002). This theory has been used to understand and predict the abundance and diversity of organisms present at carrion due to changes in inter- and intra-specific competition (Ives 1991; Kouki and Hanski 1995; Woodcock et al. 2002). In particular, stochastic colonization of carcasses and priority effects (Brundage et al. 2014) can play an important role in determining which species will dominate, and this in turn can affect rates of decomposition and the composition of the invertebrate scavenger community. A number of invertebrate species that frequent carcasses are not there to feed on the carrion. Rather, they are predators, phoronts or parasitoids of the carrion feeders, and of each other. Knowledge of the interactions among invertebrates at carrion is still poorly known (Hanski 1987b) although their role in the decomposition process is reasonably well understood (Braack 1987). A simplified trophic web is given in the chapter "Carrion Decomposition" (Fig. 1) and shows the many links between functional groups, including predators and parasitoids. These linkages illustrate the potential for many biotic factors to affect the composition of invertebrate scavenger communities.

Carrion in Aquatic Environs

In freshwater systems, aquatic organisms have evolved to tolerate a range of physical and chemical attributes. Therefore, the decomposition process is mediated through biological mechanisms such as microbial and macro invertebrate interactions, however certain physical-chemical parameters either directly or indirectly influence the pathway of decomposition. These include temperature, gas exchange, hydrology, nutrients, pH, and current/depth (Merritt and Wallace 2009; Wallace et al. 2015).

Temperature may be the most important environmental factor that influences carrion decomposition. The variation in air temperature can influence temperatures in streams and rivers and occurs spatially and temporally e.g. via geographical differences with changes in latitude, altitude, season and weather (Smith and Lavis 1974). In lentic systems, such as ponds and lakes, the size and shape can affect thermal regimes (Hutchinson 1957). Temperature influences the amount of dissolved oxygen in freshwater systems and this relationship has shaped the life histories of aquatic organisms instrumental in all stages of carcass decomposition (Burkepile et al. 2006; Sweeney and Vannote 1978; Ward and Stanford 1982). The rate of decomposition of organic matter such as carrion in both lotic and lentic systems tends to increase as a function of temperature regardless of seasonal fluctuations in temperate water bodies (Wallace et al. 2015). Higher temperatures increase aquatic insect growth and feeding rates, altering both the quality and quantity of organic matter associated with carrion (Cummins and Klug 1979). Generally, cold water temperatures whether freshwater or marine, slow decomposition rates. This especially influences microbial breakdown (Merritt and Wallace 2009).

Gas exchange, specifically, dissolved oxygen levels or the amount of oxygen in solution are intimately tied to water temperature and fluctuate daily (Addy and Green 1997). That is, oxygen solubility, or dissolved oxygen, is negatively correlated with water temperature which itself varies with current speed and turbulence (Merritt and Wallace 2009). Not only are aquatic insects tied to respiring oxygen in solution, but their role in decomposition requires oxygen. Therefore, those insects that play a significant role in the processing of carrion involved with the decomposition process such as shredder types e.g. stoneflies (**Plecoptera**: Nemouridae and Pteronarycidae) and caddisflies (**Trichoptera**: Limnephilidae) may be excluded from such activities in low oxygen environments (Merritt and Wallace 2009). Whereas, midges (**Diptera**: Chironomidae) are able to survive in oxygen depleted habitats such as ponds and swamps and may be more abundant on corpses in these environs (Erikson et al. 1996). Hence, depending on the oxygen concentration in a specific aquatic habitat, the faunal community available to colonize carrion may be quite different (Hobischak 1997). Therefore, differences in water quality as exhibited in dissolved oxygen levels may influence the diversity and abundance of aquatic invertebrates colonizing a carcass and alter the rate of decomposition (Merritt and Wallace 2009).

Physical aspects of aquatic systems e.g. flow regimes or currents in lotic (moving waters) systems such as streams are characterized temporally by monthly water discharge, frequency of high/low pulse events, duration of meteorological or anthropogenic inputs, whether events are predictable and the degree of flashiness in events (Poff et al. 1997). In lentic systems such as ponds or lakes, there are different hydrological parameters governing their influence on decomposition such as the length of the hydroperiod or period of inundation, seasonal precipitation patterns, presence or absence of inlets/outlets and water management (Stevenson and Childers 2004). The fluxes within normal hydrological processes allow for nutrient exchange and movement of aquatic organisms between systems (Osmond et al. 1995) but changes in these fluxes can impact species composition and ultimately food chain support. This may alter the ecological integrity and impact the rate of carrion decomposition in these systems (Sakaris 2013).

There are several other water chemistry parameters that may alter decomposition rates of carrion such as pH and salinity (Wallace et al. 2015). Specifically with pH, acid stress in aquatic systems may impact the diversity of organisms that initiate decomposition, such as the primary and secondary decomposers for example bacteria and invertebrates thus influencing the rate of carrion decomposition (Wallace et al. 2015). The limitation of key players such as bacteria and fungi can impact aquatic organisms such as amphipods (**Amphipoda**: Gammaridae) that have been demonstrated to be integral players in the fragmentation of vertebrate carrion in streams (Hobischak and Anderson 2002; Merritt and Wallace 2009).

Oxygen can have a major impact on colonization in the marine environment often driving decomposition. In well oxygenated waters pig carcasses have been scavenged and skeletonized within days at depths of 300 m or less (Anderson and Bell 2016) whereas in hypoxic waters, scavenging is much slower with only low-oxygen adapted species remaining at carrion (Anderson and Bell 2014). Oxygen levels at time of carcass deposition impacts decomposition. Carcasses deposited when oxygen is present, but at levels low enough to cause physiological stress, still attract large decapod crustaceans which feed and remain at a carcass despite hypoxic conditions. However, carcasses deposited when oxygen is extremely low attract very few fauna and these carcasses remain virtually intact until oxygen levels rise seasonally (Anderson and Bell 2014).

Conclusions and Future Perspectives

Invertebrates dominate most of the earth's terrestrial and aquatic biomes, and have evolved to use a vast array of resources. A wide range of invertebrate taxa have also been recorded at animal carcasses, where they may prey on or parasitize other carrion-invertebrates. Detailed observations from many different parts of the world have shown a clear succession of species, especially in terrestrial environments. An ever increasing number of case studies have also shown the important influence of various abiotic and biotic factors that influence invertebrate succession and food web dynamics at carcasses. There are fundamental differences in the ecology of invertebrate succession on carrion between terrestrial, freshwater and marine environments. For example, terrestrial carrion ecosystems include more specialized taxa than aquatic ones. However, these environments all have in common the central concept of carrion as a scarce and nutrient rich resource. Further research will help to elucidate these complex themes.

References

Addy K, Green L (1997) Oxygen and temperature. Natural resources facts. University of Rhode Island Cooperative Extension Fact Sheet No. 96-3

Anderson GS (2000) Minimum and maximum developmental rates of some forensically significant Calliphoridae (Diptera). J Forensic Sci 45(4):824–832

Anderson GS (2005) Effects of arson on forensic entomology evidence. Can Soc Forensic Sci J 38(2):49–67

Anderson GS (2009) Factors that influence insect succession on carrion. In: Byrd J, Castner E (eds) Forensic entomology: the utility of arthropods in legal investigations, 2nd edn. CRC Press, Boca Raton, pp 201–250

Anderson GS (2010) Decomposition and invertebrate colonization of cadavers in coastal marine environments. In: Amendt J, Campobasso CP, Grassberger M, Goff ML (eds) Current concepts in forensic entomology. Springer, Berlin

Anderson GS (2011) Comparison of decomposition rates and faunal colonization of carrion in indoor and outdoor environments. J Forensic Sci 56(1):136–142. https://doi.org/10.1111/j.1556-4029.2010.01539.x

Anderson GS, Bell LS (2014) Deep coastal marine taphonomy: investigation into carcass decomposition in the Saanich Inlet, British Columbia using a baited camera. PLoS One 9(10):e110710. https://doi.org/10.1371/journal.pone.0110710

Anderson GS, Bell LS (2016) Impact of marine submergence and season on faunal colonization and decomposition of pig carcasses in the Salish Sea. PLoS ONE 11(3):e0149107. https://doi.org/10.1371/journal.pone.0149107

Anderson GS, Hobischak NR (2002) Determination of time of death for humans discovered in saltwater using aquatic organism succession and decomposition rates. Canadian Police Research Centre, Technical Report., Ottawa, ON

Anderson GS, Hobischak NR (2004) Decomposition of carrion in the marine environment in British Columbia, Canada. Int J Legal Med 118(4):206–209

Anderson GS, VanLaerhoven SL (1996) Initial studies on insect succession on carrion in southwestern British Columbia. J Forensic Sci 41(4):617–625

Angioy AM, Stensmyr MC, Urru I, Puliafito M, Collu I, Hansson BS (2004) Function of the heater: the dead horse arum revisited. Proc Biol Sci 271(Suppl 3):S13–S15. https://doi.org/10.1098/rsbl.2003.0111

Anton E, Niederegger S, Beutel RG (2011) Beetles and flies collected on pig carrion in an experimental setting in Thuringia and their forensic implications. Med Vet Entomol 25(4):353–364

Archer MS (2000) Natural history observations of the native carrion beetle, *Ptomaphila lacrymosa* Schreibers (Coleoptera: Silphidae). Proc Royal Soc Vic 112:133–136

Archer MS (2002) The ecology of invertebrate associations with vertebrate carrion in Victoria, with reference to forensic entomology. University of Melbourne, Melbourne

Archer MS (2003) Annual variation in arrival and departure times of carrion insects at carcasses: implications for succession studies in forensic entomology. Austral J Zool 51(6):569–576. https://doi.org/10.1071/zo03053

Archer MS (2004) Rainfall and temperature effects on the decomposition rate of exposed neonatal remains. Sci Justice 44(1):35–41

Archer MS (2014) Comparative analysis of insect succession data from Victoria (Australia) using summary statistics *vs.* preceding mean ambient temperature models. J Forensic Sci 59:404–412

Barrios M, Wolff M (2011) Initial study of arthropods succession and pig carrion decomposition in two freshwater ecosystems in the Colombian Andes. Forensic Sci Int 212(1-3):164–172

Barton PS (2015) The role of carrion in ecosystems. In: Benbow ME, Tomberlin JK, Tarone AM (eds) Carrion ecology, evolution, and their applications. CRC Press, Boca Raton, pp 273–290. https://doi.org/10.1201/b18819-16

Barton PS, Cunningham SA, Macdonald BC, McIntyre S, Lindenmayer DB, Manning AD (2013a) Species traits predict assemblage dynamics at ephemeral resource patches created by carrion. PLoS One 8(1):e53961

Barton PS, Cunningham SA, Lindenmayer DB, Manning AD (2013b) The role of carrion in maintaining biodiversity and ecological processes in terrestrial ecosystems. Oecologia 171: 761–772

Barton PS, Cunningham SA, Manning AD, Gibb H, Lindenmayer DB, Didham RK (2013c) The spatial scaling of beta diversity. Glob Ecol Biogeogr 22(6):639–647

Barton PS, Weaver HJ, Manning AD (2014) Contrasting diversity dynamics of phoretic mites and beetles associated with vertebrate carrion. Exp Appl Acarol 63:1–13

Baz A, Cifrián B, Martín-Vega D, Baena M (2010) Phytophagous insects captured in carrion-baited traps in central Spain. Bull Insect 63:21–30

Blackith RE, Blackith RM (1990) Insect infestations of small corpses. J Nat Hist 24:699–709

Boggs CL, Dau B (2004) Resource specialization in puddling Lepidoptera. Environ Entomol 33(4):1020–1024

Bornemissza GF (1957) An analysis of arthropod succession in carrion and the effect of its decomposition on the soil fauna. Austral J Zool 5:1–12

Boyle S, Galloway A, Mason RT (1997) Human aquatic taphonomy in the Monterey Bay area. In: Haglund WD, Sorg MH (eds) Forensic taphonomy. The postmortem fate of human remains. CRC Press, Boca Raton, pp 605–613

Braack LEO (1987) Community dynamics of carrion-attendant arthropods in tropical African woodland. Oecologia 72(3):402–409

Braig HR, Perotti MA (2009) Carcasses and mites. Exp Appl Acarol 49(1-2):45–84

Britton JC, Morton B (1994) Marine carrion and scavengers. Oceanogr Mar Biol Ann Rev 32:369–434

Brodie B, Gries R, Martins A, VanLaerhoven S, Gries G (2014) Bimodal cue complex signifies suitable oviposition sites to gravid females of the common green bottle fly. Entomol Exp Appl 153(2):114–127. https://doi.org/10.1111/eea.12238

Brundage A, Benbow ME, Tomberlin JK (2014) Priority effects on the life-history traits of two carrion blow fly (Diptera, Calliphoridae) species. Ecol Entomol 39:539–547

Bucheli SR, Bytheway JA, Gangitano DA (2010) Necrophagous caterpillars provide human mtDNA evidence. J Forensic Sci 55(4):1130–1132. https://doi.org/10.1111/j.1556-4029.2010.01379.x

Burkepile DE, Parker JD, Woodson CB, Mills HJ, Kubanek J, Sobecky PA, Hay ME (2006) Chemically mediated competition between microbes and animals: microbes as consumers in food webs. Ecology 87:2821–2831

Byrd JH, Castner JL (2001) Insects of forensic importance. In: Byrd JH, Castner JL (eds) Forensic entomology: the utility of arthoprods in legal investigations. CRC Press, Boca Raton, pp 43–79. https://doi.org/10.1201/9781420036947.ch2

Carter DO, Yellowlees D, Tibbett M (2007) Cadaver decomposition in terrestrial ecosystems. Naturwissenschaften 94(1):12–24

Chaloner DT, Wipfli MS (2002) Influence of decomposing Pacific salmon carcasses on macro-invertebrate growth and standing stock in southeastern Alaska streams. N Am Benthol Soc 21(3):430–442

Chapman RF, Sankey JHP (1955) The larger invertebrate fauna of three rabbit carcasses. J Anim Ecol 24(2):395–402

Charabidze D, Bourel B, Gosset D (2011) Larval-mass effect: characterisation of heat emission by necrophageous blowflies (Diptera: Calliphoridae) larval aggregates. Forensic Sci Int 211(1-3):61–66

Chen C-P, Denlinger DL, Lee RE (1991) Seasonal VARIATION in generation time, diapause and cold hardiness in a central Ohio population of the flesh fly *Sarcophaga bullata*. Ecol Entomol 16(2):155–162

Chin HC, Marwi MA, Salleh AFM, Jeffery J, Omar B (2007) A preliminary study of insect succession on a pig carcass in a palm oil plantation in Malaysia. Trop Biomed 24(2):23–27

Colless DH, McAlpine DK (1991) Diptera. In: Naumann ID (ed) The insects of Australia, vol 2, 2nd edn. Melbourne University Press, Melbourne, pp 717–786

Colombini I, Chelazzi L (2003) Influence of marine allochthonous input on sandy beach communities. Oceanogr Mar Biol 41:115–159

Cummins KW (1974) Structure and function of stream ecosystems. Bioscience 24:631–641

Cummins KW, Klug MJ (1979) Feeding ecology of stream invertebrates. Annu Rev Ecol Syst 10:147–172
Dahlgren TG, Wiklund H, Kallstrom B, Lundalv T, Smith CR, Glover AG (2006) A shallow-water whale-fall experiment in the north Atlantic. Cah Biol Mar 47:385–389
Davies L (1999) Seasonal and spatial changes in blowfly production from small and large carcasses at Durham in lowland northeast England. Med Vet Entomol 13(3):245–251
Denno RF, Cothran WR (1976) Competitive interaction and ecological strategies of sarcophagid and calliphorid flies inhabiting rabbit carrion. Ann Entomol Soc Am 69:109–113
Deonier CC (1940) Carcass temperatures and their relation to winter blowfly activity in the Southwest. J Econ Entomol 33(1):166–170
Dillon LC (1997) Insect succession on carrion in three biogeoclimatic zones in British Columbia. M.Sc., Simon Fraser University, Burnaby, BC
Dillon LC, Anderson GS (1995) Forensic entomology: The use of insects in death investigations to determine elapsed time since death. Canadian Police Research Centre, TR-05-95, Ottawa, Ontario
Early M, Goff ML (1986) Arthropod succession patterns in exposed carrion on the island of O'ahu, Hawai'i. J Med Entomol 23:520–531
Easton AM (1966) The Coleoptera of a dead fox (*Vulpes vulpes* L.); including two species new to Britain. Entomol Mon Mag 102:205–210
Eggert A-K, Mulller JK (1997) Biparental care and social evolution in burying beetles: lesson from the larder. In: Choe JC, Crespi BC (eds) The evolution of social behaviour in insects and arachnids. Cambridge University Press, Cambridge, pp 216–236
Ellis RJ (1970) *Alloperla* stonefly nymphs: predators or scavengers on salmon eggs and alevins? Trans Am Fish Soc 4:677–683
Erikson CH, Resh VH, Lamberti BA (1996) Aquatic insect respiration. In: Merritt RW, Cummins KW (eds) An introduction to the aquatic insects of North America. Kendall/Hunt Publishing Co., Duboque, pp 29–40
Erzinclioglu YZ (1980) On the role of *Trichocera* larvae (Diptera: trichoceridae) in the decomposition of carrion in winter. Naturalist 105:133–134
Fenoglio ST, Agosta P, Cucco M (2005) Mass loss and macroinvertebrate colonisation of fish carcasses in riffles and pools of a NW Italian stream. Hydrobiologia 532:111–122
Fenoglio S, Merritt RW, Cummins KW (2014) Why do no specialized necrophagous species exist among aquatic insects? Freshwater Sci 33(3):711–715. https://doi.org/10.1086/677038
Fielding D, Newey S, van der Wal R, Irvine RJ (2014) Carcass provisioning to support scavengers: evaluating a controversial nature conservation practice. Ambio 43(6):810–819. https://doi.org/10.1007/s13280-013-0469-4
Fontaine B, Bouchet P, Vanachterberg K, Alonsozarazaga M, Araujo R, Asche M, Aspock U, Audisio P, Aukema B, Bailly N (2007) The European union's 2010 target: Putting rare species in focus. Biol Conserv 139(1-2):167–185. https://doi.org/10.1016/j.biocon.2007.06.012
Frederickx C, Dekeirsschieter J, Verheggen FJ, Haubruge E (2013) Host-habitat location by the parasitoid, nasonia vitripennis walker (Hymenoptera: Pteromalidae). J Forensic Sci 59:242–249. https://doi.org/10.1111/1556-4029.12267
Fujikura K, Fujiwara Y, Kawato M (2006) A new species of *Osedax* (Annelida : Siboglinidae) associated with whale carcasses off Kyushu, Japan. Zool Sci 23(8):733–740
Fuller ME (1934a) The insect inhabitants of carrion : a study in animal ecology. CSIRO Bull 82:5–62
Fuller ME (1934b) The early stages of *Sciadocera rufomaculata* White (Dipt. Phoridae). Proc Linnean Soc NSW 59:9–15
Fuller ME (1935) Notes on Australasian Anisopodidae (Diptera). Proc Linnaean Soc NSW 60:291–302
Gaston KJ (2000) Global patterns in biodiversity. Nature 405(6783):220–227
Gaston KJ, Lawton JH (1988) Patterns in the distribution and abundance of insect populations. Nature 331(6158):709–712

George KA, Archer MS, Toop T (2013) Abiotic environmental factors influencing blowfly colonisation patterns in the field. Forensic Sci Int 229(1-3):100–107. https://doi.org/10.1016/j.forsciint.2013.03.033

Giller PS, Malmqvist B (1998) The biology of streams and rivers. Oxford University Press, Oxford

Glover AG, Kallstrom B, Smith CR, Dahlgren TG (2005) World-wide whale worms? A new species of Osedax from the shallow north Atlantic. Proc R Soc B 272(1581):2587–2592

Glover AG, Wiklund H, Taboada S, Avila C, Cristobo J, Smith CR, Kemp KM, Jamieson AJ, Dahlgren TG (2013) Bone-eating worms from the Antarctic: the contrasting fate of whale and wood remains on the Southern Ocean seafloor. Proc Biol Sci 280(1768):20131390. https://doi.org/10.1098/rspb.2013.1390

Goddard J, Lago PK (1985) Notes on blowfly (Diptera: Calliphoridae) succession on carrion in Northern Mississippi. J Entomol Sci 20:312–317

Grassberger M, Frank C (2003) Temperature-related development of the parasitoid wasp *Nasonia vitripennis* as forensic indicator. Med Vet Entomol 17(3):257–262

Haefner JN, Wallace JR, Merritt RW (2004) Pig decomposition in lotic aquatic systems: the potential use of algal growth in establishing a postmortem submersion interval (PMSI). J Forensic Sci 49(2):330–336

Haglund WD (1993) Disappearance of soft tissue and the disarticulation of human remains from aqueous environments. J Forensic Sci 38:806–815

Hanski I (1987a) Carrion fly community dynamics: patchiness, seasonality and coexistence. Ecol Entomol 12(3):257–266

Hanski I (1987b) Nutritional ecology of dung- and carrion-feeding insects. In: Slanky F, Rodriguez J (eds) Nutritional ecology of insects, mites, and spiders. Wiley, New York, pp 837–884

Haskell NH, McShaffrey DG, Hawley DA, Williams RE, Pless JE (1989) Use of aquatic insects in determining submersion interval. J Forensic Sci 34:622–632

Hinton HE (1963) A monograph of the beetles associated with stored products, vol 1. Johnston Reprint Corporation, London

Hobischak NR (1997) Freshwater invertebrate succession and decompositional studies on carrion in British Columbia. M.P.M., Simon Fraser University, Burnaby

Hobischak NR, Anderson GS (2002) Time of submergence using aquatic invertebrate succession and decompositional changes. J Forensic Sci 47(1):142–151

Hobson RP (1932) Studies on the nutrition of blow-fly larvae. J Exp Biol 9:359–365

Hutchinson GE (1957) A treatise on limnology, vol I. Wiley, New York

Hwang C, Turner BD (2005) Spatial and temporal variability of necrophagous Diptera from urban to rural areas. Med Vet Entomol 19(4):379–391

Introna FJ, Suman TW, Smialek JE (1991) Sarcosaprophagous fly activity in Maryland. J Forensic Sci 36(1):238–243

Ives AR (1991) Aggregation and coexistence in a carrion fly community. Ecol Monogr 61:75–94

Johnson AP, Wighton SJ, Wallman JF (2014) Tracking movement and temperature selection of larvae of two forensically important blow fly species within a "maggot mass". J Forensic Sci 59(6):1586–1591. https://doi.org/10.1111/1556-4029.12472

Jones WJ, Johnson SB, Rouse GW, Vrijenhoek RC (2008) Marine worms (genus *Osedax)* colonize cow bones. Proc Biol Sci 275(1633):387–391

Kalinová B, Podskalská H, Růžička J, Hoskovec M (2009) Irresistable bouquet of death - how are burying beetles (Coleoptera: Silphidae: Nicrophorus) attracted by carcasses? Naturwissenschaften 96(8):889–899

Kavazos CRJ, Wallman JF (2012) Community composition of carrion-breeding blowflies (Diptera: Calliphoridae) along an urban gradient in south-eastern Australia. Landsc Urban Plan 106(2):183–190

Keiper JB, Chapman EG, Foote BA (1997) Midge larvae (Diptera: Chironomidae) as indicators of postmortem submersion interval of carcasses in a woodland stream: a preliminary report. J Forensic Sci 42(6):1074–1079

Kelly JA, van der Linde TC, Anderson GS (2009) The influence of clothing and wrapping on carcass decomposition and arthropod succession during the warmer seasons in Central South Africa. J Forensic Sci 54(5):1105–1112

Kouki J, Hanski I (1995) Population aggregation facilitates coexistence of many competing carrion fly species. Oikos 72:223–227

Magni PA, Venn C, Aquila I, Pepe F, Ricci P, Di Nunzio C, Ausania F, Dadour IR (2015) Evaluation of the floating time of a corpse found in a marine environment using the barnacle *Lepas anatifera* L. (Crustacea: Cirripedia: Pedunculata). Forensic Sci Int 247:e6–e10. https://doi.org/10.1016/j.forsciint.2014.11.016

Manlove JD, Disney RH (2008) The use of *Megaselia abdita* (Diptera: Phoridae) in forensic entomology. Forensic Sci Int 175(1):83–84. https://doi.org/10.1016/j.forsciint.2007.08.001

Martín-Vega D, Baz A (2011) Could the 'vulture restaurants' be a lifeboat for the recently rediscovered bone-skippers (Diptera: Piophilidae)? J Insect Conserv 15(5):747–753. https://doi.org/10.1007/s10841-011-9429-0

Martin-Vega D, Baz A, Michelsen V (2010) Back from the dead: *Thyreophora cynophila* (Panzer, 1798) (Diptera: Piophilidae) 'globally extinct' fugitive in Spain. Syst Entomol 35(4):607–613. https://doi.org/10.1111/j.1365-3113.2010.00541.x

Matthews EG (1982) A guide to the genera of beetles of South Australia. Part 2 polyphaga: staphylinoidea and hydrophiloidea. South Australia Museum, Adelaide

Matthews EG (1984) A guide to the genera of beetles of South Australia. Part 3: polyphaga: eucinetoidea, dascilloidea and scarabaeoidea. South Australia Museum, Adelaide

Matuszewski S, Bajerlein D, Konwerski S, Szpila K (2008) An initial study of insect succession and carrion decomposition in various forest habitats of Central Europe. Forensic Sci Int 180(2-3):61–69. https://doi.org/10.1016/j.forsciint.2008.06.015

Matuszewski S, Bajerlein D, Konwerski S, Szpila K (2010) Insect succession and carrion decomposition in selected forests of Central Europe. Part 1: pattern and rate of decomposition. Forensic Sci Int 194(1-3):85–93

Meierhofer I, Schwarz HH, Muller JK (1999) Seasonal variation in parental care, offspring development and reporoductive succees in the burying beetle, *Nicrophorus vespillo*. Ecol Entomol 24:73–79

Meiklejohn KA (2012) Taxonomy and systematics of the Australian Sarcophaga s.l. (Dipterea: Sarcophagidae). University of Wollongong, Wollongong

Melis C, Teurlings I, Linnell JC, Andersen R, Bordoni A (2004) Influence of a deer carcass on Coleopteran diversity in a Scandinavian boreal forest: a preliminary study. Eur J Wildl Res 50:146–149. https://doi.org/10.1007/s10344-004-0051-2

Merritt RW, Wallace JR (2009) The role of aquatic insects in forensic investigations. In: Byrd JH, Castner JL (eds) Forensic entomology: the utility of arthropods in legal investigations. CRC Press, Boca Raton, pp 272–319. https://doi.org/10.1201/NOE0849392153.ch7

Michaud JP, Schoenly KG, Moreau G (2015) Rewriting ecological succession history: did carrion ecologists get there first? Q Rev Biol 90(1):45–66

Minshall GW, Hitchcock E, Barnes JR (1991) Decomposition of rainbow trout (*Oncorhynchus mykiss*) carcasses in aforest stream ecosystem inhabited only by non-anadromous fish populations. Can J Fish Aquat Sci 48:191–195

Moore BP (1955) Notes on carrion Coleoptera in the Oxford district. Entomol Mon Mag 43:45–51

Nicola SJ (1968) Scavenging by *Alloperla* (Plecoptera: Chloroperlidae) nymphs on dead pink salmon and chum salmon embryos. Can J Zool 46:787–796

Norris KR (1965) The bionomics of blowflies. Annu Rev Entomol 10:47–68

Norris KR (1966) Daily patterns of flight activity of blowflies (Calliphoridae: Diptera) in the Canberra district as indicated by trap catches. Aust J Zool 14:835–853

O'Connor BM (2009) Astigmatid mites (Acari: Sarcoptiformes) of forensic interest. Exp Appl Acarol 49(1-2):125–133

O'Flynn MA, Moorehouse DE (1979) Species of *Chrysomya* as primary flies in carrion. J Aust Entomol Soc 19:31–32

Osmond DL, Line DE, Gale JA, Gannon RW, Knott CB, Bartenhagen KA, Turner MH et al (1995) Watersheds: water, soil and hydro-environmental decision support system
Payne JA (1965) A summer carrion study of the baby pig *Sus Scrofa* Linnaeus. Ecology 46: 592–602
Payne JA, King EW (1970) Coleoptera associated with pig carrion. Entomol Mon Mag 105:224–232
Payne JA, King EW (1972) Insect succession and decomposition of pig carcasses in water. J Georgia Entomol Soc 73:153–162
Peck SB (2001) Review of the carrion beetles of Australia and New Guinea (Coleoptera: Silphidae). Aust J Entomol 40:93–101
Perotti MA, Braig HR (2009) Phoretic mites associated with animal and human decomposition. Exp Appl Acarol 49(1-2):85–124
Pinero FS (1997) Analysis of spatial and seasonal variability of carrion beetle (Coleoptera) assemblages in two arid zones of Spain. Environ Entomol 26(4):805–814
Pitkin BR (1989) Piophilidae. In: Evenhuis NL (ed) Catalog of the Diptera of the Australasian and Oceanian regions. Bishop Museum Press, Honolulu, p 533
Poff NL, Allan JD, Bain MB, Karr JR, Prestegaard KL, Richter BD, Sparks RE, Stromberg JC (1997) The natural flow regime: a paradigm for river conservation and restoration. Bioscience 47:769–784
Pont AC (1973) Studies on Australian Muscidae (Diptera). IV a revision of the subfamilies Muscinae and Stomoxyinae. Aust J Zool Suppl Ser 21:129–296
Pont AC (1977) A revision of the Australian Fanniidae (Diptera: Calyptrata). Aust J Zool Suppl Ser 51:1–60
Putman RJ (1977) Dynamics of the blowfly, *Calliphora erythrocephala*, within carrion. J Anim Ecol 46(3):853–866
Putman RJ (1983) Carrion and dung: the decomposition of animal wastes, The institute of biology's studies in biology, vol 156. Edward Arnold, London
Richards CS, Williams KA, Villet MH (2009a) Predicting geographic distribution of seven blowfly species (Diptera: Calliphoridae) in South Africa. Afr Entomol 17(2):170–182
Richards CS, Price BW, Villet M (2009b) Thermal ecophysiology of seven carrion-feeding blowflies in Southern Africa. Entomol Exp Appl 131(1):11–19
Richter S (1993) Phoretic association between the dauerjuveniles of *Rhabditis stammeri* (Rhabditidae) and life history stages of the burying beetle *Nicrophorus vespilloides*. Nematologica 39(3):346–355
Rivers DB, Thompson C, Brogan R (2011) Physiological trade-offs of forming maggot masses by necrophagous flies on vertebrate carrion. Bull Entomol Res 101(5):599–611. https://doi.org/10.1017/S0007485311000241
Rochefort S, Giroux M, Savage J, Wheeler TA (2015) Key to forensically important piophilidae (Diptera) in the Nearctic region. Can J Arthropod Ident. https://doi.org/10.3752/cjai.2015.27
Rouse GW, Goffredi SK, Vrijenhoek RC (2004) Osedax: bone-eating marine worms with dwarf males. Science 305(5684):668–671
Sakaris P (2013) A review of the effects of hydrologic alteration on fisheries and biodiversity and the management and conservation of natural resources in regulate driver systems. In: Bradley PM (ed) Environmental sciences: current perspectives in contaminant hydrology and water resources sustainability. Intech, London. https://doi.org/10.5772/55963
Santos WE, Carneiro LT, Alves ACF, Creao-Duarte AJ, Martins CF (2014) Stingless bees (Hymenoptera: Apidae: Meliponini) attracted to animal carcasses in the brazilian dry forest and implications for forensic entomology. Sociobiology 61(4):490–493
Schoenly K, Reid W (1987) Dynamics of heterotrophic succession in carrion-arthropod assemblages: discrete series or a continuum of change? Oecologia 73:191–202
Seastedt TR, Mameli L, Gridley K (1981) Arthropod use of invertebrate carrion. Am Midl Nat 105(1):124–129
Seevers CH, Herman LH (1978) A generic and tribal version of the North American Aleocharinae (Coleoptera: Staphylinidae). Fieldiana Zool 71:1–289

Shalaby OA, deCarvalho LML, Goff ML (2000) Comparison of patterns of decomposition in a hanging carcass and a carcass in contact with soil in a xerophytic habitat on the Island of Oahu, Hawaii. J Forensic Sci 45(6):1267–1273

Silveira OT, Esposito MC, dos Santos JN, Gemaque FE (2005) Social wasps and bees captured in carrion traps in a rainforest in Brazil. Entomol Sci 8(1):33–39

Simmons P (1925) The ham beetle, *Necrobia rufipes* De Geer. J Agric Res 30(9):845–863

Slone DH, Gruner SV (2007) Thermoregulation in larval aggregations of carrion-feeding blow flies (Diptera: Calliphoridae). J Med Entomol 44(3):516–523

Smith KGV (1986) A manual of forensic entomology. Trustees of The British Museum (Nat. Hist.)/Cornell University Press, London

Smith CR, Baco AR (2003) Ecology of whale falls at the deep-sea floor. Oceanogr Mar Biol Annu Rev 41:311–354

Smith K, Lavis ME (1974) Environmental influences on the temperature of a small upland stream. Oikos 26:228–236

Stevenson C, Childers DL (2004) Hydroperiod and seasonal effects on fish decomposition in an oligotrophic Everglades marsh. Wetlands 24(3):529–537

Sweeney BW, Vannote RL (1978) Size variation and the distribution of hemimetabolous aquatic insects: two thermal equilibrium hypotheses. Science 200:444–446

Szpila K, Madra A, Jarmusz M, Matuszewski S (2015) Flesh flies (Diptera: Sarcophagidae) colonising large carcasses in Central Europe. Parasitol Res 114(6):2341–2348. https://doi.org/10.1007/s00436-015-4431-1

Taboada S, Bas M, Leiva C, Garriga M, Sardá R, Avila C (2016) Life after death: shallow-water Mediterranean invertebrate communities associated with mammal bones. Mar Ecol 37(1):164–178. https://doi.org/10.1111/maec.12257

Tomberlin JK, Adler PH (1998) Seasonal colonization and decomposition of rat carrion in water and on land in an open field in South Carolina. J Med Entomol 35(5):704–709

Vance GM, Vandyk JK, Rowley WA (1995) Device for sampling aquatic insects associated with carrion in water. J Forensic Sci 40(3):479–482

VanLaerhoven SL, Anderson GS (1999) Insect succession on buried carrion in two biogeoclimatic zones of British Columbia. J Forensic Sci 44(1):32–43

Voss SC, Forbes SL, Dadour IR (2008) Decomposition and insect succession on cadavers inside a vehicle environment. Forensic Sci Med Pathol 4(1):22–32

Voss SC, Spafford H, Dadour IR (2009) Hymenopteran parasitoids of forensic importance: host associations, seasonality, and prevalence of parasitoids of carrion flies in Western Australia. J Med Entomol 46(5):1210–1219

Voss SC, Spafford H, Dadour I (2010) Temperature-dependent development of the parasitoid *Tachinaephagus zealandicus* on five forensically important carrion fly species. Med Vet Entomol 24:189–198

Wallace JR (2015) Aquatic vertebrate carrion decomposition. In: Benbow ME, Tomberlin JK, Tarone AM (eds) Carrion ecology, evolution, and their applications. CRC Press, Boca Raton, pp 247–272. https://doi.org/10.1201/b18819-15

Wallace JR, Merritt RW, Kimbirauskas RK, Benbow ME, McIntosh M (2008) Caddisflies assist with homicide case: determining a postmortem submersion interval using aquatic insects. J Forensic Sci 53(1):219–221

Wallace JR, Byrd JH, LeBlanc HN, Cervenka VJ (2015) North America. In: Tomberlin JK, Benbow ME (eds) Forensic entomology. Contemporary topics in entomology. CRC Press, Boca Raton, pp 187–202. https://doi.org/10.1201/b18156-23

Wallman JF, Leys R, Hogendoorn K (2005) Molecular systematics of Australian carrion-breeding blowflies (Diptera : Calliphoridae) based on mitochondrial DNA. Invertebr Syst 19(1):1–15. https://doi.org/10.1071/is04023

Ward JV, Stanford JA (1982) Thermal responses in the evolutionary ecology of aquatic insects. Annu Rev Entomol 27:97–117

Warren JA, Anderson GS (2013) Effect of fluctuating temperatures on the development of a forensically important blow fly, *Protophormia terraenovae* (Diptera: Calliphoridae). Environ Entomol 42(1):167–172. https://doi.org/10.1603/EN12123

Wipfi MS, Hudson J, Caouette J (1998) Influence of salmon carcasses on stream productivity: response of biofilm and benthic macroinvertebrates in southeastern Alaska, U.S.A. Can J Fish Aquat Sci 55:1503–1511

Woodcock BA, Watt AD, Leather SR (2002) Aggregation, habitat quality and coexistence: a case study on carrion fly communities in slug cadavers. J Anim Ecol 71(1):131–140

Zhu J, Chaudhury MF, Tangtrakulwanich K, Skoda SR (2013) Attractants of the secondary screwworm, *Cochliomyia macellaria* (F.) released from Rotten chicken liver. J Chem Ecol 39(11-12):1407–1414

Zwick P (1979) Contributions to the knowledge of Australian Cholevidae (Catoptidae auct.; Coleoptera). Aust J Zool Suppl Ser 70:1056

Vertebrate Scavenging Communities

Nuria Selva, Marcos Moleón, Esther Sebastián-González, Travis L. DeVault, Maria Martina Quaggiotto, David M. Bailey, Sergio A. Lambertucci, and Antoni Margalida

Contents

N. Selva (✉)
Institute of Nature Conservation, Polish Academy of Sciences, Kraków, Poland

M. Moleón
Departamento de Biología Aplicada, Universidad Miguel Hernández, Alicante, Spain

Departamento de Biología de la Conservación, Estación Biológica de Doñana (EBD-CSIC), Sevilla, Spain

Department of Zoology, University of Granada, Granada, Spain

E. Sebastián-González
Departamento of Applied Biology, Miguel Hernández University, Alicante, Spain

T. L. DeVault
U.S. Department of Agriculture, Animal and Plant Health Inspection Service, Wildlife Services, National Wildlife Research Center, Ohio Field Station, Sandusky, OH, USA
e-mail: Travis.L.DeVault@usda.gov

M. M. Quaggiotto · D. M. Bailey
Institute of Biodiversity, Animal Health and Comparative Medicine, College of Medical, Veterinary and Life Sciences, University of Glasgow, Glasgow, UK
e-mail: david.bailey@glasgow.ac.uk

S. A. Lambertucci
ECOTONO Lab, INIBIOMA (CONICET-Comahue National University), Bariloche, Argentina
e-mail: slambertucci@comahue-conicet.gob.ar

A. Margalida
Institute for Game and Wildlife Research IREC, CSIC-UCLM-JCCM, Ciudad Real, Spain

Division of Conservation Biology, University of Bern, Bern, Switzerland
e-mail: a.margalida@csic.es

P. P. Olea et al. (eds.), *Carrion Ecology and Management*, Wildlife Research Monographs 2, https://doi.org/10.1007/978-3-030-16501-7_4

Facultative *Versus* Obligate Scavengers

In 1877, Sanborn Tenney, a professor of natural history at Williams College in Massachusetts, USA, wrote about scavengers as those animals using or removing what others have ignored or discarded (Tenney 1877). Scavenging is likely to have been common since the very first origins of vertebrates. Many theropod dinosaurs, which dominated terrestrial ecosystems for more than 160 million years, likely gained an important part of their energetic requirements by feeding on carrion (Kane et al. 2016). In the Pleistocene, when mammalian megafauna was abundant (Martin 1989; Ruxton and Houston 2003), scavenging was common among carnivores, and even among early hominins, which were better scavengers than hunters (Blumenschine and Cavallo 1992; Moleón et al. 2014a).

Later on, with large carcasses becoming scarcer (Barnosky et al. 2004), the temporal patchiness of carrion availability might have inhibited evolution towards strict specialisation for scavenging in most vertebrates. As a result, obligate scavengers, i.e. species that rely entirely or near entirely on carrion as food resource, are rare. In the recent past, no reptiles or mammals have evolved into obligate scavengers (DeVault et al. 2003). Some benthic scavengers (e.g. hagfish: family Myxinidae) rely widely on carrion and may be considered obligate scavengers (Smith and Baco 2003; Beasley et al. 2015). However, among terrestrial vertebrates, only vultures (families Accipitridae and Cathartidae) have evolved into obligate scavengers (DeVault et al. 2003; Beasley et al. 2015; Moleón et al. 2014b).

Even so, facultative scavenging is fairly common and scavenging activity appears to be pervasive across the animal kingdom (DeVault et al. 2003; Wilson and Wolkovich 2011; Beasley et al. 2015; Mateo-Tomás et al. 2015). Facultative scavengers are animals that scavenge at variable rates but that can subsist on other food resources in the absence of carrion. All mammalian predators (e.g., raccoons *Procyon lotor*, wolves *Canis lupus*, bears -family Ursidae-, foxes *Vulpes* spp., hyenas *Hyaena* spp., lions *Panthera leo*), birds of prey (e.g., kites, eagles, buzzards-family Accipitridae) and corvids (e.g., ravens, crows-family Corvidae) are facultative scavengers (DeVault et al. 2003; Selva 2004; Wilson and Wolkovich 2011; Moleón et al. 2014b; Pereira et al. 2014). Some species, typically regarded as strict predators, also scavenge frequently, like the northern goshawk (*Accipiter gentilis*), the weasel (*Mustela nivalis*, Selva 2004) and the European otter (*Lutra lutra*, O'Sullivan et al. 1992). Some omnivores which are not strict predators (e.g. wild boar *Sus scrofa*) also consume carrion frequently (Selva et al. 2005). Woodpeckers and tits feed on carcasses during winter, and other passerines exploit resources provided by carcasses, like maggots or hairs to build nests (Moreno-Opo and Margalida 2013). An amazing spectrum of animals will consume carrion if the opportunity arises, including the hippopotamus (*Hippopotamus amphibius*), white-tailed deer (*Odocoileus virginianus*), and many snake species (DeVault and Krochmal 2002; Beasley et al. 2015). Carrion is by far the most important food for some reptiles (e.g. the lace monitor, *Varanus varius*, Guarino 2001).

Morphological, Physiological and Behavioral Adaptations to Scavenging

Terrestrial Scavengers

The adaptations shown by vultures includes morphological characters facilitating the dismembering of and access to the interior of large corpses, like long necks, strong beaks and scarce feather coverage on head and neck to prevent fouling (Kruuk 1967; Houston 1975, 1979). Vultures have excellent visual abilities, and some species like vultures from the genus *Cathartes* also possess a well-developed sense of smell to locate carrion (Houston 1986; Ruxton and Houston 2004). Their large body size allows vultures to cope with periods of food scarcity. In addition, vultures reduce the potential lethal effects of toxins and pathogens due to a number of physiological and immunological adaptations, including an acid digestive pH (Houston and Cooper 1975). Vultures have also developed foraging strategies to exploit randomly distributed resources, i.e. to explore large areas by gliding, and the transmission, active or passive, of information (Houston 1974; Buckley 1996; Ruxton and Houston 2004; Jackson et al. 2008). These adaptations make vultures very efficient scavengers compared to mammalian carnivores, which show a more limited mobility due to territoriality and the higher energetic costs of walking (Houston 1979; Ruxton and Houston 2004; Moleón et al. 2014b). Similar adaptations are shown by avian facultative scavengers. Corvids, and common ravens *Corvus corax*, in particular, actively recruit each other to carcasses, and share information about carrion location at their communal roosts (Heinrich 1988; Marzluff and Heinrich 1991; Heinrich et al. 1993; Marzluff et al. 1996; Wilmers et al. 2003a; Wright et al. 2003).

A scavenging lifestyle often involves watching and following others, both at the intra- and inter-specific levels. The transfer of social information within the scavenging guild is quite fluid (Moleón et al. 2014b). Avian scavengers may follow mammalian predators to feed on their kills. Vultures have been observed circling over large carnivores before hunting and being attracted to sounds of hyaenas (Attwell 1963; Kruuk 1967, 1972). Ravens associate with wolves (Stahler et al. 2002), polar bears *Ursus maritimus* (Gilchrist and Robertson 2000), and Siberian tigers *Panthera tigris altaica* (Matyushkin 1974) to feed on their kills. Ravens also react to gunshots and have learned to exploit gut piles left by hunters (White 2005). Mesopredators also follow large predators to feed on their kills, e.g. red *Vulpes vulpes* and arctic foxes *Vulpes lagopus* are reported to follow wolves and polar bears, respectively (Stirling and Archibald 1977; Jędrzejewska and Jędrzejewski 1998). Large scavenging birds, and particularly vultures and eagles, observe the activities of smaller birds, like crows or magpies, which are usually first to discover carcasses (König 1974; Prior and Weatherhead 1991; Vogeley 1999). Flocks of corvids are an unequivocal sign of carrion presence and often attract eagles and other raptors (Knight and Knight 1983; Selva 2004). Raptors often act as producers of information, while vultures are information scroungers (Kane et al. 2014). Griffon vultures *Gyps fulvus* use socially acquired information and they follow vultures that sink vertically to land to increase carcass location success (Cortés-Avizanda et al. 2014). Mammals also respond to aggregations of scavenging birds; hyenas and jackals react to alighting vultures by running towards the place where they have come down (Kruuk 1967).

Mammalian scavengers, *Cathartes* vultures and reptiles use the sense of smell to locate carcasses. In the case of small carrion items, chemical cues are the dominant stimuli used by scavengers to locate them (DeVault and Rhodes 2002). In snakes, olfaction also seems to be the principal sensory modality to detect carrion (DeVault and Krochmal 2002). They often consume carcasses of small animals like rodents, birds, fish, frogs and other snakes, which are inaccessible to avian and large mammalian scavenging competitors. Many snake species are attracted to decomposing odors and carrion (DeVault and Krochmal 2002). Scavenging is also widespread among crocodiles. They detect carrion, even concealed carcasses and/or at large distances, by chemical cues in the water and air, and promote olfaction by gular pumping (Weldon et al. 1990; Weldon and Ferguson 1993). Like vultures, crocodiles also possess a low stomach pH that inhibits bacteria from putrefying meat (Diefenbach 1975, in Platt et al. 2007) and can survive long periods without food.

Many major scavengers are social species. Vultures and corvids aggregate in flocks at carcasses and are commensal with other species. Some reptiles, like the Komodo dragon *Varanus komodoensis*, often feed communally at large carcasses (Bull et al. 2010). Within avian scavenger guilds, interspecific competition has been reduced by divergent eco-morphological characters and dominance hierarchies that lead to an ordered resource partitioning (Kruuk 1967; Wallace and Temple 1987; Houston 1988; Hertel 1994). The costs of adaptation to obligate scavenging are high, as evidenced by the small portion of terrestrial vertebrates relying exclusively on carrion (DeVault et al. 2003). Probably the most critical adaptations are related to detection (including the costs of locomotion) and detoxification of carrion (Shivik 2006). However, scavenging seems highly beneficial as an occasional energy acquisition strategy, given the fact that a large number of terrestrial vertebrates are facultative scavengers (DeVault et al. 2003; Wilson and Wolkovich 2011).

Fig. 1 Examples of scavengers in different ecosystems (from top to bottom): polar bears congregate at a whale carcass in the Arctic (photo by Alexander Gruzdev); wolves are often followed by raven flocks to feed on their kills (photo by Dan Hartman); the scavenger community of the African savanna is one of the richest (photo by Sarah C. Jones); condors and other vulture species rely often on livestock carcasses in Patagonian steppes (photo by Manuel de la Riva)

Aquatic Scavengers

There are numerous fish species which scavenge. Across a depth range of 800–5000 m, around 22% of the fish assemblage is thought to be scavengers and this proportion is relatively consistent across this range (Yeh and Drazen 2011). Scavenging requires a different set of sensory and swimming performance characteristics than are required for herbivorous and predatory fish. Food resources are less predictable, requiring some degree of endurance to allow fish to reach their next meal, but can be large (e.g. whale carcasses) when obtained (see chapter "Carrion Availability in Space and Time"). Carrion does not attempt to escape and so there is no requirement for burst swimming performance, though animals may need to compete for, or defend, carrion. Unlike on land, carrion may move with the current, but in general scavengers must detect this resource without the aid of prey movement as a signal. The characteristics of marine carrion shape the physiology, behavior and morphology of swimming scavengers. Scavenging fish have been studied in the deep sea for decades, in spite of the difficulty of working in this environment. Observations of scavenging fish in the abyss by Isaacs and Schwartzlose (1975) changed forever the scientific view of the deep ocean by showing that "active animals" were present even in this environment with few food resources.

There is strong evidence that carrion is usually detected from the odor plume emitted from the carrion. Observations at bait show fish arriving from downstream of the bait (Wilson and Smith 1984), indicating that the fish had either followed a trail of chemicals released from the carrion (chemotaxis) or swum generally up-current (rheotaxis) after detecting that carrion was in their vicinity. Because most studies of scavengers are conducted by still or video photography of bait, scientists seldom have information on how the scavengers behaved prior to arrival. Løkkeborg (1998) tracked cod *Gadus morhua* in fjords and observed fish turning towards bait and swimming to it. These results indicated that fish were active prior to bait detection at ranges of up to 700 m. Deep sea researchers also fed transponders to fish inside small bait packages at a photographic lander (a free-falling autonomous vehicle) to track them for periods of 3–9 h, at ranges of up to 500 m from the camera lander (Bagley et al. 1994, 2004). Photography at the lander allowed the species and size of the scavenger to be determined, while a current meter measured the direction and speed of currents. Evidence from the tracks of the fish suggested that abyssal grenadiers (*Coryphaenoides* spp.) swum perpendicular to the current direction. This behavior would maximize their chance of crossing an odor plume being emitted by a carrion item on the seafloor (Bagley et al. 2004).

Bailey and Priede (2002) used a simple mathematical model to predict the arrival time patterns of fish using different search strategies. These were "cross-current" scavenging, where fish swam across the current until detecting a plume and then turned upstream to swim to the bait; "sit and wait", where the scavenger remained stationary unless an odor plume was detected; and "drifting", where scavengers went with the current until coming close enough to the carrion to hear scavengers feeding or detect other evidence of activity (e.g. stimulated bioluminescence).

Using a range of assumptions, the arrival patterns predicted that the "cross current" model was the best fit to real data for abyssal fish.

All models assumed that once in the vicinity of the bait the fish would localize it using sound, hydrodynamic disturbance or bioluminescence, but at that time the process had not been studied. Using video at bait showed deep sea fish "pole-dancing" around the lander, repeatedly circling the legs of the vehicle and other structures in the field of view. A combination of video analysis and anatomical investigation showed that the fish were using a chin barbell to taste their way around the vicinity of the bait. It appears that for these fish the odor plume got them close to the bait, after which the signal was not clear enough for them to locate it directly. Tasting the seabed until the carrion was obtained seemed to be the solution (Bailey et al. 2007).

In general, we know little about how fish respond to carrion, which is surprising as the scavenging behavior by fish is the basis of the long-line fishing and fish trapping industries (Stoner 2004). Whereas deep sea fish are often surprisingly tolerant of competitors around carrion, given how sparse these resources might be, this is definitely not the case for other taxa (Stoner 2004). Wolffish *Anarhichas denticulatus* have been observed aggressively defending bait (Godø et al. 1997) and preventing cod and haddock Melanogrammus aeglefinus from obtaining it. Aggression between shallow-water fish clearly reduces the numbers of fish able to feed on carrion (Dunlop et al. 2015) by scaring smaller individuals away from this food source. The extent of aggression around carrion was density dependent, with fish being more likely to interact aggressively at high densities. Carrion-defending and aggressive behaviors probably affect the importance of scavenging across fish assemblages, with some taxa and size ranges dominating the resource. Density-dependent aggression of the type observed by Dunlop et al. (2015) will tend to reduce the difference in catches between areas (and times) of high and low population density. As a result, the effects of spatial management, or trends with time, could be hidden by differences in fish behavior. This is a very practical example of how an understanding of scavenger ecology is essential to management.

Scavengers Across the World

Tundra and Polar Regions

The tundra is a relatively simple ecosystem inhabited by many endemic species adapted to low productivity and cold temperatures, like the Arctic fox and the polar bear. Tundra food webs are marked by complex temporal fluctuations; the best known example is the cycles of voles and lemmings (Killengreen et al. 2011 and references therein). In high latitudes, carrion is a key food resource used to survive periods of food scarcity and overcome harsh environmental conditions. Carrion resources, mostly of marine origin, are associated with seabird and seal colonies, carcasses of marine mammals washed ashore and other marine subsidies, like dead fish (Fig. 1).

Giant petrels *Macronectes halli* and *M. giganteus* are the principal scavenging seabirds in the Antarctic ecosystems, and feed mainly on penguin and seal carrion. Seal carrion is highly important for giant petrels during the breeding season, particularly for males. These species also scavenge on floating carcasses during pelagic trips (Hunter 1985; González-Solís et al. 2002). Giant petrels are also frequent scavengers at penguin colonies, together with kelp gulls *Larus dominicanus*, lesser sheathbills *Chionis minor* and sub-Antarctic skuas *Catharacta lonnbergi* (Hunter 1991). The populations of some seal species in the Antarctic have expanded, promoting an increase in carrion and placenta availability, which has benefited scavengers like the skuas (Phillips et al. 2004).

In the northern polar regions, carrion may come additionally from carcasses of large herbivores, like the reindeer *Rangifer tarandus* or muskox *Ovibos moschatus*, and from polar bear kills. Polar bears kill mostly ringed seals *Phoca hispida* and bearded seals *Erignathus barbatus*, and leave behind a substantial portion of the carcasses, which are scavenged afterwards by Arctic foxes and eventually also by other polar bears (Stirling and Archibald 1977). Polar bears can also kill bigger prey, like beluga whales *Delphinapterus leucas*, even in high numbers when they are entrapped in ice. Lowry et al. (1987) documented the killing of more than 40 belugas in an ice entrapment and observed 15 polar bears feeding on them. Twelve days later, carrion was distributed widely and at least 30 polar bears were recorded feeding. Such bear aggregations have been recorded scavenging on large whale carcasses (Bentzen et al. 2007; Fig. 1). Ivory gulls *Pagophila eburnea* and glaucous gulls *Larus hyperboreus* are important avian scavengers in Svalbard (Gjertz and Lydersen 1986). In the Canadian Arctic, ravens, gulls, arctic foxes, and even wolves have been also recorded scavenging on polar bear kills (Stirling and Archibald 1977).

The Arctic fox is the mammalian scavenger per excellence in the tundra and polar habitats. Their use of carrion and marine subsidies is highest during winter and lemming population declines (Roth 2002). Carrion storage, and food caching in general, is particularly common among Arctic foxes as a response to trophic pulses and fluctuating environments (Careau et al. 2007; Samelius et al. 2007). With the shrinking ice cover and tundra under global warming, the influx of predator-scavengers from neighboring ecosystems (e.g. red foxes from boreal forests) is increasing competition and eventually could exclude Arctic foxes from carcasses (Killengreen et al. 2011, 2012).

Temperate and Boreal Forests

Temperate and boreal forests are complex ecosystems where scavenging is widespread. Although there are very few obligate scavengers (e.g. turkey vultures *Cathartes aura*), all predators are scavengers to some extent. Most important are avian and mammalian scavengers. For instance, predators such as the goshawk and the weasel, or insectivores like shrews and woodpeckers also consume carrion when the opportunity arises. More than 30 species of birds and mammals were recorded consuming carrion in the forests of Bialowieza, northeastern Poland (Selva 2004). Typically, the bulk of the carrion found in these forests are ungulates, mainly provided by large predators such as wolves, by harsh winters, or by human-caused mortality like hunting (Houston 1978; Wilmers et al. 2003a, b; Selva and Fortuna 2007; Mateo-Tomás et al. 2015). Carrion is a very important food resource for vertebrates in winter and early spring, and support hibernating species like the brown bear *Ursus arctos* (Green et al. 1997). Scavengers, and particularly the brown and black bear *Ursus americanus*, also rely on carrion pulses from spawning salmon that arrive every year in some rivers of these forests (Fortin et al. 2007; Levi et al. 2015, see chapter "Ecological Functions of Vertebrate Scavenging"). Caching food (i.e. storing it in locations hidden from the sight of conspecifics and other species) is common among the main scavengers like corvids and canids.

The most efficient and frequent scavenger in northern forests is probably the common raven (Wilmers et al. 2003b; Selva et al. 2005; Wikenros et al. 2013; Fig. 1). Other corvids like jays, magpies, crows and nutcrackers are frequent visitors to carcasses (Wilmers et al. 2003b; Selva et al. 2005; Wikenros et al. 2013; Ray et al. 2014). Among raptors, the common buzzard *Buteo buteo*, as well as eagles (e.g. white-tailed eagle *Haliaeetus albicilla*, golden eagle *Aquila chrysaetos* and bald eagle *Haliaeetus leucocephalus*) rely on carrion during some periods and efficiently exploit hunter-kills, often concentrated in space and time and exposed in more open areas within the forest (Wilmers et al. 2003a; Selva et al. 2005; Mateo-Tomás et al. 2015; Fig. 1).

Mammalian scavengers in temperate and boreal forests are also well represented. As in other biomes, canids such as red foxes, coyotes *Canis latrans*, raccoon dogs *Nyctereutes procyonoides*, and wolves are among the most frequent scavengers (Selva et al. 2005; Wikenros et al. 2013). Wolves not only subsidize other scavengers with carrion, but also facilitate scavenging of large ungulate carcasses by opening them (Selva et al. 2003). Some mustelids, like martens *Martes* spp. and wolverines *Gulo gulo*, are quite dependent on carrion, particularly in years when alternative sources of prey (e.g. rodents or hares) are scarce (Jedrzejewski et al. 1993; Koskela et al. 2013). Felids are not frequent users of carrion, and often suffer kleptoparasitism by larger species like the brown bear, the wild boar, and the wolverine, with a limited or significant effect on lynx predation rates depending on the area (Selva 2004; Mattisson et al. 2011; Krofel et al. 2012).

Mediterranean Regions

Scavengers in these regions have been traditionally associated with cattle farming. Carrion has been mostly supplied through livestock carcasses, hunting of wild ungulates, and more recently, artificial feeding stations or vulture restaurants and dumps (Oro et al. 2013). The scavenger species at these carrion supplies are the same, but the number of individuals of these species, as well as their time of detection and consumption is very different (Cortés-Avizanda et al. 2010, 2012, 2014, Moreno-Opo et al. 2015a, b, 2016). Within the Mediterranean region, Spain is home to >90% of the obligate European scavengers (Eurasian griffon vulture *Gyps fulvus*, Egyptian vulture *Neophron percnopterus*, bearded vulture *Gypaetus barbatus* and cinereous vulture *Aegypius monachus*, Fig. 1) and of a rich community of facultative scavengers which includes some threatened species like the Spanish imperial eagle *Aquila adalberti*, the red kite *Milvus milvus,* and the Iberian lynx *Lynx pardinus*. The guild of obligate avian scavengers in other Mediterranean countries, like Italy, France, Portugal or Greece, constitutes less than 10% of the vulture population in the European Union (Deinet et al. 2013; Margalida et al. 2014). This has promoted several reintroductions to recover the scarce populations of some avian scavengers, like the bearded vulture in the Alps (Austria, Italy, France and Switzerland) and Massif Central (France), the Eurasian griffon vulture in Italia, Israel and France or the cinereous vulture in France (see reviews in Terrasse 2001; Margalida and Heredia 2005; Eliotout 2007; Donázar et al. 2009).

Among avian scavengers, Eurasian griffon vultures and common ravens are the most habitual species scavenging carrion resources in the Mediterranean (Cortés-Avizanda et al. 2010; Mateo-Tomás et al. 2015; Moreno-Opo et al. 2015a, 2016). In this sense, griffon vultures play important ecosystem services eliminating ca. 10 thousand tons of meat per year, and, in the case of bearded vultures, 200 tons of bones (Margalida and Colomer 2012; Margalida et al. 2012). Other corvids that frequently visit carrion resources are the carrion crow *Corvus corone*, magpie *Pica pica* and Eurasian jay *Garrulus glandarius*. Among raptors other than vultures, the common buzzard, the red and black kite, the golden and Spanish imperial eagles and, more occasionally goshawks, obtain additional food resources from carrion subsidies (Sánchez et al. 2009; Sánchez-Zapata et al. 2010; Mateo-Tomás et al. 2015; Moreno-Opo et al. 2016; Margalida et al. 2017). In particular, non-adult individuals of golden and Spanish imperial eagles are prone to use scavenging as a feeding strategy in winter, coinciding with low levels of their main natural prey and the occupancy of low quality habitats (Sánchez et al. 2009; Sánchez-Zapata et al. 2010; Margalida et al. 2017). With respect to mammalian scavengers in Mediterranean areas, the most frequent are the red fox, the stone marten *Martes foina* and the wild boar (Mateo-Tomás et al. 2015).

Savannas

The earliest studies on scavengers were mostly conducted in the African savanna, which encompasses the majority of the savanna regions on the planet. This is not surprising because this continent is home of probably the most emblematic scavengers: vultures and hyaenas (Fig. 1). Classic studies by Petrides (1959), Attwell (1963), Kruuk (1966, 1967, 1972) and Houston (1974, 1975, 1979) on these species have inspired, and still do, many scavenging ecologists worldwide. The scavenger communities of savanna regions of South America, Asia and Australia are poorly known, so here we focus on African scavengers. However, the general characteristics of the scavenging fauna are probably pervasive among continents, with the exception that there are no obligate scavengers in Australia.

The savanna regions of the African continent support the richest community of vultures in the world, including ten resident and one wintering species, eight of which are endemic or near-endemic (Del Hoyo 1994). This species array ranges from consumers of meat and other soft tissues (e.g. Rüppell's vulture *Gyps rüppellii*) to bone eaters (bearded vulture) and even one species that rarely scavenges and has evolved to exploit palm-nuts and capture small prey (palm-nut vulture *Gypohierax angolensis*). Foraging habits among African vultures are also variable, including species that forage solitarily (e.g. white-headed vulture *Aegypius occipitalis*) and species that forage in pairs (e.g. lappet-faced vulture *A. tracheliotos*) and in groups (e.g. white-headed vulture *Gyps africanus*). The morphological, ecological and behavioral diversity of African vultures, together with dominance hierarchies at carcasses, results in an ordered resource partitioning that likely reduces competition for carrion (Kruuk 1967; Hertel and Lehman 1998; Kendall et al. 2012; Kendall 2013). Such a variety has been related to the high diversity of large mammalian herbivores, whose carcasses provide major feeding resources for scavengers. Indeed, Houston (1983) suggested that the evolutionary origin of vultures was closely related to the presence of long-distance migratory ungulates, which still survive in a few African regions.

The community of facultative scavengers inhabiting the African savanna is also diverse and includes omnivorous species such as corvids (e.g. white-necked raven *Corvus albicollis*), medium-sized predators such as birds of prey (e.g. tawny eagle *Aquila rapax*) and mammalian carnivores (e.g. side-striped jackal *Canis adustus*), and apex predators such as hyaenas and large cats (Fig. 1). In areas with resident herbivore populations, spotted hyaenas (*Crocuta crocuta*) and lions can outcompete vultures and consume most of the ungulate carrion biomass (Moleón et al. 2015). The brown hyena (*Parahyaena brunnea*), i.e. the mammal species most specialized to use carrion (up to 90% of its diet; Skinner and Chimimba 2005), is found in the driest savannas of Southern Africa.

Steppes

Steppes are located in several regions of the world, mainly Eurasia, North America and South America. This cold, dry grassland is an important biome for scavengers.

In the Americas there are seven species of obligate scavengers, all of them from the family Cathartidae (known as vultures from the New Word). These vultures occur from high mountains to the coastline and from woodlands to the steppes. Apart from them, other facultative scavengers, both mammals and birds, aggregate and compete in carcasses found in those open areas.

In South American steppes, three are the species of obligate scavengers, the Andean condor *Vultur gryphus*, the turkey vulture and the black vulture *Coragyps atratus* (Fig. 1). Other bird species that facultatively consume carcasses are the southern *Caracara plancus* and chimango *Milvago chimango* caracaras, and the black-chested buzzard eagle *Geranoaetus melanoleucus* (Travaini et al. 1998; Lambertucci et al. 2009a). Mostly during the dark hours, carcasses are also consumed by mammals such as pumas *Puma concolor*, foxes *Lycalopex* spp., skunks *Conepatus* spp., and armadillos *Zaedyus pichiy*, and also by passerines and lizards (Travaini et al. 1998; Elbroch and Wittmer 2012). Currently, all these species scavenge mainly on carcasses from domestic animals since steppes have been historically used for extensive livestock raising, which has produced a sharp decline in the populations of native herbivores (Fig. 1). Scavengers also feed on carcasses of exotic wild species, such as the red deer *Cervus elaphus* and the European hare *Lepus europaeus*. Native species such as guanacos *Lama guanicoe*, and rheas *Rhea pennata* that once were the main supply of carrion are currently much less consumed in many areas due to their low numbers which brought them to an ecological extinction (Novaro et al. 2000; Lambertucci et al. 2009b).

As in other biomes, competition for carrion is important and seems to relate to the size of the species. Generally, bigger species (condors) dominate over the carcasses (Wallace and Temple 1987). However, when the number of individuals of other species strongly increases, the size-related hierarchy could be alternated and smaller species can dominate at carcasses (Carrete et al. 2010). In South American steppes this is the case of black vultures, that can reach very high densities in the surroundings of human settlements and might oust Andean condors at carcasses where generally they would be dominant (Carrete et al. 2010; Barbar et al. 2015).

Scavenger composition is also influenced by the size of the carcass, with few individuals of few species feeding on hare carcasses (Travaini et al. 1998) and more species in larger numbers feeding on larger carcasses like sheep, guanaco or horse (e.g. Lambertucci et al. 2009a). The place where the carcass is located can modify the order of arrival and consumption patterns. Carcasses close to roads are mainly used by smaller bird scavengers as caracaras, while larger species as condors and turkey vultures are the first to eat at carcasses far from roads (Lambertucci et al. 2009a). Moreover, bird scavenger composition can also be affected by weather conditions, and for instance, in gusty or very windy days carcasses are mainly used by larger species (Shepard and Lambertucci 2013). This is particularly relevant in Patagonian steppes which have cold-temperate climate with strong winds.

Regarding mammals, pumas, the largest carnivore in the area, can be observed scavenging, but they are more important as suppliers of carcasses for other species. In fact, they can provide large amounts of meat which are used by at least 12 species in the Patagonian steppes. Pumas need to increase their kill rate to compensate for the losses due to scavengers (Elbroch and Wittmer 2013). Two species of fox (*Lycalopex cualpaeus* and *Lycalopex griseus*) are commonly observed displacing smaller species in order to feed on puma kills.

Tropical Forests

In tropical forests Cathartidae and Accipitridae vultures constitute important members of the scavenger community (Houston 1979), but they differ substantially in their distribution and the habitats they occupy. For example, in the tropical forests of Asia and Africa, obligate vulture species are rare when compared to neotropical forests, where species such as Cathartidae vultures are well represented. Two New World vulture species, the turkey vulture and the greater yellow-headed vulture *Cathartes melambrotus*, are the most common vultures of neotropical forests. In these forests, mammalian carnivores are rather inefficient at locating carcasses. This is a consequence of their limitations in the daily foraging areas covered and the trade-off between the energy invested to search for food and the distance travelled to find temporal and unpredictable food resources. Thus, mammalian carnivores are probably outcompeted by avian scavengers and consume carrion just when the chance appears. Avian scavengers, which use little energy in soaring (Pennycuick 1983) and can quickly cover large areas of forest, are more efficient carrion foragers (Houston 1984, 1986).

With respect to obligate vulture species, the consumption of carrion in forest habitats is possible thanks to the well-developed olfactory lobe and sense of smell which is used for finding food in forested areas (Chapman 1929; Houston 1984, 1986). The three Cathartes species—the turkey, the greater yellow-headed and the lesser yellow-headed *Cathartes burrovianus*—differ only in their wing and tail shape and in the color of their heads, having the remarkable ability to locate food by smell (Houston 2001). American black vultures and king vultures *Sarcoramphus papa* are also frequent on tropical forests but have no sense of smell; they often rely on Cathartes vultures behavior to find food.

Agricultural and Other Human-Modified Landscapes

Despite human domination over much of the planet (Waters et al. 2016), scavenging ecology often has been studied in relatively natural environments (e.g., Houston 1974; Wilmers et al. 2003b; DeVault et al. 2004; Selva et al. 2005). However, studies conducted in agricultural and other human-dominated environments suggest that vertebrate scavengers are also very active in such habitats. In this section, we briefly review studies investigating carrion use by vertebrates in heavily-altered landscapes, focusing on agricultural habitats, a major land use across the world where diversity is often reduced and vertebrate communities are dominated by a few generalist species (Swihart et al. 2003).

In an intensively farmed region of northern Indiana, USA, small carrion items, such as mouse carcasses, were scavenged extensively by two species of generalist mesopredators: raccoons and Virginia opossums *Didelphis virginiana* (DeVault et al. 2011). The high scavenging efficiency by vertebrates in the agricultural region was attributed to elevated densities of raccoons and Virginia opossums, and the reduced search area for carrion in the sparsely distributed forested habitat islands (DeVault et al. 2011). For larger carcasses in the same northern Indiana agricultural area, Olson et al. (2016) found that 87% of raccoon, Virginia opossum, and rabbit *Oryctolagus* spp. carcasses were fed upon by 16 species of vertebrate scavengers, and included more bird scavengers (seven species) than was found for the smaller mouse carcasses in the earlier study (DeVault et al. 2011). Other studies have investigated scavenging efficiency by vertebrates on carcasses (usually birds) found in agricultural fields themselves. Scavenging efficiency ranged from 38% to 92% for small bird carcasses (Balcolm 1986; Tobin and Dolbeer 1990; Linz et al. 1997; Kostecke et al. 2001) and was 96% for large birds (mallards *Anas platyrhynchos*; Peterson et al. 2001).

Wide-ranging scavengers that use highly altered, agricultural habitats as well as more natural habitats have been shown to adjust foraging strategies based upon available resources. For example, in the USA, turkey vultures fed extensively on domestic farm animal carcasses in an agricultural region (Coleman and Fraser 1987), whereas their diet was almost exclusively wild vertebrates in a heavily forested area (Kelly et al. 2007). Likewise, in Bulgaria, the diet of the facultative scavenging golden jackal *Canis aureus* included a large percentage of domestic animals in an agricultural area, whereas in a forested region wild ungulates were the dominant food items (Raichev et al. 2013).

Scavenging efficiency by vertebrates varies widely across human-dominated landscapes, and appears to be influenced by many factors including habitat, composition of the vertebrate and invertebrate community, and climate. Broadly, it appears that carcass use by vertebrates in agricultural regions is equivalent to, or in some cases higher than that found in more natural habitats. Smaller carcasses produced in agricultural habitats are usually removed by mammalian scavengers, whereas bird scavengers often use larger carcasses.

Deserts

In arid and semiarid environments, water availability is often the limiting factor in the decomposition process. Carcasses can desiccate before they are completely decomposed in some circumstances (Payne 1965; Parmenter and MacMahon 2009), and dried-out carcasses are generally less attractive to vertebrate scavengers than those that retain some water (Parmenter and MacMahon 2009). Water loss from carcasses is strongly influenced by surface area-to-volume ratios, with smaller carcasses having higher ratios and, thus, greater water loss rates (Parmenter and MacMahon 2009).

Despite water limitations, vertebrate scavengers are active in arid environments and usually feed upon available carcasses, especially when temperatures are cool (Stoddart 1970; Travaini et al. 1998; Parmenter and MacMahon 2009). Furthermore, carcass use by facultative scavengers can actually increase following water shortages because of a relative lack of alternative food resources. Brown et al. (2006) showed that carcass use by vertebrate scavengers increased following a period of below-average rainfall coincident with the El Niño Southern Oscillation in semi-arid southeastern Australia.

Like in other ecosystems, a variety of mammal and bird species are prolific scavengers in arid and semiarid environments. Stoddart (1970) studied the fate of 45 black-tailed jackrabbit *Lepus californicus* carcasses in northwestern Utah, USA. Of these, 60% were scavenged by birds (species undetermined). In arid southern Australia, corvids (*Corvus coronoides* and *C. bennetti*), wedge-tailed eagles *Aquila audax* and red foxes fed extensively on kangaroo offcuts left in the field after butchering (Read and Wilson 2004). Considering smaller carcasses, Rogers et al. (2014) found that common ravens, kit foxes *Vulpes macrotis*, coyotes, and greater roadrunners *Geococcyx californianus* found and fed upon 20 out of 25 carcasses of small migrating birds that died from colliding with power lines in the Sonoran Desert (California, USA).

Coastal Systems

The world's coastline is an estimated 1,634,701 km long, ranging across all the climatic zones, from the polar, through the temperate, to the tropical regions. Coastal ecosystems represent the interface between land and sea and provide carrion mainly via colonies of mammals, seabirds and reptiles or through deposition of carcasses originated at sea (Polis and Hurd 1996; Moss 2017). Vertebrate scavengers inhabiting the coastal surface and exploiting carcasses provided by pinniped and seabird colonies include mostly birds and mammals. The procellariids southern and northern giant petrels opportunistically scavenge on dead seals and penguins in Antarctica (Bruyn and Cooper 2005). In Southern Chile, black vultures, kelp gull and feral dogs *Canis familiaris* compete for carrion occurring in a sea lion *Otaria flavescens* colony (Pavés et al. 2008). At temperate latitudes, the great black-backed gull *Larus marinus* and the red fox consume placenta and dead pups at grey seal *Halichoerus grypus* breeding colonies (Twiss et al. 2003; Culloch 2012; Quaggiotto et al. 2018; Fig. 1). In Namibia (Africa) brown hyenas restrict their range to the coastal area to scavenge on carrion produced during pupping season of the Cape fur seal *Arctocephalus pusillus* (Kuhn et al. 2008).

Similarly, marine inputs deposited on the coastline of California attracted coyotes to the shore, causing a numerical response in the population density which increased between 2.4 and 13.7 times in relation to the adjacent desert (Rose and Polis 1998). In experimental conditions, fish carcasses on a sandy beach in Australia determined the formation of consumers' aggregation around the carrion patches. In this sub-tropical coastal environment not only the opportunistic Torresian crows *Corvus orru* and silver gull *Chroicocephalus novaehollandiae*, but also the birds of prey like the white-bellied sea eagle *Haliaeetus leucogaster*, whistling kite *Haliastur sphenurus* and brahminy kite *Haliastur indus*, and the lizard lace monitor were observed scavenging on the fish carcasses (Huijbers et al. 2015). Marine carrion is crucial in environments such as deserts, sandy beaches and tundra, where productivity is low (Spiller et al. 2010).

On urbanized beaches in Australia the scavenging community feeding on fish carcasses included rats *Rattus* spp., domestic cats *Felis catus*, dogs and red foxes, but not raptors. Raptors occurred only in rural areas, where they completely removed carrion within 24 h (Huijbers et al. 2013). Mammals scavenging at urban locations did not cover the functional role of raptors, as they could not compensate the extent of scavenging activity of the avian counterpart (Huijbers et al. 2015). By changing species composition of the coastal scavenging community, urbanization has also an impact on the scavenging processes occurring in these systems. Furthermore, the effect of invasive species, such as the red fox in Australia, can be detrimental to the endemic scavengers. Red foxes, for instance, can remove a large number of carcasses during the night and before the diurnal raptors even start foraging (Brown et al. 2016).

Carrion consumption by coastal scavengers may vary because of human activity (e.g. Lambertucci et al. 2018). Ronconi et al. (2014) found that great black-backed gulls living on Sable Island (Canada) rely more on seal carrion as food source than forty years ago. An increase of the seal population, changes in fish availability, reduction of fishery discards and alterations in the surrounding ecosystem could all have led to this shift in gull's diet.

Marine and Freshwater Systems

In marine systems, carrion is exploited by many invertebrates at any depth, while vertebrate scavenging species are mostly found closer to the deep-sea floor. In shallow waters, fish observed visiting experimental marine mammal carcasses were found to be feeding on the amphipods colonising the carcass (Jones et al. 1998). Whale falls in the bathyal zone attract a large number of sleeper sharks *Somniosus pacificus*, hagfish *Eptatretus deani* and *Mixine circifrons*, and macrourids which scavenge the soft tissue of the whale carcass. These species characterize the first stage of the whale consumption, called mobile-scavenger stage, which lasts from months to years. Around 4400 m depth, grenadiers *Coryphaenoides armatus* were observed approaching experimental cetacean carcasses, feeding on the amphipods at the carcass, rather than scavenging (Jones et al. 1998). Floating whale carcasses attract scavenging elasmobranchs, such as the great white shark *Carcharodon carcharias*, the tiger shark *Galeocerdo cuvier* and the blue shark *Prionace glauca* (Long and Jones 1996; Dudley et al. 2000; Curtis et al. 2006). Although it is thought that scavenging behavior in sharks is common, this was occasionally observed and rarely quantified (Fallows et al. 2013). These large predatory fish congregate or solitarily arrive at carcasses attracted by the olfactory cues released by the decaying whale and transported by wind (Fallows et al. 2013). Juvenile individuals also scavenge on dead whales (Dicken 2008). However, adults are the most frequent attendants of such banquet, which are hypothesized to trigger mating between mature individuals (Fallows et al. 2013). Scavenging by sharks is common at whale calving grounds where both carrion and predatory opportunities increase (Bonfil et al. 2010). There is evidence that tiger sharks also scavenge on bird dies-offs at sea (Gallagher et al. 2011).

Notoriously, killer whales *Orcinus orcas*, together with sharks, were the main vertebrate scavengers of whale carcasses and their remains produced by the whaling industry. It is likely that killer whales adopted this opportunistic behavior and specialized in scavenging on such nutritious resources when these were patchy and temporally predictable. Before (and possibly after) the whaling era, instead, scavenging represented one of the foraging alternatives used opportunistically by these apex predators (Whitehead and Reeves 2005). Scavenging seabirds, such as kelp gulls, Antarctic skuas *Catharacta antarctica*, black browed albatrosses *Thalassarche melanophris* and different species of petrels often associate with hunting killer whales to easily gain some pieces of food from the remains of the killed prey (Williams et al. 1990).

Carrion in freshwater ecosystems is mainly derived from fish or aquatic birds, and it is used by a diverse array of scavengers, even Eurasian otters (O'Sullivan et al. 1992). Some freshwater systems experience a carrion pulse every year derived from fish migrating to their spawning areas, like salmon (see chapter "Ecological Functions of Vertebrate Scavenging"). Such food pulses support high-density populations of brown bears (i.e. Hilderbrand et al. 1999) and other mammals such as wolves, marten, mink, and coyotes, and a diverse array of avian scavengers including bald eagles, ravens, jays, mergansers, gulls, and even owls (Levi et al. 2015). Some of these predator-scavengers of salmon, like the glaucous-winged gulls (*Larus glaucescens*) and brown bears shifted their spatial distribution together with salmon, thus closely tracking the phenology of their main food resource (Schindler et al. 2013). Mobile scavengers, like gulls, by transporting nutrients derived from fish carrion to distant locations, have an important impact on local aquatic communities (Payne and Moore 2006).

Structure of Scavenger Communities

Interactions among obligate scavengers have always been recognized as highly organized and driven by competition (Kruuk 1967; Wallace and Temple 1987), whereas scavenging has traditionally been viewed as a more random and opportunistic process for facultative scavengers (Hiraldo et al. 1991). In the last decade, however, many studies have demonstrated that vertebrate scavenger communities can show complex patterns that deviate from random processes (Selva and Fortuna 2007). We can consider that an assemblage is structured if it is more ordered than expected by chance. In random communities, interactions are the result of opportunistic encounters (i.e. a facultative scavenger finding a carcass without an active search for it). Structured communities show identifiable patterns and are a consequence of one or several interacting processes shaping which interactions are possible and which more likely (Selva and Fortuna 2007).

One of the patterns that has been used to study the structure of vertebrate scavenger communities is nestedness. A community of interacting species is nested if the specialist species interact with a subset of the species that interact with the more generalist species (Bascompte et al. 2003). This concept has been widely identified in interacting communities, especially mutualistic interactions (Bascompte and Jordano 2007), with important implications for the stability and persistence of mutualistic networks (Bastolla et al. 2009; Rohr et al. 2014). A nested pattern in scavenger communities emerges when the species feeding on carcasses visited by few consumers are subsets of the assemblage of species feeding on carcasses visited by more consumers (Fig. 2). Selva and Fortuna (2007) were the first authors that identified nestedness in a scavenger assemblage. Since

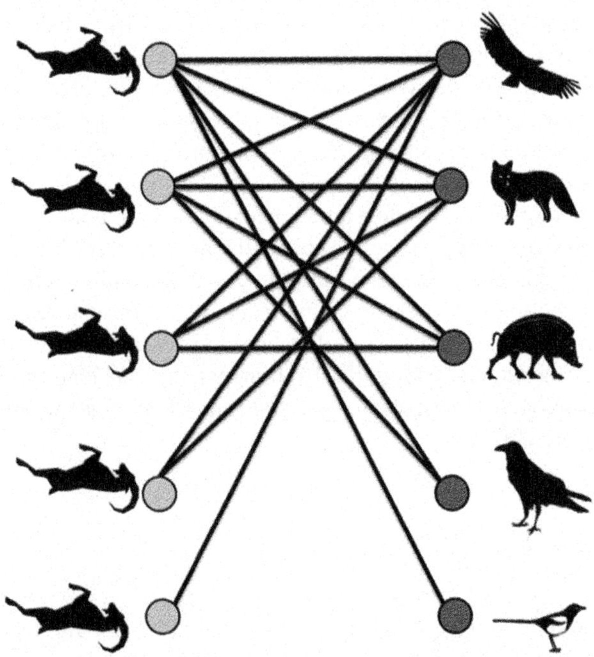

Fig. 2 Schematic representation of a perfectly nested scavenger community. Light gray circles represent carcasses, dark grey circles represent scavengers. A carcass is linked to a scavenger if it is consumed by that species. Scavengers are from top to bottom: vulture, fox, wild boar, raven and magpie. In this example, the species feeding on carcasses visited by few consumers (i.e. vulture) are subsets of the assemblage of species feeding on carcasses visited by more consumers (i.e. first carcass on the top, consumed by all the scavengers). Picture performed by E. Sebastián-González. Silhouettes from The Noun Project https://thenounproject.com, authors: S. Laing, A. Bearne, M. Turan Ercan, H. Richir, Bluetip Design and P. Lehmann

then, we know that nestedness is not universal (i.e. it does not appear in all the scavenger communities), and that it depends on many interacting factors, such as the type of carcass, composition of the scavenger assemblage, and the competition degree of the community (Sebastián-González et al. 2013; Allen et al. 2014; Moleón et al. 2015).

The type of carcass is one of the factors affecting the structure of scavenger communities. For example, the contribution to nestedness (i.e. how much a species or a carcass follows the nested pattern) was high in carcasses originated from human hunting, it was medium for carcasses from predator kills and lower from animals dead from disease or starvation (Selva and Fortuna 2007). This pattern is related to the spatio-temporal distribution and predictability of the resources. Hunter kills are normally aggregated in an area (i.e. hunting spots) and in time (i.e. hunting season). This can lead to a super-saturation of the resource, allowing the exploitation of carcasses by more consumers (Wilmers et al. 2003a, but see Cortés-Avizanda et al. 2012 for a different perspective of a similar problem). In contrast, death by disease or starvation is a much more random process that can occur at any time and location, and these carcasses are mainly consumed by carrion specialists. Other carcass characteristic found to be very important for assemblage structure is size. Scavenger communities consuming larger carcasses (>10 kg) were organized under a nested pattern, while small carcasses showed a more random organization (Sebastián-González et al. 2013; Moleón et al. 2015). Larger carcasses may be partitioned by a higher number of species, promoting both competitive and facilitative processes, which can lead to more structured communities. Small carcasses, however, are often consumed by a single individual, increasing the likelihood of opportunistic encounters and the randomness in the interactions.

Independently of the characteristics of the carcass, the composition of the scavenger community also has important implications for the structure of the assemblage. Birds presented a higher contribution to nestedness than mammals, indicating that they are the main responsible for the nested structure of the community (Selva and Fortuna 2007). Moreover, Sebastián-González et al. (2016) studied scavenger communities in Spain and found that those assemblages including obligate scavengers (i.e. vultures) were nested, while those lacking vultures presented a more random organization (see also Sebastián-González et al. 2013). Two facilitation mechanisms may be behind the role of key species in the structure of the community: carcass location and skin opening. Avian scavengers in general, and vultures in particular, are more adapted to scavenging because of their low energy-demanding search strategies (Selva et al. 2005; Jackson et al. 2008). Thus, birds are able to encounter and consume most of the carcasses available in an area and can act as indicators of the presence of the carcass for other species. If the ability to locate the resource differs among scavenger species, the facilitation of carcass location may lead to an increase in the nestedness of the assemblage. Vultures can also facilitate carcass use to other species by opening thick-skin animals. Many scavengers are unable to consume carcasses from intact (i.e. unopened) animals because they lack the tools (e.g.

strong beak) or the strength to break the skin. The same mechanism has been proposed for hyenas (Moleón et al. 2015), pumas (Allen et al. 2014) and wolves (Selva et al. 2003). Limitation of the trophic resources by competitive interactions can also increase nestedness through interspecific dominance hierarchies. For example, the presence of black bears in a Californian scavenger community increased nestedness by decreasing complementary use of carrion by other species (Allen et al. 2014).

Nestedness has been found to be high under stressful conditions in several animal communities (Worthen et al. 1998; Baber et al. 2004; Cook et al. 2004). Theoretical studies on mutualistic interaction networks suggest that nestedness can reduce competition among species and enhance coexistence (Bastolla et al. 2009; Rohr et al. 2014). Scavenger communities seem to mirror other animal communities in their relationship among nestedness and competition. Selva and Fortuna (2007) found that the scavenger community in the cold season, when food is scarcer and weather is more extreme, was more nested than at the warm season. Moreover, both species richness and nestedness are characteristic of scavenger networks that were robust to high levels of interspecific competition for carrion (Sebastián-González et al. 2016). Thus, nestedness may favor the coexistence of interacting species by counterbalancing the effects of high competition.

Thus, we conclude that scavenger assemblages show non-random nested organizations that can emerge when carcasses are not monopolized by a single consumer (e.g. because they are small). In general, nestedness in scavenger communities is the consequence of three interacting processes: (1) predictability in the availability of the resource, (2) facilitative processes provided by key species, and (3) differences in the abilities of the scavenger species to compete for the carrion.

Conclusions and Future Perspectives

Although there are few obligate scavengers, carrion consumption is a foraging strategy widely spread among vertebrates in all biomes. The scavenger species, the number of obligate scavengers, and even the adaptations to locate and consume carrion differ among ecosystems. However, everywhere scavenging is a profitable strategy to acquire energy, even if sporadic, at the individual, population and community level, and an important path for energy and nutrient cycling.

Recent advances in technology, such as automatic cameras or satellite telemetry, have exponentially improved the amount and quality of collected data, which have greatly contributed to obtain a complete picture of many scavenger communities and a better understanding of their functioning. This has been particularly important for ecosystems where direct observations have been difficult or impossible, like polar or abyssal regions. Although important achievements in scavenging ecology have been made in the last decades, there are still pending tasks, such as the identification of all scavenger species in some ecosystems and their inter- and intra-specific interactions at carrion, or the assessment of the role of carrion in the ecology and population dynamics of many obligate and facultative scavenger species.

Box 1 Vultures in Tibet and Sky Burials

The Himalayan range is the mountain region in the world where more species of vultures exist. However, the information on species population numbers and ecology for that area is rather limited. Most of the research in southern Asia comes from India, and is particularly focused on the vulture crisis suffered there (Green et al. 2004). As much as nine species of vultures are found in the southern portion of the Himalayan range. Eight of those species are present in China, seven of which are present in the Tibetan plateau as well, one of the richest vulture regions in the world (MaMing and Xu 2015).

In the Tibetan plateau, which is over 4000 m asl, vultures are the main scavengers and feed not only on wildlife and livestock carrion, but also on human corpses. A common cultural practice in this area are the sky burials. In this type of funeral the dead human body is placed on a platform on a mountain from where scavengers (mainly vultures) consume it. The remains are disposed for scavengers in as generous a way as possible (Fig. 3). The Himalayan griffon vulture *Gyps himalayensis* is the most common species at sky burials, and can be counted at hundreds in a single funeral; cinereous and bearded vultures are also observed at sky burials (MaMing et al. 2016). Tibetans believe that the cessation of breathing is only a first stage of death. Tibetan Buddhists view death as the journey from this life to the next one, and consider vultures as sacred animals aiding humans in this journey. This makes vultures highly important for both their environment and culture.

Fig. 3 Vultures in the Tibet feed on human corpses at the called "sky burials". Photo credit: Lama Rinczen

Acknowledgements The authors sincerely thank Alexander Gruzdev, Dan Hartman, Sarah C. Jones, Lama Rinczen and Manuel de la Riva for their pictures.

References

Allen ML, Elbroch LM, Wilmers CC et al (2014) Trophic facilitation or limitation? Comparative effects of pumas and black bears on the scavenger community. PLoS ONE 9(7):e102257

Attwell RIG (1963) Some observations on feeding habits, behavior and inter-relationships of Northern Rhodesian vultures. Ostrich 34:235–247

Baber MJ, Fleishman E, Babbitt KJ et al (2004) The relationship between wetland hydroperiod and nestedness patterns in assemblages of larval amphibians and predatory macroinvertebrates. Oikos 107:16–27

Bagley P, Smith A, Priede I (1994) Tracking movements of deep demersal fishes in the Porcupine seabight, north-east Atlantic Ocean. J Mar Biol Assoc 74:473–480

Bagley P, Priede I, Amieson A et al (2004) Lander techniques for deep-ocean biological research. Int J Soc Underw Technol 26:3–12

Bailey D, Priede I (2002) Predicting fish behaviour in response to abyssal food falls. Mar Biol 141:831–840

Bailey D, Wagner H, Jamieson A et al (2007) A taste of the deepsea: the roles of gustatory and tactile searching behaviour in the grenadier fish *Coryphaenoides armatus*. Deep Sea Res Part I Oceanogr Res Pap 54:99–108

Balcolm R (1986) Songbird carcasses disappear rapidly from agricultural fields. Auk 103:817–820

Barbar F, Werenkraut V, Morales JM et al (2015) Emerging ecosystems change the spatial distribution of top carnivores even in poorly populated areas. PLoS One 10(3):e0118851

Barnosky AD, Koch PL, Feranec RS et al (2004) Assessing the causes of late Pleistocene extinctions on the continents. Science 306:70–75

Bascompte J, Jordano P (2007) Plant-animal mutualistic networks: the architecture of biodiversity. Annu Rev Ecol Evol Syst 38:567–593

Bascompte J, Jordano P, Melián CJ et al (2003) The nested assembly of plant-animal mutualistic networks. Proc Natl Acad Sci U S A 100:9383–9387

Bastolla U, Fortuna MA, Pascual-García A et al (2009) The architecture of mutualistic networks minimizes competition and increases biodiversity. Nature 458:1018–1020

Beasley JC, Olson ZH, DeVault TL (2015) Ecological role of vertebrate scavengers. In: Benbow ME, Tomberlin J, Tarone A (eds) Carrion ecology, evolution, and their application. CRC Press, Boca Raton, pp 107–128

Bentzen TW, Follman EH, Amstrup SC et al (2007) Variation in winter diet of southern Beaufort Sea polar bears inferred from stable isotope analysis. Can J Zool 85:596–608

Blumenschine RJ, Cavallo JA (1992) Scavenging and human evolution. Sci Am 267:70–76

Bonfil R, Francis MP, Duffy C et al (2010) Large-scale tropical movements and diving behavior of white sharks *Carcharodon carcharias* tagged off New Zealand. Aquat Biol 8:115–123

Brown OJF, Field J, Letnic M (2006) Variation in the taphonomic effect of scavengers in semi-arid Australia linked to rainfall and the El Niño Southern Oscillation. Int J Osteoarchaeol 16:165–176

Brown M, Schlacher T, Schoeman D et al (2016) Invasive carnivores alter ecological function and enhance complementarity in scavengers assemblages on ocean beaches. Ecology 96: 2715–2725

Bruyn PJN, Cooper J (2005) Who's the boss? Giant petrel arrival times and interspecific interactions at a seal carcass at sub-Antarctic Marion Island. Polar Biol 28:571–573

Buckley NJ (1996) Food finding and the influence of information, local enhancement, and communal roosting on foraging success of North American vultures. Auk 113:473–488

Bull JJ, Jessop TS, Whiteley M (2010) Deathly drool: evolutionary and ecological basis of septic bacteria in Komodo dragon mouths. PLoS ONE 5(6):e11097
Careau V, Giroux JF, Berteaux D (2007) Cache and carry: hoarding behaviour of Arctic fox. Behav Ecol Sociobiol 62:87–96
Carrete M, Lambertucci SA, Speziale K et al (2010) Winners and losers in human-made habitats: interspecific competition outcomes in two Neotropical vultures. Anim Conserv 13:390–398
Chapman FM (1929) My tropical air castle. D Appleton-Century Co, New York
Coleman JS, Fraser JD (1987) Food habits of black and turkey vultures in Pennsylvania and Maryland. J Wildl Manag 51:733–739
Cook RR, Angermeier PL, Finn DS et al (2004) Geographic variation in patterns of nestedness among local stream fish assemblages in Virginia. Oecologia 140:639–649
Cortés-Avizanda A, Carrete M, Donázar JA (2010) Managing supplementary feeding for avian scavengers: guidelines for optimal design using ecological criteria. Biol Conserv 143:1707–1715
Cortés-Avizanda A, Jovani R, Carrete M et al (2012) Resource unpredictability promotes species diversity and coexistence in an avian scavenger guild: a field experiment. Ecology 93:2570–2579
Cortés-Avizanda A, Jovani R, Donázar JA et al (2014) Bird sky networks: how do avian scavengers use social information to find carrion? Ecology 95:1799–1808
Culloch R (2012) Short note: observations from video footage of red fox (*Vulpes vulpes*) activity within a grey seal (*Halichoerus grypus*) breeding colony on the UK Mainland. Aquat Mamm 38:81–85
Curtis TH, Kelly JT, Menard KL et al (2006) Observations on the behavior of white sharks scavenging from a whale carcass at Point Reyes, California. Calif Fish Game 92:113–124
da Diefenbach CO (1975) Gastric function in Caiman crocodilus (Crocodylia: Reptilia)—I. Rate of gastric digestion and gastric motility as a function of temperature. Comp Biochem Physiol A 51(2):259–265
Deinet S, Ieronymidou C, McRae L et al (2013) Wildlife comeback in Europe: the recovery of selected mammal and bird species. Final report to Rewilding Europe by ZSL, BirdLife International and the European Bird Census Council, London
Del Hoyo J (1994) In: Elliott A, Sargatal J (eds) New world vultures to guineafowl. Handbook of the birds of the world, vol 2. Lynx Edicions, Barcelona
DeVault TL, Krochmal AR (2002) Scavenging by snakes: an examination of the literature. Herpetologica 58:429–436
DeVault TL, Rhodes OE (2002) Identification of vertebrate scavengers of small mammal carcasses in a forested landscape. Acta Theriol 47:185–192
DeVault TL, Rhodes OE Jr, Shivik JA (2003) Scavenging by vertebrates: behavioral, ecological, and evolutionary perspectives on an important energy transfer pathway in terrestrial ecosystems. Oikos 102:225–234
DeVault TL, Brisbin IL, Rhodes OE (2004) Factors influencing the acquisition of rodent carrion by vertebrate scavengers and decomposers. Can J Zool 82:502–509
DeVault TL, Olson ZH, Beasley JC et al (2011) Mesopredators dominate competition for carrion in an agricultural landscape. Basic Appl Ecol 12:268–274
Dicken ML (2008) First observations of young of the year and juvenile great white sharks (*Carcharodon carcharias*) scavenging from a whale carcass. Mar Freshw Res 59:596–602
Donázar JA, Margalida A, Campión D (eds) (2009) Vultures, feeding stations and sanitary legislation: a conflict and its consequences from the perspective of conservation biology, Munibe 29 (Suppl). Sociedad de Ciencias Aranzadi, San Sebastián
Dudley SF, Anderson-Reade MD, Thompson GS et al (2000) Concurrent scavenging off a whale carcass by great white sharks, *Carcharodon carcharias*, and tiger sharks, *Galeocerdo cuvier*. Fish Bull Ocean Atmos Adm 98:646–649
Dunlop K, Scott M, Parsons D et al (2015) Do agonistic behaviours bias baited remote underwater video surveys of fish? Mar Ecol 36:810–818
Elbroch LM, Wittmer HU (2012) Table scraps: inter-trophic food provisioning by pumas. Biol Lett 8:776–779

Elbroch LM, Wittmer HU (2013) Nuisance ecology: do scavenging condors exact foraging costs on pumas in Patagonia? PLoS ONE 8(1):e53595
Eliotout B (2007) Le vautour fauve. Delachaux et niestlé, Lausanne
Fallows C, Gallagher A, Hammerschlag N (2013) White sharks (*Carcharodon carcharias*) scavenging on whales and its potential role in further shaping the ecology of an apex predator. PLoS ONE 8(4):e60797
Fortin JK, Farley SD, Rode KD et al (2007) Dietary and spatial overlap between sympatric ursids relative to salmon use. Ursus 18:19–29
Gallagher A, Jackson T, Hammerschlag N (2011) Occurrence of tiger shark (*Galeocerdo cuvier*) scavenging on avian prey and its possible connection to large-scale bird die-off in the Florida Keys. Fla Sci 4:264–269
Gilchrist HG, Robertson GJ (2000) Observations of marine birds and mammals wintering at polynyas and ice edges in the Belcher Islands, Nunavut, Canada. Arctic 53:61–68
Gjertz I, Lydersen C (1986) Polar bear predation on ringed seals in the fast-ice of Hornsund, Svalbard. Polar Res 4:65–68
Godø O, Huse I, Michalsen K (1997) Bait defence behaviour of wolffish and its impact on long-line catch rates. ICES J Mar Sci 54:272–275
González-Solís J, Croxall JP, Briggs DR (2002) Activity patterns of giant petrels, *Macronectes* spp., using different foraging strategies. Mar Biol 140:197–204
Green GI, Mattson DJ, Peek JM (1997) Spring feeding on ungulate carcasses by grizzly bears in Yellowstone National Park. J Wildl Manag 61:1040–1055
Green RE, Newton I, Shultz S et al (2004) Diclofenac poisoning as a cause of vulture population declines across the Indian subcontinent. J Appl Ecol 41:793–800
Guarino F (2001) Diet of a large carnivorous lizard, *Varanus varius*. Wildl Res 28:627–630
Heinrich B (1988) Winter foraging at carcasses by three sympatric corvids, with emphasis on recruitment by the raven, *Corvus corax*. Behav Ecol Sociobiol 23:141–156
Heinrich B, Marzluff JM, Marzluff CS (1993) Common ravens are attracted by appeasement calls of food discoverers when attacked. Auk 110:247–254
Hertel F (1994) Diversity in body size and feeding morphology within past and present vulture assemblages. Ecology 75:1074–1084
Hertel F, Lehman N (1998) A randomized nearest-neighbor approach for assessment of character displacement: the vulture guild as a model. J Theor Biol 190:51–61
Hilderbrand GV, Schwartz CC, Robbins CT et al (1999) The importance of meat, particularly salmon, to body size, population productivity, and conservation of North American brown bears. Can J Zool 77:132–138
Hiraldo F, Blanco JC, Bustamante J (1991) Unspecialized exploitation of small carcasses by birds. Bird Study 38:200–207
Houston DC (1974) The role of griffon vultures *Gyps africanus* and *Gyps rüpellii* as scavengers. J Zool 172:35–46
Houston DC (1975) Ecological isolation of African scavenging birds. Ardea 63:55–64
Houston DB (1978) Elk as winter-spring food for carnivores in northern Yellowstone National Park. J Appl Ecol 15:653–661
Houston DC (1979) The adaptations of scavengers. In: Sinclair ARE, Griffiths MN (eds) Serengeti, dynamics of an ecosystem. University of Chicago Press, Chicago, pp 263–286
Houston DC (1983) The adaptive radiation of griffon vultures. In: Wilbur SR, Jackson JA (eds) Vulture biology and management. University of California Press, Berkeley, pp 135–152
Houston DC (1984) Does the king vulture *Sarcoramphus papa* use a sense of smell to locate food? Ibis 126:67–69
Houston DC (1986) Scavenging efficiency of turkey vultures in tropical forest. Condor 88:318–323
Houston DC (1988) Competition for food between neotropical vultures in forest. Ibis 130:402–417
Houston DC (2001) Vultures and condors. Clin Baxter Photography Ltd, Grantown-on-Spey
Houston DC, Cooper JE (1975) The digestive tract of the white-back griffon vulture and its role in disease transmission among wild ungulates. J Wildl Dis 11:306–313

Huijbers CM, Schlacher TA, Schoeman DS et al (2013) Urbanisation alters processing of marine carrion on sandy beaches. Landsc Urban Plan 119:1–8

Huijbers CM, Schlacher TA, Schoeman DS et al (2015) Limited functional redundancy in vertebrate scavenger guilds fails to compensate for the loss of raptors from urbanized sandy beaches. Divers Distrib 21:55–63

Hunter S (1985) The role of giant petrels in the Southern Ocean ecosystem. In: Siegfried WR, Laws RM, Condy PR (eds) Antarctic nutrient cycles and food webs. Springer, New York, pp 534–542

Hunter S (1991) The impact of avian predator-scavengers on king penguin *Aptenodytes patagonicus* chicks at Marion Island. Ibis 133:343–350

Isaacs J, Schwartzlose R (1975) Active animals of the deep-sea floor. Sci Am 233:85–91

Jackson AL, Ruxton GD, Houston DC (2008) The effect of social facilitation on foraging success in vultures: a modelling study. Biol Lett 4:311–313

Jędrzejewska B, Jędrzejewski W (1998) Predation in vertebrate communities. The Białowieża Primeval Forest as a case study. Springer, Berlin

Jędrzejewski W, Zalewski A, Jędrzejewska B (1993) Foraging by pine marten *Martes martes* in relation to food resources in Białowieża National Park, Poland. Acta Theriol 38: 405–426

Jones E, Collins M, Bagley P et al (1998) The fate of cetacean carcasses in the deep sea: observations on consumption rates and succession of scavenging species in the abyssal north-east Atlantic Ocean. Proc R Soc B 265:1119–1127

Kane A, Jackson AL, Ogada DL et al (2014) Vultures acquire information on carcass location from scavenging eagles. Proc R Soc B 281:20141072

Kane A, Healy K, Ruxton GD et al (2016) Body size as a driver of scavenging in theropod dinosaurs. Am Nat 187:706–716

Kelly NE, Sparks DW, DeVault TL et al (2007) Diet of black and turkey vultures in a forested landscape. Wilson J Ornithol 119:267–270

Kendall C (2013) Alternative strategies in avian scavengers: how subordinate species foil the despotic distribution. Behav Ecol Sociobiol 67:383–393

Kendall C, Virani MZ, Kirui P et al (2012) Mechanisms of coexistence in vultures: understanding the patterns of vulture abundance at carcasses in Masai Mara National Reserve, Kenya. Condor 114:523–531

Killengreen ST, Lecomte N, Ehrich E et al (2011) The importance of marine vs. human induced subsidies in the maintenance of an expanding mesocarnivore in the Arctic tundra. J Anim Ecol 80:1049–1060

Killengreen ST, Strømseng E, Yoccoz NG et al (2012) How ecological neighbourhoods influence the structure of the scavenger guild in low arctic tundra. Divers Distrib 18:563–574

Knight SK, Knight RL (1983) Aspects of food finding by wintering bald eagles. Auk 100: 477–484

König C (1974) On the behaviour of vultures on carcasses in Spain. J Ornithol 115:289–320

Koskela A, Kojola I, Aspi J et al (2013) Effect of reproductive status on the diet composition of wolverines (*Gulo gulo*) in boreal forests of eastern Finland. Ann Zool Fenn 50:100–106

Kostecke RM, Linz GM, Bleier WJ (2001) Survival of avian carcasses and photographic evidence of predators and scavengers. J Field Ornithol 72:439–447

Krofel M, Kos I, Jerina K (2012) The noble cats and the big bad scavengers: effects of dominant scavengers on solitary predators. Behav Ecol Sociobiol 66:1297–1304

Kruuk H (1966) Clan-system and feeding habits of spotted hyaenas (*Crocuta crocuta* Erxleben). Nature 209:1257–1258

Kruuk H (1967) Competition for food between vultures in East Africa. Ardea 55:171–193

Kruuk H (1972) The spotted hyaena. A study of predation and social behavior. University of Chicago Press, Chicago

Kuhn B, Wiesel I, Skinner J (2008) Diet of brown hyaenas (*Parahyaena brunnea*) on the Namibian coast. Trans R Soc S Afr 63:1–8

Lambertucci SA, Speziale KL, Rogers TE et al (2009a) How do roads affect the habitat use of an assemblage of scavenging raptors? Biodivers Conserv 18:2063–2074
Lambertucci SA, Trejo A, Di Martino S et al (2009b) Spatial and temporal patterns in the diet of the Andean condor: ecological replacement of native fauna by exotic species. Anim Conserv 12:338–345
Lambertucci SA, Navarro J, Sanchez Zapata JA, Hobson KA, Alarcón PAE, Wiemeyer G, Blanco G, Hiraldo F, Donázar JA (2018) Tracking data and retrospective analyses of diet reveal the consequences of loss of marine subsidies for an obligate scavenger, the Andean condor. Proc R Soc B Biol Sci 285(1879):20180550
Levi T, Wheat RE, Allen JM et al (2015) Differential use of salmon by vertebrate consumers: implications for conservation. PeerJ 3:e1157
Linz GM, Bergman DL, Bleier WJ (1997) Estimating survival of song bird carcasses in crops and woodland. Prairie Nat 29:7–13
Løkkeborg S (1998) Feeding behaviour of cod, *Gadus morhua*: activity rhythm and chemically mediated food search. Anim Behav 56:371–378
Long D, Jones R (1996) White shark predation and scaveging on cetaceans in the Eastern North Pacific Ocean. In: Kimley AP, Ainley DG (eds) Great white sharks: the biology of *Carcharodon carcharias*. Academic, San Diego, pp 293–307
Lowry LF, Burns JJ, Nelson RR (1987) Polar bear, *Ursus maritimus*, predation on belugas, *Delphinapterus leucas*, in the Bering and Chukchi Seas. Can Field Nat 101:141–146
MaMing R, Xu G (2015) Status and threats to vultures in China. Vulture News 68:3–24
MaMing R, Lee L, Yang X, Buzzard P (2016) Vultures and sky burials on the Qinghai-Tibet plateau. Vulture News 71:22–35
Margalida A, Colomer MA (2012) Modelling the effects of sanitary policies on European vulture conservation. Sci Rep 2:753
Margalida A, Heredia R (eds) (2005) Biología de la conservación del quebrantahuesos (*Gypaetus barbatus*) en España. Organismo Autónomo Parques Nacionales, Madrid
Margalida A, Carrete M, Sánchez-Zapata JA et al (2012) Good news for European vultures. Science 335:284
Margalida A, Bogliani G, Bowden C et al (2014) One health approach to use of pharmaceuticals. Science 346:1296–1298
Margalida A, Colomer MA, Sánchez R et al (2017) Behavioral evidence of hunting and foraging techniques by a top predator suggests the importance of scavenging for pre-adults. Ecol Evol 7:4192–4199
Martin PS (1989) Prehistoric overkill: the global model. In: Martin PS, Klein RG (eds) Quaternary extinctions. University of Arizona Press, Tucson, pp 354–403
Marzluff JM, Heinrich B (1991) Foraging by common ravens in the presence and absence of territory holders: an experimental analysis of social foraging. Anim Behav 42:755–770
Marzluff JM, Heinrich B, Marzluff CS (1996) Raven roosts are mobile information centres. Anim Behav 51:89–103
Mateo-Tomás P, Olea PP, Moleón M et al (2015) From regional to global patterns in vertebrate scavenger communities subsidized by big game hunting. Divers Distrib 21:913–924
Mattisson J, Andrén H, Persson J et al (2011) Influence of intraguild interactions on resource use by wolverines and Eurasian lynx. J Mammal 92:1321–1330
Matyushkin EN (1974) Large carnivores and scavengers of the middle Sikhote-Alin. Bulleten Moskovskogo Obshchestva Ispytatelei Prirody 79:5–21
Moleón M, Sánchez-Zapata JA, Margalida A et al (2014a) Humans and scavengers: the evolution of interactions and ecosystem services. Bioscience 64:394–403
Moleón M, Sánchez-Zapata JA, Selva N et al (2014b) Inter-specific interactions linking predation and scavenging in terrestrial vertebrate assemblages. Biol Rev 89:1042–1054
Moleón M, Sánchez-Zapata JM, Sebastián-González E et al (2015) Carcass size shapes the structure and functioning of an African scavenging assemblage. Oikos 124:1391–1403
Moreno-Opo R, Margalida A (2013) Carcasses provide resources not exclusively to scavengers: patterns of carrion exploitation by passerine birds. Ecosphere 4:105

Moreno-Opo R, Trujillano A, Arredondo A et al (2015a) Manipulating size, amount and appearance of food inputs to optimize supplementary feeding programs for European vultures. Biol Conserv 181:27–35

Moreno-Opo R, Trujillano A, Margalida A (2015b) Optimization of supplementary feeding programs for European vultures depend on environmental and management factors. Ecosphere 6:127

Moreno-Opo R, Trujillano A, Margalida A (2016) Behavioral coexistence and feeding efficiency drive niche partitioning in European avian scavengers. Behav Ecol 27:1041–1052

Moss B (2017) Marine reptiles, birds and mammals and nutrient transfers among the seas and the land: an appraisal of current knowledge. J Exp Mar Biol Ecol 492:63–80

Novaro AJ, Funes MC, Walker RS (2000) Ecological extinction of native prey of a carnivore assemblage in Argentine Patagonia. Biol Conserv 92:25–33

O'Sullivan WM, Sleeman DP, Murphy DM (1992) Otters *Lutra lutra* feeding on carrion. Ir Nat 24:140–143

Olson ZH, Beasley JC, Rhodes OE (2016) Carcass type affects local scavenger guilds more than habitat connectivity. PLoS ONE 11(2):e0147798

Oro D, Genovart M, Tavecchia G et al (2013) Ecological and evolutionary implications of food subsidies from humans. Ecol Lett 16:1501–1514

Parmenter RR, MacMahon JA (2009) Carrion decomposition and nutrient cycling in a semiarid shrub-steppe ecosystem. Ecol Monogr 79:637–661

Pavés H, Schlatter R, Espinoza C (2008) Scavenging and predation by black vultures *Coragyps atratus* at a South American sea lion breeding colony. Vulture News 58:4–15

Payne JA (1965) A summer carrion study of the baby pig *Sus scrofa* Linnaeus. Ecology 46: 592–602

Payne LX, Moore JW (2006) Mobile scavengers create hotspots of freshwater productivity. Oikos 115:69–80

Pennycuick CJ (1983) Thermal soaring compared in three dissimilar tropical bird species, *Fregata magnificens*, *Pelecanus occidentalis* and *Coragyps atratus*. J Exp Biol 102:307–325

Pereira LM, Owen-Smith N, Moleón M (2014) Facultative predation and scavenging by mammalian carnivores. Mammal Rev 44:44–55

Peterson CA, Lee SL, Elliott JE (2001) Scavenging of waterfowl carcasses by birds in agricultural fields of British Columbia. J Field Ornithol 72:150–159

Petrides GA (1959) Competition for food between five species of East African vultures. Auk 76:104–106

Phillips RA, Phalan B, Forster IP (2004) Diet and long-term changes in population size and productivity of brown skuas *Catharacta antarctica lonnbergi* at Bird Island, South Georgia. Polar Biol 27:555–561

Platt SG, Rainwater TR, Snider S et al (2007) Consumption of large mammals by *Crocodylus moreletii*: field observations of necrophagy and interspecific kleptoparasitism. Southwest Nat 52:310–317

Polis GA, Hurd SD (1996) Linking marine and terrestrial food webs: allochthonous input from the ocean supports high secondary productivity on small islands and coastal land communities. Am Nat 147:396–423

Prior KA, Weatherhead PJ (1991) Competition at the carcass: opportunities for social foraging by turkey vultures in southern Ontario. Can J Zool 69:1550–1556

Quaggiotto M-M, Barton PS, Morris C et al (2018) Seal carrion is a predictable resource for coastal ecosystems. Acta Oecol 88:41–51

Raichev EG, Tsunoda H, Newman C et al (2013) The reliance of the golden jackal (*Canis aureus*) on anthropogenic foods in winter in central Bulgaria. Mamm Study 38:19–27

Ray R–R, Seibold H, Heurich M (2014) Invertebrates outcompete vertebrate facultative scavengers in simulated lynx kills in the Bavarian Forest National Park, Germany. Anim Biodivers Conserv 37(1):77–88

Read JL, Wilson D (2004) Scavengers and detritivores of kangaroo harvest offcuts in arid Australia. Wildl Res 31:51–56

Rogers AM, Gibson MR, Pockette T et al (2014) Scavenging of migratory bird carcasses in the Sonoran desert. Southwest Nat 59:544–549

Rohr RP, Saavedra S, Bascompte J (2014) On the structural stability of mutualistic systems. Science 345:1253497

Ronconi RA, Steenweg RJ, Taylor PD et al (2014) Gull diets reveal dietary partitioning and ecosystem changes at a remote colony. Mar Ecol Prog Ser 514:247–261

Rose M, Polis G (1998) The distribution and abundance of coyotes: the effects of allochthonous food subsidies from the sea. Ecology 79:998–1007

Roth JD (2002) Temporal variability in arctic fox diet as reflected in stable carbon isotopes; the importance of sea ice. Oecologia 133:70–77

Ruxton GD, Houston DC (2003) Could *Tyrannosaurus rex* have been a scavenger rather than a predator? An energetics approach. Proc R Soc Lond B 270:731–733

Ruxton GD, Houston DC (2004) Obligate vertebrate scavengers must be large soaring fliers. J Theor Biol 228:431–436

Samelius G, Alisauskas RT, Hobson KA et al (2007) Prolonging the arctic pulse: long-term exploitation of cached eggs by Arctic foxes when lemmings are scarce. J Anim Ecol 76:873–880

Sánchez R, Margalida A, González LM et al (2009) Temporal and spatial differences in the feeding ecology of the Spanish imperial eagle *Aquila adalberti* during the non-breeding season: effects of the rabbit population crash. Acta Ornithol 44:53–58

Sánchez-Zapata JA, Eguía S, Blázquez M et al (2010) Unexpected role of ungulate carcasses in the diet of golden eagles *Aquila chrysaetos* in Mediterranean mountains. Bird Study 57:352–360

Schindler DE, Armstrong JB, Bentley KT et al (2013) Riding the crimson tide: mobile terrestrial consumers track phenological variation in spawning of an anadromous fish. Biol Lett 9:20130048

Sebastián-González E, Sánchez-Zapata JA, Donázar JA et al (2013) Interactive effects of obligate scavengers and scavenger community richness on lagomorph carcass consumption patterns. Ibis 155:881–885

Sebastián-González E, Moleón M, Gibert JP et al (2016) Nested species-rich networks of scavenging vertebrates support high levels of interspecific competition. Ecology 97:95–105

Selva N (2004) The role of scavenging in the predator community of Białowieża Primeval Forest (E Poland). PhD thesis, University of Sevilla, Spain

Selva N, Fortuna MA (2007) The nested structure of a scavenger community. Proc R Soc B 274:1101–1108

Selva N, Jędrzejewska B, Jędrzejewski W et al (2003) Scavenging on European bison carcasses in Białowieża Primeval Forest (eastern Poland). Ecoscience 10:303–311

Selva N, Jedrzejewska B, Jedrzejewski W et al (2005) Factors affecting carcass use by a guild of scavengers in European temperate woodland. Can J Zool 83:1590–1601

Shepard EL, Lambertucci SA (2013) From daily movements to population distributions: weather affects competitive ability in a guild of soaring birds. J R Soc Interface 10:20130612

Shivik JA (2006) Are vultures birds, and do snakes have venom, because of macro- and microscavenger conflict? Bioscience 56:819–823

Skinner JD, Chimimba CT (2005) The mammals of the Southern African subregion. Cambridge University Press, Cambridge

Smith CR, Baco AR (2003) Ecology of whale falls at the deep-sea floor. Oceanogr Mar Biol 41:311–354

Spiller DA, Piovia-Scorr J, Wright AN et al (2010) Marine subsidies have multiple effects on coastal food webs. Ecology 91:1424–1434

Stahler D, Heinrich B, Smith D (2002) Common ravens, *Corvus corax*, preferentially associate with grey wolves, *Canis lupus*, as a foraging strategy in winter. Anim Behav 64:283–290

Stirling I, Archibald WR (1977) Aspects of predation of seals by polar bears. J Fish Res Board Can 34:1126–1129

Stoddart LC (1970) A telemetric method for detecting jackrabbit mortality. J Wildl Manag 34:501–507

Stoner A (2004) Effects of environmental variables on fish feeding ecology: implications for the performance of baited fishing gear and stock assessment. J Fish Biol 65:1445–1471
Swihart RK, Gehring TM, Kolozsvary MB et al (2003) Response of 'resistant' vertebrates to habitat loss and fragmentation: the importance of niche breadth and range boundaries. Divers Distrib 9:1–18
Tenney S (1877) A few words about scavengers. Am Nat 11:129–135
Terrasse JF (2001) Le gypaète barbu. Delachaux et niestlé, Lausanne
Tobin ME, Dolbeer RA (1990) Disappearance and recoverability of songbird carcasses in fruit orchards. J Field Ornithol 61:237–242
Travaini A, Donazar JA, Rodriguez A et al (1998) Use of European hare (*Lepus europaeus*) carcasses by an avian scavenging assemblage in Patagonia. J Zool 246:175–181
Twiss SD, Duck C, Pomeroy PP (2003) Grey seal (*Halichoerus grypus*) pup mortality not explained by local breeding density on North Rona, Scotland. J Zool 259:83–91
Vogeley W (1999) Use of carcasses by Cape griffons *Gyps coprotheres* and food competition – observations at a vulture restaurant in south-eastern Botswana. Ornithol Beobachter 96:13–23
Wallace MP, Temple SA (1987) Competitive interactions within and between speies in a guild of avian scavengers. Auk 104:290–295
Waters CN, Zalasiewicz J, Summerhayes C et al (2016) The Anthropocene is functionally and stratigraphically distinct from the Holocene. Science 351:aad2622
Weldon PJ, Ferguson MWJ (1993) Chemoreception in crocodilians: anatomy, natural history, and empirical results. Brain Behav Evol 41:239–245
Weldon PJ, Swanson DJ, Olson JK et al (1990) The American alligator detects food chemicals in aquatic and terrestrial environments. Ethology 85:191–198
White C (2005) Hunters ring dinner bell for ravens: experimental evidence of a unique foraging strategy. Ecology 86:1057–1060
Whitehead H, Reeves R (2005) Killer whales and whaling: the scavenging hypothesis. Biol Lett 1:415–418
Wikenros C, Sand H, Ahlqvist P et al (2013) Biomass flow and scavengers use of carcasses after re-colonization of an apex predator. PLoS ONE 8(10):e77373
Williams AJ, Dyer BM, Randall RM, Komen J (1990) Killer whales *Orcinus orca* and seabirds "play" predation and association. Mar Ornithol 18:37–41
Wilmers CC, Stahler DR, Crabtree RL et al (2003a) Resource dispersion and consumer dominance: scavenging at wolf- and hunter- killed carcasses in Greater Yellowstone, USA. Ecol Lett 6:996–1003
Wilmers CC, Crabtree RL, Smith DW et al (2003b) Trophic facilitation by introduced top predators: grey wolf subsidies to scavengers in Yellowstone National Park. J Anim Ecol 72:909–916
Wilson R, Smith K (1984) Effect of near-bottom currents on detection of bait by the 4 abyssal grenadier fishes *Coryphaenoides* spp., recorded in situ with a video 5 camera on a free fall vehicle. Mar Biol 84:83–91
Wilson EE, Wolkovich EM (2011) Scavenging: how carnivores and carrion structure communities. Trends Ecol Evol 26:129–135
Worthen WB, Jones MT, Jettojn RM (1998) Community structure and environmental stress: desiccation promotes nestedness in mycophagous fly communities. Oikos 81:45–54
Wright J, Stone RE, Brown N (2003) Communal roosts as structured information centers in the raven, *Corvus corax*. J Anim Ecol 72:1003–1014
Yeh J, Drazen J (2011) Baited-camera observations of deep-sea megafaunal scavenger ecology on the California slope. Mar Ecol Prog Ser 424:145–156

Carrion Decomposition

Philip S. Barton and Joseph K. Bump

Contents

P. S. Barton (✉)
Fenner School of Environment and Society, Australian National University,
Canberra, ACT, Australia
e-mail: philip.barton@anu.edu.au

J. K. Bump
Forest Wildlife Research and Education, Department of Fisheries, Wildlife, and Conservation
Biology, University of Minnesota, St. Paul, MN, USA
e-mail: bump@umn.edu

P. P. Olea et al. (eds.), *Carrion Ecology and Management*, Wildlife Research
Monographs 2, https://doi.org/10.1007/978-3-030-16501-7_5

Introduction

Life concentrates biologically limiting resources. Primary producers harness relatively diffuse solar energy and earth elements to produce tissues that feed the vast array of herbivores globally. Herbivore tissue in turn, as a resource, increases in nutritional value and feeds a diversity of predators across multiple trophic levels. In death, animal tissues are consumed or re-enter the cycle via decomposition. In this way, death disperses biologically limiting resources. This simplification of global resource cycling via biological pathways highlights the fundamental importance of dead animal matter (carrion; e.g. vertebrate carcasses) in maintaining biodiversity and ecological processes, especially in terrestrial ecosystems. Carrion occurs in all biomes and can be conceptualized as a bottleneck in the vital flow of energy and limiting nutrients with profound ecological and evolutionary consequences. Until mortality, consumers temporarily concentrate and sequester energy, nutrients, and moisture in their standing biomass. As a result, carrion constitutes a critical resource for a variety of communities and it strongly affects scavenger guilds, the maintenance of biodiversity, and life-sustaining ecological processes (Barton et al. 2013a).

Carrion decomposition is a fundamental ecological process involving the breakdown of dead animals and the recycling of their embodied nutrients and energy through other organisms and their environment. The decomposition process involves a combination of physical and biological degradation and breakdown of a carcass into smaller fragments and its constituent biochemical components. This occurs through the combined action of intrinsic (e.g. enzymatic and internal microbial processes), and extrinsic processes (e.g. fragmentation and consumption by organisms). All of these processes are strongly moderated by a wide variety of abiotic (e.g. temperature, moisture), and biotic factors such as vegetation and inter-specific competition among scavengers. In this chapter, we have taken a broad perspective of carrion decomposition to include all biotic and abiotic processes that affect the breakdown of an animal carcass. Subsequent chapters give greater detail on the specific taxa (see chapter "Invertebrate Scavenging Communities") and processes (see chapter "Ecological Functions of Vertebrate Scavenging"). We outline the decomposition process, the role of different groups of organisms, and key moderating abiotic and environmental factors. We conclude with a discussion of some future research directions that highlight how different technologies and interdisciplinary collaboration can yield greater understanding of this important process.

Carrion Decomposition

Animal Carrion as a Distinct Resource

Carrion is a distinctive form of detritus in ecosystems when compared with other forms of detritus, such as animal dung, fruiting fungal bodies, or the many kinds of plant detritus, like litter, dead wood, or fruit (Barton et al. 2013a; Finn 2001). It is important to highlight three key attributes of carrion that affects its decomposition. First, the remains of animals are generally rare and comprise only a fraction of the organic biomass in ecosystems (Barton et al. 2013a; Parmenter and MacMahon 2009). In contrast, plant litter is ubiquitous in many ecosystems, from terrestrial grasslands to marine kelp forests. Although the occurrence of different forms of plant detritus can vary (e.g. leaves vs. branches), the availability of plant detritus is generally not as limited as carrion. Second, carrion is very rich in nutrients, whereas plant litter is comparatively nutrient-poor with low concentrations of key macronutrients such as nitrogen and phosphorus (Moore et al. 2004; Swift et al. 1979). The nitrogen content of carrion can be up to five times higher, and moisture up to 10 times higher, than several kinds of plant litter (Carter et al. 2007). Third, carrion decomposes orders of magnitude faster than plant litter (Parmenter and MacMahon 2009), and is only available as a resource to other organisms for a brief window of time. These factors combine to make carrion spatially patchy in occurrence, very nutrient rich, and ephemeral. These distinctive features of carrion not only make it a unique resource in ecosystems, but also strongly affects how it decomposes.

The Decomposition Process

The processes responsible for the decomposition of an individual carcass have been described as creating a 'cadaver decomposition island' (sensu Carter et al. 2007). This decomposition island is a very localized and intense 'hot spot' and 'hot moment' of chemical and biological activity that ultimately leads to the breakdown and recycling of the energy and nutrients in a dead animal through other organisms and the wider environment. This occurs through both chemical and biological processes. The intrinsic chemical processes, such as autolysis and putrefaction, are not covered in detail in this chapter. Briefly, after death of the animal, cells no longer receive oxygen or nutrients, and they are unable to maintain normal functioning. This leads to uncontrolled enzymatic and biochemical activity, which leads to cell death, autolysis and the breakdown and leakage of cell membranes. A more in-depth coverage of this aspect of decomposition can be found elsewhere (e.g. Carter et al. 2007; Dent et al. 2004; Forbes and Carter 2015). The extrinsic biological drivers of decomposition are largely due to other organisms, and is initiated by the rapid proliferation of bacteria present in the digestive tract and on the epidermal surface. These microbes release their own enzymes that further break down the animal's cells. Together with the intrinsic chemical activity, the early stages of decomposition are characterized by the release of gases, which act as cues that attract early insect colonizers of carrion that specialize on carrion as a food resource (see chapter "Invertebrate Scavenging Communities"), and in turn begins the physical break down of the animal carcass through consumption of internal fluids and tissues. Further physical decomposition of carrion is facilitated by vertebrate scavengers that disarticulate carcasses and consume large portions of tissue (see chapter "Vertebrate Scavenging Communities"). This results in fragmentation of the carcass into smaller parts as pieces are physically moved apart and scattered by animals. The net product of these chemical, biological and physical processes leads to the decomposition and disintegration of a carcass.

Temporal Progression of Carrion Decay

A key aspect of carrion decomposition is its temporal component, involving the progressive breakdown of tissues through time. In their detailed review, Michaud et al. (2015) noted that carrion decomposition is often viewed as a series of steps or stages by many researchers, and this step-wise description is pervasive in both the entomological and forensic literature. This view, of course, serves as a useful way to differentiate between assemblages of species or qualitative features of a carcass at different points in time. This approach is further perpetuated by sampling and observation methods, which are often necessarily restricted to points in time, thus leading to clear distinctions when making comparisons among samples and observations. Nevertheless, it is important to reiterate (with the above-mentioned methodological constraints in mind) that little evidence exists of substantive step-changes in decomposition, and that the process (mostly) occurs gradually through time. The description of decay 'stages' should therefore be considered as phases of decomposition through which a typical carcass passes gradually and not abruptly, and care should be taken in focusing on discrete stages where there is a continuum of change.

A typical classification scheme of temporal decay stage includes fresh, bloat, active, advance, and dry decay. A *fresh* carcass lasts for only a few minutes to hours after death, depending on temperature, before it enters a bloat stage whereby internal bacteria proliferate inside the gut, and leak into the lymphatic system and other tissues. The *bloating* is caused by the gases released by the bacteria and decaying tissues, and include methane, hydrogen sulfide, and carbon dioxide (Forbes and Carter 2015). The release of gases during the bloat stage is often the trigger for colonisation by carrion insects, such as blowflies, with gases acting as an important cue for species to search and locate the carcass (see chapter "Invertebrate Scavenging Communities"). *Active decay* of an animal carcass is characterized by the putrefaction and liquefaction of carcass tissues, and the release of a complex array of volatile organic compounds (Forbes and Carter 2015). A rapid loss of carcass mass occurs due to the consumption of tissues by scavenger organisms, both vertebrate and invertebrate (see chapters "Invertebrate Scavenging Communities" and "Vertebrate Scavenging Communities"). Moisture is also lost from the carcass into the environment via evaporation and leakage of fluids into the soil. Vertebrate scavengers can play a role in disarticulating and moving larger carcasses, and local weather conditions such as wind and rain, may play a role in spreading the components of a carcass such as bones, skin fragments and fur/feathers. Progression into *advanced decay* involves the final breakdown of soft tissues and the appearance of the skeleton. *Dry decay* occurs over a much longer time frame, and depending on the environment, the recalcitrant parts of a carcass, particularly the ligaments, nails, hair, and skeleton, may persist for many months and years.

The Contribution of Organisms to the Decomposition Process

A diverse range of taxa are involved with the decomposition process, with several being critical to the rate and/or completeness of decay through either the direct consumption of carrion, or preying on other organisms that do. Experimental exclusion or removal of different components of the necrobiome have also demonstrated their importance in accelerating mass loss from a carcass (Lauber et al. 2014; Pechal et al. 2014a). Below we outline the role of three broad groups of organisms—microbes, invertebrates and vertebrates—in the decomposition process. We focus only on the most important taxa, with further and more comprehensive details on succession and species-specific roles of invertebrates and vertebrates given in other chapters.

Microbial Decomposers

Microbial decomposers (bacteria, archaea, fungi, protists) have a critical role in the decomposition of carrion (Carter et al. 2008; Crippen et al. 2015; Lauber et al. 2014; Metcalf et al. 2015). The diversity of the microbial community found at carrion is impressive, with many thousands of taxa found both in and on carcasses (Crippen et al. 2015; Pechal et al. 2013, 2014b). The primary mode of action of both bacterial and fungal decomposers is the secretion of enzymes to breakdown complex organic molecules into smaller ones that are then metabolized for their growth and rapid proliferation. Microbes, primarily bacteria, are present in great numbers both in the gut of animals and on the surface of the skin. Gut bacteria have a particularly important role in the decomposition of animals, and can influence the rate of decay (Lauber et al. 2014), as well as the appearance and odour of carcasses. The main aerobic bacteria include several taxa from the Firmicutes and Bacteroidetes phyla, and anaerobic bacteria include the commonly known *Lactobacillus*, *Streptococcus*, and *Staphylococcus* that play a role in the fermentation of various organic compounds in carrion (Forbes and Carter 2015). Fungi may become more abundant on older carcasses as conditions become drier and more hostile for bacteria (Carter and Tibbett 2003). Importantly, microbial communities play a critical role in accelerating decomposition rates, and can drive the decomposition process in the absence of all other eukaryotic organisms. It has been established through experimental work, for example, that decomposition rates of mice carcasses occurred at twice the rate in soil with intact microbial communities compared with sterile soil (Lauber et al. 2014). The importance of microbes to decomposition lies in their ability to produce a broad array of degradative enzymes, and their ability to colonise and use a broad range of carrion substrates, including all internal tissues and organs, as well as keratinous hair and skin, and even bone.

Arthropods

Arthropods comprise both primary and secondary consumer trophic roles at carcasses, and their presence at a carcass can accelerate the decomposition process through consumption of the different components of a carcass. Surprisingly few experimental tests have been performed to directly assess the role of arthropods, but these few studies have identified important contributions (Farwig et al. 2014; Payne 1965; Pechal et al. 2014a). A detailed overview of the different arthropod taxa present at carrion is given in chapter "Invertebrate Scavenging Communities", and here we focus on their functional roles.

Braack (1987) provided a useful classification scheme of the functional role of different arthropod groups by synthesizing his research on the arthropod fauna at carcasses in Africa. Not all functional groups may be present at a carcass in other parts of the world, or indeed at different carcasses in the same biome. For example, the rumen contents and large horns of an antelope attract a different set of fauna than a small carnivore carcass where these elements are absent. The relationship between the taxa that make up these functional groups (Fig. 1) shows that some taxa may feed on more than one carcass component, and therefore perform more than one functional role. This also applies to adults and larvae of the same species, which may each specialize on a different component. Such variation highlights varying

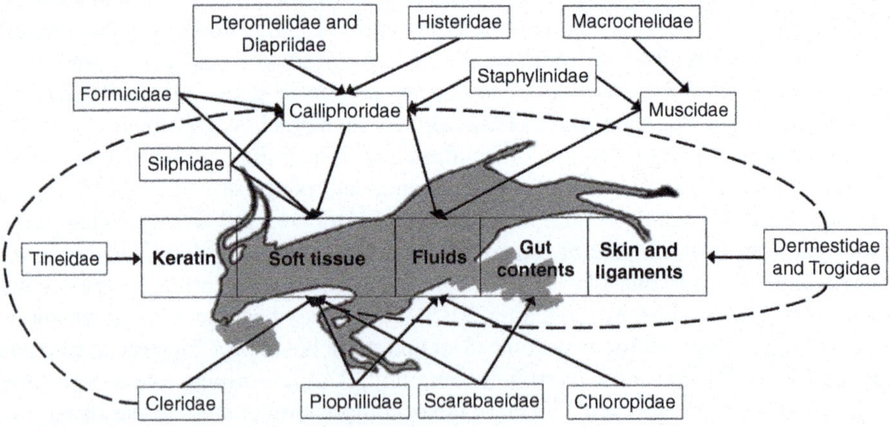

Fig. 1 A food web of the main arthropod taxa associated with an impala carcass in southern Africa (re-drawn from Braack 1987). Different carrion-associated taxa specialize on different components of the carcass (shaded grey), but there can be overlap between components of the carcass. Credits: the chapter authors

degrees of functional redundancy and complementarity among species at different carcasses in different biomes. The *sarcophagous* component feeds on carcass soft tissues, including muscle and body fluids, and is dominated by various fly taxa (e.g. Calliphoridae, Muscidae). Several taxa have a minor role, including Diptera: Piophilidae, Chloropidae, Coleoptera: Silphidae, Cleridae, Scarabaeidae, or Hymenoptera: Formicidae. The *coprophagous* component are most relevant to herbivore carcasses and are attracted to and feed on the gut contents. Typically dominated by the Scarabaeidae, but other taxa typically found at dung pads will also be found at carcasses. The *dermatophagous* component feed on the skin, and is dominated by the beetle taxa Dermestidae and Trogidae. The *keratophagous* component is a highly specialized fauna that feed on horns, nails, hair and hooves, and include the Tineidae and Trogidae. The *detritivore* component is a more generalized fauna that feed on various organic material, but is distinct from the Sarcophagous component by being opportunistic rather than specialists. The detritivore component includes many taxa, including those species that specialize on other components, such as the Formicidae, Dermestidae and Trogidae. The *predator* component is a large and diverse group of taxa that feed predominantly on the abundant fly larvae and eggs, but may also feed opportunistically on other arthropods. The predator component includes the Formicidae, the beetle taxa Histeridae, Silphidae, Cleridae, Staphylinidae, and the mite taxa Macrochelidae and Parasitidae. Spiders will also opportunistically prey upon arthropods on carcass. The *parasitic* component includes several wasp species that are parasitoids of fly larvae. Many mites at carcasses disperse via phoresy, but are technically not parasites of their fly or beetle hosts. The predator and parasitic functional groups may play a role in decomposition by moderating the abundance or activity of other consumer groups.

Vertebrates

Vertebrates involved with decomposition are typically termed 'scavengers' (see chapter "Vertebrate Scavenging Communities"). They play a key role in the consumption and fragmentation of large carcasses, and can be responsible for consuming the majority of carrion in some ecosystems (DeVault et al. 2003; Wilson and Wolkovich 2011). Small carcasses can be entirely consumed by scavengers, but larger carcasses may only be partially consumed, resulting in disarticulation and dispersal of carcass material through the landscape. The importance of vertebrates in carrion removal varies among ecosystems, but it is clear that vertebrates generally accelerate carrion decomposition rates where they occur (DeVault et al. 2003; Ogada et al. 2012; Parmenter and MacMahon 2009). This may depend, in part, on the size and composition of the scavenger guild (Sebastián-González et al. 2016; Moleón et al. 2015). Some scavenger guilds are more diverse than others, with some guilds dominated by only a few large facultative scavenger vertebrates, such as dingos and raptors in arid Australia (Read and Wilson 2004), whereas other ecosystems have a relatively large and diverse guild with both facultative and obligate scavengers, such as lions, hyenas, jackals and several vulture species in the Serengeti of Africa (Hunter et al. 2007). Importantly, many vertebrate predators will opportunistically scavenge on animal remains, and so also fill an important role in carrion consumption and carcass removal (Mateo-Tomás et al. 2015; Moleón et al. 2014; Moreno-Opo and Margalida 2013).

Factors Affecting Decomposition

Temperature and Moisture

Temperature and moisture are the two most important abiotic factors affecting the rate of decomposition of carrion. Temperature is critical in determining rates of chemical reactions driving autolysis and putrefaction, the metabolic activity of microbes, and development rates of fly larvae. Moisture is also critical in providing a substrate for effective enzymatic activity and bacterial proliferation, as well as preventing desiccation of carrion tissues and larvae of flies. The temperature-dependence of many processes in the natural world has led to the development of the 'accumulated degree days' (ADD) principle. This principle uses variation in daily temperature and the total time taken for a process to occur (Simmons et al. 2010). ADD has been examined widely in the forensic sciences (Megyesi et al. 2005; Michaud and Moreau 2009) to generate predictions for post-mortem interval. This is often based on the developmental rates of key insect taxa such as flies. The development of fly larvae (and other holometabolous insects) through their instar stages requires a minimum threshold temperature. Above this threshold (and if sufficient food resources are available), higher temperatures will result in faster development and faster progression through instar stages until an upper threshold is reached and survival is reduced. By summing daily minimum and maximum temperatures at a carcass, and relating this to larval instar stage, it is possible to accurately determine the time the larvae were first deposited at the carcass, and therefore the age (or time of death) of the carcass. Temperature and moisture are closely linked, with very low temperatures preserving carcasses for extended periods by reducing moisture availability. For example Musk ox (*Ovibos moschatus*) carcasses in the arctic circle have been shown to have extended effects on the localized environment, perhaps for over 10 years and due in part to lower temperatures (Danell et al. 2002). Conversely, extreme high temperatures can also reduce moisture availability, and can increased likelihood of mummification of carrion tissues, thus slowing decomposition.

Vegetation and Habitat

Vegetation can affect decomposition by moderating the abiotic environment (i.e. through provision of shade), as well as by providing habitat to different species involved with decomposition. Many studies have compared arthropod communities from different habitats and found important differences. For example, differences in carrion arthropods between grassland or forest habitats (de Almeida and Louzada 2009), and urban, farmland and forest habitats (Kavazos and Wallman 2012) have been reported. These differences in species occurrence have the capacity to alter decomposition rates.

Vegetation structure can also impede access by large vertebrate scavengers. For example, consider that carrion initially attracts scavengers from a relatively local area. If the amount of carrion is small it will be consumed by competitively dominant scavengers that arrive first. However, if the amount of carrion is greater than is swiftly consumed locally, then it is reasonable to expect scavengers to continue to recruit from increasingly distant areas. Some scavengers, especially avian species, are more capable of recruiting to or discovering carrion more rapidly than others (Wilmers et al. 2003). However, the increased feeding radii of avian scavengers is dependent upon the visibility of carrion. If carrion is sufficient obscured from visual detection by vegetative or other habitat characteristics then it is reasonable to expect that differences in decomposition can occur due to varying degrees of carrion use and disarticulation.

Burial and Soil

Burial can alter decomposition by moderating temperature and moisture, limiting the external microbial fauna that comes into contact with the carcass, and preventing access by carrion-feeding arthropods and vertebrates (Carter and Tibbett 2008; Payne et al. 1968; VanLaerhoven and Anderson 1999). Burial therefore typically slows decomposition. Large carcasses may also produce an adipocere—a fatty layer that forms around a carcass following biochemical decomposition and hydrolysis of tissues. This can act as a barrier to nutrient and microbial movement away from the carcass and may prolong the decomposition process (Dent et al. 2004; Forbes et al. 2005).

Submersion

The decomposition of carrion in aquatic and marine environments is fundamentally different from that in terrestrial systems (Beasley et al. 2012; Parmenter and Lamarra 1991; Wallace 2015). The vast majority of research on carrion decomposition has been conducted in terrestrial environments rather than aquatic environments (freshwater or marine), and this has influenced much of our understanding of the process. Some key points of difference between decomposition in terrestrial and aquatic environments were synthesized by Beasley et al. (2012). These include the overarching effect of water acting as a medium through which organisms and nutrients must move through to go towards or away from the decomposing carcass. This can have an effect on which animals are attracted to a carcass via movement of chemical cues downstream or with currents. Another critical effect of water is the moderation of temperature, which is a major determinant of decomposition rate. Carrion in shallow water can experience fluctuations in temperature extremes, whereas in deep water temperature is more constant. In marine systems, ocean depth can determine the composition of the scavenger community and the decomposition process. Deep sea abyssal biomes (Smith and Baco 2003), for example, have very different scavenger assemblages than shallow marine environments (Moore and Howarth 1996), and oxygen concentrations may have a role in which species are able to colonize carcasses (Anderson and Bell 2014). The decomposition of carcasses in marine and freshwater environments is also affected by faunal colonization and succession (see chapter "Invertebrate Scavenging Communities"). Key points of difference between freshwater and marine systems is the absence of insects in the marine environment (replaced largely by crustacean arthropods), and the comparative lack of carrion specialists in freshwater environments. Whether a carcass is completely submerged, floating at the surface, or in contact with the bed/ floor of the water column, can affect what taxa is able to colonize and how rapidly the carcass may be consumed.

Carcass Size

The size of a carcass has important implications for how it decomposes, including its rate of decomposition and what organisms might use the carcass as a food resource. The size of a carcass affects how it moderates internal and ambient temperatures, which is critical to decomposition rate. Large carcasses can contain more than 210,000 fly larvae (Braack 1987), and this generates heat that increases internal temperatures. This can accelerate microbial and intrinsic decay processes relative to smaller carcasses that cannot accommodate such large maggot masses. For example, Hewadikaram and Goff (1991) showed that 15 kg pig carcasses had higher internal temperature relative to 8 kg carcasses. Further, they showed that the internal temperature of small pig carcasses more closely matched that of ambient temperatures.

The ecological study of carrion has been conducted on many kinds of carcasses (Barton et al. 2013a), including bison (Towne 2000), pigs (Mądra et al. 2015), kangaroos (Macdonald et al. 2014), rabbits (De Jong and Chadwick 1999), rats (Carter et al. 2008), and even slugs (Woodcock et al. 2002) to name but a few. This has resulted in a wide range of separate studies involving different carcasses. Due to the requirements of proper scientific method, samples are typically replicated across the same kind of animal carcass. This has meant that few studies have explicitly compared decomposition among different carcass types to see how this attracts different sets of organisms (but see Moleón et al. 2015; Parmenter and MacMahon 2009). One way to understand the effect of carcass size is to view each carcass as a food particle (Moleón et al. 2015). When viewed this way, the size of a carcass will determine what organisms will use it as a food resource, and therefore the fate of the carcass and its nutrients. To illustrate this point, carcasses of very small animals (i.e. insects, small rodents) are typically consumed whole by scavenging animals such as birds or other small vertebrates (DeVault and Rhodes 2002). Very large carcasses (e.g. ungulates) can often be larger than any of their predators or potential scavengers. In these instances, each carcass may support multiple species of scavenger organism (Selva et al. 2005), and this introduces new factors such as species interactions that may affect decomposition. A study by Moleón et al. (2015) found that scavenger assemblages were more species rich at larger carcasses, and that this was largely due to the presence of facultative scavengers (i.e. mammal carnivores) rather than obligate scavengers (i.e. vultures). A consequence of this size-driven assemblage difference was that decomposition rate, expressed as percentage biomass consumed, was negatively associated with carcass size, indicating that rate of decay was slower for bigger carcasses.

Ecological Theory Underpinning Carrion Decomposition

The study of carrion decomposition has been informed by several important theories relating to populations and communities of carrion-associated organisms. Most work on carrion decomposition has been dedicated to case studies and detailed field observations, with less emphasis on hypothesis testing and development of theory (Michaud et al. 2015). The ecological attributes of carrion are well known, and include its (relatively) predictable decay pattern (Payne 1965; Schoenly and Reid 1987) and its unpredictable patchiness and ephemeral occurrence (Barton et al. 2013a; Braack 1987; Doube 1987; Finn 2001). These ideas capture much of the theory describing faunal communities involved with carrion decomposition, including succession, competition, and coexistence and aggregation. Other processes link carrion to ecosystems at larger scales, including carrion-based subsidies of communities and resource pulse dynamics (Oro et al. 2013; Polis and Hurd 1996; Yang 2006), as well as predators decoupling carrion distribution from live-prey distribution (Bump et al. 2009a).

Succession

The succession (i.e. the sequenced arrival and departure) of different organisms at a carcass through the decomposition process is very well documented (Bornemissza 1957; Matuszewski et al. 2011; Mégnin 1894; Payne et al. 1968; Schoenly and Reid 1987). Succession theory is one of the most widely used theories relating to carrion decomposition, especially with regard to the study of the diversity and structure of arthropod communities (Barton et al. 2014; Benbow et al. 2013; Hobischak et al. 2006; Richards and Goff 1997), and its application to the estimation of postmortem interval for forensic purposes (Archer 2014; Pechal et al. 2014b; Schoenly 1992;

Schoenly et al. 1992). Indeed, there is evidence that the concept of succession was first formalized by forensic entomologists working with human cadavers in the late 1800s (Mégnin 1894), which contrasts with the history of succession theory usually attributed to plant ecologists in the early 1900s (Michaud et al. 2015). Succession theory as it relates to carrion, however, differs somewhat from how it is applied to other ecological phenomena such as plant colonization and community development following disturbance (Turner et al. 1997). This is because carrion decomposition does not lead towards a 'climax' community where species and resources reach equilibrium. Rather, carrion is a finite resource that rapidly reduces in quality and quantity until depleted.

The processes driving succession relate to the change in carrion quality through time, including changes in the moisture and nutrient content, as well as the mix of volatile organic compounds released from the carcass. A fresh carcass is fundamentally different to dry remains in this regard, and this determines which specialists, generalist scavengers, and higher predators, will be attracted to a carcass at different points in time. Indeed, the temporal change in the 'signature' of chemicals released from a carcass may be an important mechanism driving insect succession, although his has not been explicitly tested (Michaud et al. 2015). Another key mechanism behind succession is *facilitation* (Connell and Slatyer 1977). It occurs when species present at a carcass have a role in modifying the resource after its arrival, thus preparing the carcass for colonization by another species. Although suggested by some (Schoenly and Reid 1987; Smith and Baco 2003), it has not been explicitly tested (Michaud et al. 2015).

Competition

The rare occurrence of carrion, and its high nutrient content, produces intense competition among species for the limited resources available at carcasses (Braack 1987; Kneidel 1984). This has led to the selection of key traits of species to enable them to rapidly disperse, locate, colonize, and feed on the carrion for growth and development of offspring (Tomberlin et al. 2011). Any species that can perform these tasks faster or more efficiently than others can potentially monopolize the resource and have a competitive advantage. Direct competition, however, should lead to the evolved specialization of species to reduce overlap in their resource requirements. It is evident that niche specialization exists among scavenger communities, especially for diverse arthropod assemblages (Bessa et al. 2014; Braack 1987) and obligate vertebrate guilds (DeVault et al. 2003). However, there are other mechanisms that also underpin competitive dynamics, such as resource partitioning, population aggregation, and chemically mediated interactions.

Partitioning of carrion resources among species can occur within and among individual carcasses. Different combinations of species of arthropods and vertebrates have been shown to utilize different kinds and sizes of carcasses (Hewadikaram and Goff 1991; Moleón et al. 2015; Sebastián-González et al. 2016) and this represents a kind of resource partitioning *among* different carcasses. Nestedness in species-rich vertebrate scavenger assemblages has been reported from carcasses in Spain (Sebastián-González et al. 2016) and in Poland (Selva and Fortuna 2007), and is thought to be a consequence of competition among species-rich assemblages. This non-random occurrence of species among carcasses is evidence of partitioning among carcasses. However, partitioning is best known *within* carcasses where different species utilize different parts of a carcass (e.g., Fig. 1). For some species, competition may have resulted in niche differentiation, such as among flies competing for moist tissues and fluids (e.g. Calliphoridae, Muscidae; (Hanski 1987b). For other species, partitioning may represent an adaptation to an available niche where there is low competition. The tough, keratinous horns and hooves of some animals, for example, is exploited by tineid moths or trogid beetles and few other arthropods (Braack 1987), likely due to low competition for this challenging substrate.

Interactions between competitively superior and inferior species can affect levels of coexistence among species seemingly sharing the same patchy resource. The 'aggregation model of coexistence' for a community of species sharing a single resource proposes that if the distribution of a competitively superior species is spatially aggregated, i.e. clumped at patches, then this frees other patches of resource to be colonized by competitively inferior species (Atkinson 1985). This model has been examined in detail for communities of carrion flies (Hanski 1987a; Ives 1991), and has shown a general tendency for increasing aggregation to lead to increasing levels of coexistence among species competing for the same carrion resource. Factors such as carcass size and habitat quality have also been shown to affect species aggregation and levels of species coexistence (Woodcock et al. 2002). Another consequence of competition among species for a patchy resource is the evolved capacity for rapid location, colonization, and exploitation of the resource patches (Barton et al. 2013b; Ruxton and Houston 2004). The early arrival of a species prior to a competitor, may confer a competitive advantage by providing a window of opportunity to consume the resource unencumbered by other species. Such 'priority effects' are another potential mechanism underpinning patterns of succession and coexistence among species competing for the same resource (Alford and Wilbur 1985; Brundage et al. 2014).

Competitive interactions among decomposers and scavengers can be mediated by chemicals. For example, the interactions between microbial decomposers and vertebrate carrion consumers has been known for some time (DeVault et al. 2003; Janzen 1977). However, it is only more recently that examples have been documented for the competitive exclusion of arthropods by bacteria (Burkepile et al. 2006; Tomberlin et al. 2012) and bacteria by arthropods (Hoback et al. 2004; Rozen et al. 2008). This competitive dynamic might have significant implications for rates of carrion decomposition if rapid consumption by flies, for example, is prevented. Further implications for ecosystems may include diversion of nutrients into the soil decomposer system where they are metabolized in the soil by bacteria and fungi at the site of a carcass.

Top-Down Versus Bottom-Up Effects on Carrion Communities

The study of foodwebs often focuses on a plant-herbivore-predator perspective of trophic interactions and community structure, but this overlooks the importance of detritus (both plant and animal) as an alternative resource base (Barton 2015; Moore et al. 2004). The role of carrion resources in structuring scavenger communities is poorly understood, although this is growing for some vertebrates (DeVault et al. 2003; Wilson and Wolkovich 2011). Further details on vertebrate scavenger communities is given in chapter "Vertebrate Scavenging Communities". Theory suggests that carrion resources may underpin a larger amount of energy and nutrient flow than consumption of live prey (Wilson and Wolkovich 2011).

Carrion can have an important role in structuring food-webs via top-down (i.e. predator-based) regulation of lower trophic levels and bottom-up (i.e. resource-based) regulation of higher trophic levels (DeVault et al. 2003; Dyer and Letourneau 2003; Mateo-Tomás et al. 2015). Although the top-down vs. bottom-up dichotomy is a useful way to conceptualize trophic interactions, recent evidence has shown that predators often have multiple links to other components in a foodweb. For example, many predators (arthropods and vertebrates) will consume carrion when the opportunity arises (Moleón et al. 2014; von Berg et al. 2012; Wilson and Wolkovich 2011). This suggests that a compartmentalized view of trophic levels is simplistic, and that the role of predators in structuring carrion-centered food webs are more complicated and warrant further study.

There are several examples of bottom-up control of scavenger communities, particularly in island or cave ecosystems (Polis and Hurd 1996; Schneider et al. 2011). Such systems often lack primary production, and may be entirely dependent on external detrital resources, including carrion, as a source of energy and nutrients. For example, work by Polis and Hurd (1995) examined the role of carrion on oceanic islands. They revealed a causal relationship between marine-derived carrion and terrestrial foodweb dynamics of islands. Another example involves work by Schneider et al. (2011) and the manipulation of amounts of carrion in a series of caves. They showed that carrion-associated arthropod communities in these caves were profoundly affected by carrion addition and removal, thus revealing the extent to which these cave biota are dependent upon outside carrion subsidies (Schneider et al. 2011).

Decomposition Linkages Between Ecosystems

The movement of carrion within and across ecosystems can underpin the functioning of some food webs and ecosystems by subsidizing detritus resources (Polis et al. 1997). For example, many small or young islands have very limited autochthonous biomass production due to limited soils or unfavorable growing conditions. In these environments, carrion from nesting birds, or fish, bird and seal carcasses washed in from the sea, can supplement the resource base for the island food web. This has been demonstrated by Polis and Hurd (1995) on islands in the Gulf of California, where they measured invertebrate densities on a range of islands of different size. Carrion inputs were shown to be critical to these ecosystems, and supported high abundances of scavenger invertebrates and their spider predators.

The movement of carrion across ecosystems can also occur through the migration of live animals. A good example of this is the migration of anadromous salmon (e.g. *Salmo* spp. and *Oncorhynchus* spp.) up streams from the ocean as part of their annual spawning runs (Hocking and Reynolds 2011). This migration from marine to freshwater ecosystems, and the subsequent die-off of thousands of individuals results in the mass transfer of substantial quantities of salmon biomass, with benefits for many scavenger invertebrates (Hocking et al. 2009). A similar phenomenon has been documented for the mass emergence of aquatic insects and their dispersal to nearby riparian and terrestrial habitat (Dreyer et al. 2012, 2015). Although insect carcasses are small, this is made up for by their sheer quantity and high densities. Such carcass-mediated transfers of carbon and nitrogen can be substantial across ecosystem boundaries, and has the potential to affect ecological processes and nutrient cycling.

Patch Dynamics and Landscape Heterogeneity

The spatial and temporal patchiness of animal carcasses in the environment has been repeatedly found to influence the dynamics of species populations and assemblages at carrion (Barton et al. 2013a). Much of this work, however, has been at small scales with a focus on individual carcasses. This has helped to consolidate the dominant paradigm of the 'ephemeral resource patch', which provides an effective general framework for how resource patchiness can affect the spatio-temporal dynamics of species populations or communities (Finn 2001). It also draws conceptual parallels between carrion resources and other spatially discrete and temporary resources, such as dung pads (Doube 1987), fungal fruiting bodies (Heard 1998) and plant fruits (Sevenster and VanAlphen 1996). By contrast, scaling up this work to understand the role of carrion in landscapes and ecosystems has been rare (Bump et al. 2009a). Further, an outward perspective that questions the larger-scale role of carrion in nutrient cycling and landscape heterogeneity is lacking, despite there being some high-profile evidence of the large-scale impact of carrion (Hocking and Reynolds 2011; Yang 2004).

The patchiness of carrion can affect landscape heterogeneity in terms of biodiversity and ecological processes. For example, two carcasses at *different* decay stages will support a greater range of species than two carcasses at the *same* decay stage, and this has consequences for supporting overall biodiversity within landscapes (Barton et al. 2013a). Carcasses can also drive soil nutrient heterogeneity (Bump et al. 2009a), and can provide establishment opportunities for grasses (Barton et al. 2013b) and for trees (Barton et al. 2013b; Bump et al. 2009b). When in sufficient quantities, and multiplied across space and over time, the geographic location and turnover of carcasses can have significant effects on whole ecosystems (Bump et al. 2009a).

It was previously thought that well-consumed carrion did not contribute to ecological heterogeneity. We have learned, however, that in some cases even well consumed carrion can cause significant physical disturbances and nutrient pulses that alter mosaics of biogeochemical complexity, nutrient availability, microbial biomass, activity, and composition, and soil biodiversity. Consumptive activities disperse carrion resources beyond a discrete area. Carcass sites also concentrate other materials resulting from the activity of consumers. Initially, carcass sites receive resource inputs from predators and scavengers, including feces and urine. Enhanced plant quality and growth at carcass sites can attract herbivores and their predators, which also results in excretory inputs (Danell et al. 2002, 2006; Towne 2000). Insects and birds may also contribute significant inputs of chitin and keratin, respectively. Cumulatively, these inputs can extend the duration and quantity of energy and nutrient exchange at carcass sites, thereby producing positive feedbacks that prolong elevated levels of heterotrophic activity and ecological heterogeneity. These changes affect above ground biomass, community composition, foliar nutrient levels, and tree seedling establishment and sapling growth (Bump et al. 2009a, b; Towne 2000). The ability of carrion to alter steady-state edaphic and ecological properties represents a strong, reciprocal feedback between above and belowground communities. This occurs via mortality of above ground organisms, which leads to multiple effects on belowground communities that then alter aboveground ecological properties (Bardgett and Wardle 2010; Dreyer et al. 2015; Schmitz et al. 2010). As a consequence, this work is highly cross-disciplinary, spanning multiple scales of ecological inquiry.

Conclusions and Future Perspectives

Knowledge of carrion decomposition will continue to inform many aspects of ecology, biodiversity conservation, natural resource management, as well as the forensic medical sciences. The decomposition of carrion results from a complex series of chemical, biological, and physical processes that ultimately recycles the energy and nutrients embodied in an animal through other organisms and its environment. Understanding of carrion decomposition has been developed by specialists in chemistry, medical pathology, microbiology, soil science, entomology, and ecology, and the future of decomposition research will need to foster deeper interdisciplinary collaboration among these specialists.

Key emerging areas of carrion decomposition research will stem from advances in technology and the rapid reduction in costs of various forms of molecular analysis. Further advances will also come from multi-disciplinary research that integrates molecular and ecological perspectives. Emerging areas and advances in carrion decomposition will also benefit from increased spatial variance sampling that will allow researchers to determine under what circumstances carrion decompositions contributes to or diminishes environmental heterogeneity at multiple scales. At the macro scale, continued comparative investigation of controls (e.g. weather, disease, predation) of carrion distribution in space and time are necessary to predict how climate change will likely affect carrion availability and decomposition in ecosystems.

Molecular Tools and the Microbial Frontier

Rapid developments in technology, including improved instrumentation and computational power, is generating ever increasing amounts of biomolecular data. This is true for all aspects of the life sciences, and there is clear potential to apply this to understanding carrion decomposition. One example of this is the emergence of ecogenomics and the rapid assessment of diversity, composition and function of hyperdiverse microbial communities (Metcalf et al. 2015; Pechal and Benbow 2016; Pechal et al. 2014b). A particularly exciting area of research is the role of the microbiome in decomposition, and how this is influenced by other carrion-feeding organisms such as blowflies (Pechal et al. 2013). Although each carcass has its own community of microbes in its gut and on its skin, the effect of colonizing insects in bringing new bacteria to a carcass, and potentially altering decomposition patterns, remains poorly understood. The diversity and composition of the microbiome is likely to have important effects on carrion decomposition, and the rapid and accurate assessment of microbial diversity using molecular tools will be critical to this future research area (Crippen et al. 2015; Metcalf et al. 2015; Pechal et al. 2014b). Further, there is the unknown possibility of carrion gut microbiota 'seeding' soils, potentially increasing localised soil microbial diversity and broader spatial microbial heterogeneity.

Community Interactions

There is a renewed focus on understanding the dynamics and interactions among species associated with carrion, and which combine to form the 'necrobiome' (sensu Benbow et al. 2013). This includes all the microbial, arthropod, and vertebrate taxa that are part of the carrion foodweb and are responsible for the decomposition of carrion. An exciting and emerging area is the interaction between microbes and insects. As noted above, the microbiome of a carcass has a hugely important role in its decomposition (Metcalf et al. 2015), yet interactions among insect species and among insects and bacteria also have important implications for decomposition (Pechal et al. 2013; Tomberlin et al. 2012). Research into the complex interactions among species, within and among carcasses, and at multiple spatial scales, is emerging as critical to understanding the decomposition process (Benbow et al. 2015), and its role in driving biodiversity dynamics, nutrient cycling, and ecosystem functioning more broadly (Barton et al. 2013a).

Particular recent interest in vertebrate scavengers (e.g. Moleón et al. 2014; Wilson and Wolkovich 2011) stems from the decline of apex predators in many ecosystems worldwide (Estes et al. 2011) and subsequent functional replacement by a range of other predators. Apex predators, such as lions or wolves, generate carrion by leaving the un-consumed remains of kills (Bump et al. 2009a), which is used by other smaller predators and scavengers. However, many predator species increase their scavenging of carrion during times of low prey availability (Wilson and Wolkovich 2011). This can result in changed structure of food webs, and altered pathways of nutrient and energy flow through ecosystems. A research priority is therefore understanding how changes to carrion food webs, and their species interactions, might translate to changes in ecosystem functioning (Barton et al. 2013a; Tomberlin et al. 2011; Wilson and Wolkovich 2011). A significant barrier to this work, however, is the lack of studies that have compared communities of multiple taxa such as microbes, arthropods, birds and carnivores (but see Parmenter and MacMahon 2009; Read and Wilson 2004; Selva and Fortuna 2007; Tomberlin et al. 2012).

Despite the recent recognition of the ecological importance of carrion, the spatial and temporal dynamics of carrion distribution via large vertebrate carcasses is comparatively unexamined. Yet, it is of fundamental ecological importance because large vertebrate carcasses can facilitate niche provisioning and specialization, and alter resource competition dynamics for other species. For example, carcass-derived nutrients can shift competitive relationships among primary producers (Barton et al. 2013b; Bugalho et al. 2011; Towne 2000), serve as a disturbance mechanism (Danell et al. 2006), and carcass sites constitute specialized, critical habitat for obligate carrion feeders and reproducers (Selva and Fortuna 2007).

Decomposition and Global Change

Many aspects of global change have the potential to alter carrion dynamics and the decomposition process. This includes changes to land use and the habitat that supports carrion scavenger and decomposer communities (DeVault et al. 2011; Klein 1989), and the transmission of disease and the timing and magnitude of animal mass mortality events (Fey et al. 2015). The drivers of land use change include agricultural expansion and intensification, as well as urbanization. When changes to land use occur, some consumers of carrion may be lost, or their spatial or temporal distribution changed, with potential implications for decomposition processes and rates (Kavazos and Wallman 2012; Klein 1989). For example, loss of species may result in reduced functional complementarity and redundancy among carrion consumers. It is already known that many vertebrates will opportunistically scavenge on carrion (see chapter "Vertebrate Scavenging Communities"), and there is evidence that some intra-guild compensation occurs when dominant scavengers are removed or lost due to land use changes (Olson et al. 2012). Similarly, several species of blow fly (Diptera: Calliphoridae) typically co-occur at carcasses, but recent evidence suggests that urbanization affects fly community composition (Kavazos and Wallman 2012), and this is likely to change levels of inter-specific competition, and perhaps even carrion decomposition rates.

Far less is known about the potential impacts of global climate change on carrion decomposition. The strong temperature dependence of the decomposition would suggest this important process is highly vulnerable to temperature changes, with higher temperatures likely accelerating rates of decay. However, the impact of climate change on species of microbe or insect critical to decomposition, or their interactions, is unknown. An expected effect includes range shifts due to changes in optimal developmental envelopes (Chen et al. 2011; Walther et al. 2002), and this is likely to move species into new areas, and out of existing ones. This has implications for both the ecology of carrion decomposition and its application to forensic problems that depend on knowledge of species presence and development rates.

Acknowledgements PSB was funded by the Australian Research Council (DE150100026).

References

Alford RA, Wilbur HM (1985) Priority effects in experimental pond communities: competition between *Bufo* and *Rana*. Ecology 66:1097–1105

Anderson GS, Bell LS (2014) Deep coastal marine taphonomy: investigation into carcass decomposition in the saanich inlet, British Columbia Using a Baited Camera. PLoS ONE 9:e110710

Archer M (2014) Comparative analysis of insect succession data from Victoria (Australia) using summary statistics versus preceding mean ambient temperature models. J Forensic Sci 59:404–412

Atkinson WD (1985) Coexistence of Australian rainforest Diptera brooding in fallen fruit. J Anim Ecol 54:507–518

Bardgett RD, Wardle DA (2010) Aboveground-belowground linkages. Biotic interactions, ecosystem processes, and global change. Oxford University Press, Oxford

Barton PS (2015) The role of carrion in ecosystems. In: Benbow EM, Tomberlin JK, Tarone AM (eds) Carrion ecology, evolution, and their applications. CRC Press, Boca Raton, p 512

Barton PS, Cunningham SA, Lindenmayer DB, Manning AD (2013a) The role of carrion in maintaining biodiversity and ecological processes in terrestrial ecosystems. Oecologia 171:761–772

Barton PS, Cunningham SA, Macdonald BCT, McIntyre S, Lindenmayer DB, Manning AD (2013b) Species traits predict assemblage dynamics at ephemeral resource patches created by carrion. PLoS ONE 8:e53961

Barton PS, Weaver HJ, Manning AD (2014) Contrasting diversity dynamics of phoretic mites and beetles associated with vertebrate carrion. Exp Appl Acarol 63:1–13

Beasley JC, Olson ZH, DeVault TL (2012) Carrion cycling in food webs: comparisons among terrestrial and marine ecosystems. Oikos 121:1021–1026

Benbow ME, Lewis AJ, Tomberlin JK, Pechal JL (2013) Seasonal necrophagous insect community assembly during vertebrate carrion decomposition. J Med Entomol 50:440–450

Benbow ME, Pechal JL, Mohr RM (2015) Community and landscape ecology of carrion. In: Benbow ME, Tomberlin JK, Tarone AM (eds) Carrion ecology, evolution, and their applications. CRC Press, Boca Raton, pp 151–185

Bessa F, Baeta A, Marques JC (2014) Niche segregation amongst sympatric species at exposed sandy shores with contrasting wrack availabilities illustrated by stable isotopic analysis. Ecol Indic 36:694–702
Bornemissza GF (1957) An analysis of arthropod succession in carrion and the effect of its decomposition on the soil fauna. Aust J Zool 5:1–12
Braack LEO (1987) Community dynamics of carrion-attendant arthropods in tropical African woodland. Oecologia 72:402–409
Brundage A, Benbow ME, Tomberlin JK (2014) Priority effects on the life-history traits of two carrion blow fly (Diptera, Calliphoridae) species. Ecol Entomol 39:539–547
Bugalho MN, Lecomte X, Goncalves M, Caldeira MC, Branco M (2011) Establishing grazing and grazing-excluded patches increases plant and invertebrate diversity in a Mediterranean oak woodland. For Ecol Manag 261:2133–2139
Bump JK, Peterson RO, Vucetich JA (2009a) Wolves modulate soil nutrient heterogeneity and foliar nitrogen by configuring the distribution of ungulate carcasses. Ecology 90: 3159–3167
Bump JK, Webster CR, Vucetich JA, Peterson RO, Shields JM, Powers MD (2009b) Ungulate carcasses perforate ecological filters and create biogeochemical hotspots in forest herbaceous layers allowing trees a competitive advantage. Ecosystems 12:996–1007
Burkepile DE et al (2006) Chemically mediated competition between microbes and animals: microbes as consumers in food webs. Ecology 87:2821–2831
Carter DO, Tibbett M (2003) Taphonomic mycota: fungi with forensic potential. J Forensic Sci 48:168–171
Carter DO, Tibbett M (2008) Does repeated burial of skeletal muscle tissue (*Ovis aries*) in soil affect subsequent decomposition? Appl Soil Ecol 40:529–535
Carter DO, Yellowlees D, Tibbett M (2007) Cadaver decomposition in terrestrial ecosystems. Naturwissenschaften 94:12–24
Carter DO, Yellowlees D, Tibbett M (2008) Temperature affects microbial decomposition of cadavers (*Rattus rattus*) in contrasting soils. Appl Soil Ecol 40:129–137
Chen IC, Hill JK, Ohlemüller R, Roy DB, Thomas CD (2011) Rapid range shifts of species associated with high levels of climate warming. Science 333:1024–1026
Connell JH, Slatyer RO (1977) Mechanisms of succession in natural communities and their role in community stability and organization. Am Nat 111:1119–1144
Crippen TL, Benbow ME, Pechal JL (2015) Microbial interactions during carrion decomposition. In: Benbow ME, Tomberlin JK, Tarone AM (eds) Carrion ecology, evolution, and their applications. CRC Press, Boca Raton, pp 31–63
Danell K, Berteaux D, Brathen KA (2002) Effect of muskox carcasses on nitrogen concentration in tundra vegetation. Arctic 55:389–392
Danell K, Bergström R, Duncan P, Pastor J (2006) Large herbivore ecology, ecosystem dynamics and conservation. Cambridge University Press, Cambridge
de Almeida SDP, Louzada JNC (2009) Community structure of scarabaeinae (Scarabaeidae: Coleoptera) in Brazilian savannah phytophysiognomies and its importance for conservation. Neotrop Entomol 38:32–43
De Jong GD, Chadwick JW (1999) Decomposition and arthropod succession on exposed rabbit carrion during summer at high altitudes in Colorado, USA. J Med Entomol 36:833–845
Dent BB, Forbes SL, Stuart BH (2004) Review of human decomposition processes in soil. Environ Geol 45:576–585
DeVault TL, Rhodes OE (2002) Identification of vertebrate scavengers of small mammal carcasses in a forested landscape. Acta Theriol 47:185–192
DeVault TL, Rhodes OE, Shivik JA (2003) Scavenging by vertebrates: behavioral, ecological, and evolutionary perspectives on an important energy transfer pathway in terrestrial ecosystems. Oikos 102:225–234
DeVault TL, Olson ZH, Beasley JC, Rhodes OE (2011) Mesopredators dominate competition for carrion in an agricultural landscape. Basic Appl Ecol 12:268–274

Doube BM (1987) Spatial and temporal organization in communities associated with dung pads and carcasses. In: Gee JHR, Giller PS (eds) Organization of communities past and present. Blackwell Scientific Publications, Oxford

Dreyer J, Hoekman D, Gratton C (2012) Lake-derived midges increase abundance of shoreline terrestrial arthropods via multiple trophic pathways. Oikos 121:252–258

Dreyer J, Townsend PA, Hook JC, Hoekman D, Vander Zanden MJ, Gratton C (2015) Quantifying aquatic insect deposition from lake to land. Ecology 96:499–509

Dyer LA, Letourneau D (2003) Top-down and bottom-up diversity cascades in detrital vs. living food webs. Ecol Lett 6:60–68

Estes JA et al (2011) Trophic downgrading of planet earth. Science 333:301–306

Farwig N, Brandl R, Siemann S, Wiener F, Mueller J (2014) Decomposition rate of carrion is dependent on composition not abundance of the assemblages of insect scavengers. Oecologia 175:1291–1300

Fey SB et al (2015) Recent shifts in the occurrence, cause, and magnitude of animal mass mortality events. Proc Natl Acad Sci 112:1083–1088

Finn JA (2001) Ephemeral resource patches as model systems for diversity-function experiments. Oikos 92:363–366

Forbes SL, Carter DO (2015) Processes and mechanisms of death and decomposition of vertebrate carrion. In: Benbow EM, Tomberlin JK, Tarone AM (eds) Carrion ecology, evolution, and their applications. CRC Press, Boca Raton, p 512

Forbes SL, Dent BB, Stuart BH (2005) The effect of soil type on adipocere formation. Forensic Sci Int 154:35–43

Hanski I (1987a) Carrion fly community dynamics: patchiness, seasonality and coexistence. Ecol Entomol 12:257–266

Hanski I (1987b) Nutritional ecology of dung- and carrion-feeding insects. In: Slansky F, Rodriguez J (eds) Nutritional ecology of insects, mites, and spiders. Wiley, New York, pp 837–884

Heard SB (1998) Resource patch density and larval aggregation in mushroom-breeding flies. Oikos 81:187–195

Hewadikaram KA, Goff ML (1991) Effect of carcass size on rate of decomposition and arthropod succession patterns. Am J Forensic Med Pathol 12:235–240

Hoback WW, Bishop AA, Kroemer J, Scalzitti J, Shaffer JJ (2004) Differences among antimicrobial properties of carrion beetle secretions reflect phylogeny and ecology. J Chem Ecol 30:719–729

Hobischak NR, VanLaerhoven SL, Anderson GS (2006) Successional patterns of diversity in insect fauna on carrion in sun and shade in the Boreal Forest Region of Canada, near Edmonton, Alberta. Can Entomol 138:376–383

Hocking MD, Reynolds JD (2011) Impacts of salmon on riparian plant diversity. Science 331:1609–1612

Hocking MD, Ring RA, Reimchen TE (2009) The ecology of terrestrial invertebrates on Pacific salmon carcasses. Ecol Res 24:1091–1100

Hunter JS, Durant SM, Caro TM (2007) Patterns of scavenger arrival at cheetah kills in Serengeti National Park Tanzania. Afr J Ecol 45:275–281

Ives AR (1991) Aggregation and coexistence in a carrion fly community. Ecol Monogr 61:75–94

Janzen DH (1977) Why fruits rot, seeds mold, and meat spoils. Am Nat 111:691–713

Kavazos CRJ, Wallman JF (2012) Community composition of carrion-breeding blowflies (Diptera: Calliphoridae) along an urban gradient in south-eastern Australia. Landsc Urban Plan 106:183–190

Klein BC (1989) Effects of forest fragmentation on dung and carrion beetle communities in central amazonia. Ecology 70:1715–1725

Kneidel KA (1984) Competition and disturbance in communities of carrion breeding Diptera. J Anim Ecol 53:849–865

Lauber CL, Metcalf JL, Keepers K, Ackermann G, Carter DO, Knight R (2014) Vertebrate decomposition is accelerated by soil microbes. Appl Environ Microbiol 80:4920–4929

Macdonald BCT, Farrell M, Tuomi S, Barton PS, Cunningham SA, Manning AD (2014) Carrion decomposition causes large and lasting effects on soil amino acid and peptide flux. Soil Biol Biochem 69:132–140

Mądra A, Frątczak K, Grzywacz A, Matuszewski S (2015) Long-term study of pig carrion entomofauna. Forensic Sci Int 252:1–10

Mateo-Tomás P et al (2015) From regional to global patterns in vertebrate scavenger communities subsidized by big game hunting. Divers Distrib 21:913–924

Matuszewski S, Bajerlein D, Konwerski S, Szpila K (2011) Insect succession and carrion decomposition in selected forests of Central Europe. Part 3: succession of carrion fauna. Forensic Sci Int 207:150–163

Mégnin P (1894) La faune des cadavres application de l'entomologie à la médecine légale. In: Encylopédie Scientifique des Aide-Mémoire, Paris

Megyesi MS, Nawrocki SP, Haskell NH (2005) Using accumulated degree-days to estimate the postmortem interval from decomposed human remains. J Forensic Sci 50:618–626

Metcalf JL et al (2015) Microbial community assembly and metabolic function during mammalian corpse decomposition. Science 351(6269):158–162

Michaud J-P, Moreau G (2009) Predicting the visitation of carcasses by carrion-related insects under different rates of degree-day accumulation. Forensic Sci Int 185:78–83

Michaud JP, Schoenly KG, Moreau G (2015) Rewriting ecological succession history: did carrion ecologists get there first? Q Rev Biol 90:45–66

Moleón M, Sanchez-Zapata JA, Selva N, Donazar JA, Owen-Smith N (2014) Inter-specific interactions linking predation and scavenging in terrestrial vertebrate assemblages. Biol Rev 89:1042–1054

Moleón M, Sánchez-Zapata JA, Sebastián-González E, Owen-Smith N (2015) Carcass size shapes the structure and functioning of an African scavenging assemblage. Oikos 124: 1391–1403

Moore PG, Howarth J (1996) Foraging by marine scavengers: effects of relatedness, bait damage and hunger. J Sea Res 36:267–273

Moore JC et al (2004) Detritus, trophic dynamics and biodiversity. Ecol Lett 7:584–600

Moreno-Opo R, Margalida A (2013) Carcasses provide resources not exclusively to scavengers: patterns of carrion exploitation by passerine birds. Ecosphere 4:108

Ogada DL, Torchin ME, Kinnaird MF, Ezenwa VO (2012) Effects of vulture declines on facultative scavengers and potential implications for mammalian disease transmission. Conserv Biol 26:453–460

Olson ZH, Beasley JC, DeVault TL, Rhodes OE (2012) Scavenger community response to the removal of a dominant scavenger. Oikos 121:77–84

Oro D, Genovart M, Tavecchia G, Fowler MS, Martínez-Abraín A (2013) Ecological and evolutionary implications of food subsidies from humans. Ecol Lett 16:1501–1514

Parmenter RR, Lamarra VA (1991) Nutrient cycling in a freshwater marsh: the decomposition of fish and waterfowl carrion. Limnol Oceanogr 36:976–987

Parmenter RR, MacMahon JA (2009) Carrion decomposition and nutrient cycling in a semiarid shrub-steppe ecosystem. Ecol Monogr 79:637–661

Payne JA (1965) A summer carrion study of the baby pig *Sus scrofa* Linnaeus. Ecology 46:592–602

Payne JA, King EW, Beinhart G (1968) Arthropod succession and decomposition of buried pigs. Nature 219:1180

Pechal JL, Benbow ME (2016) Microbial ecology of the salmon necrobiome: evidence salmon carrion decomposition influences aquatic and terrestrial insect microbiomes. Environ Microbiol 18:1511–1522

Pechal JL, Crippen TL, Tarone AM, Lewis AJ, Tomberlin JK, Benbow ME (2013) Microbial community functional change during vertebrate carrion decomposition. PLoS ONE 8:e79035

Pechal JL, Benbow ME, Crippen TL, Tarone AM, Tomberlin JK (2014a) Delayed insect access alters carrion decomposition and necrophagous insect community assembly. Ecosphere 5:1–21

Pechal JL, Crippen TL, Benbow ME, Tarone AM, Dowd S, Tomberlin JK (2014b) The potential use of bacterial community succession in forensics as described by high throughput metagenomic sequencing. Int J Legal Med 128:193–205

Polis GA, Hurd SD (1995) Extraordinarily high spider densities on islands: flow of energy from the marine to terrestrial food webs and the absence of predation. Proc Natl Acad Sci U S A 92:4382–4386

Polis GA, Hurd SD (1996) Linking marine and terrestrial food webs: allochthonous input from the ocean supports high secondary productivity on small islands and coastal land communities. Am Nat 147:396–423

Polis GA, Anderson WB, Holt RD (1997) Toward an integration of landscape and food web ecology: the dynamics of spatially subsidized food webs. Annu Rev Ecol Syst 28:289–316

Read JL, Wilson D (2004) Scavengers and detritivores of kangaroo harvest offcuts in arid Australia. Wildl Res 31:51–56

Richards EN, Goff ML (1997) Arthropod succession on exposed carrion in three contrasting tropical habitats on Hawaii island, Hawaii. J Med Entomol 34:328–339

Rozen DE, Engelmoer DJP, Smiseth PT (2008) Antimicrobial strategies in burying beetles breeding on carrion. Proc Natl Acad Sci U S A 105:17890–17895

Ruxton GD, Houston DC (2004) Obligate vertebrate scavengers must be large soaring fliers. J Theor Biol 228:431–436

Schmitz OJ, Hawlena D, Trussell GC (2010) Predator control of ecosystem nutrient dynamics. Ecol Lett 13:1199–1209

Schneider K, Christman MC, Fagan WF (2011) The influence of resource subsidies on cave invertebrates: results from an ecosystem-level manipulation experiment. Ecology 92:765–776

Schoenly K (1992) A statistical analysis of successional patterns in carrion arthropod assemblages: implications for forensic entomology and determination of the postmortem interval. J Forensic Sci 37:1489–1513

Schoenly K, Reid W (1987) Dynamics of heterotrophic succession in carrion arthropod assemblages: discrete series or a continuum of change? Oecologia 73:192–202

Schoenly K, Goff ML, Early M (1992) A basic algorithm for calculating the postmortem interval from arthropod successional data. J Forensic Sci 37:808–823

Sebastián-González E et al (2016) Nested species-rich networks of scavenging vertebrates support high levels of interspecific competition. Ecology 97:95–105

Selva N, Fortuna MA (2007) The nested structure of a scavenger community. Proc R Soc B Biol Sci 274:1101–1108

Selva N, Jedrzejewska B, Jedrzejewski W, Wajrak A (2005) Factors affecting carcass use by a guild of scavengers in European temperate woodland. Can J Zool 83:1590–1601

Sevenster JG, VanAlphen JJM (1996) Aggregation and coexistence. 2. A neotropical Drosophila community. J Anim Ecol 65:308–324

Simmons T, Adlam RE, Moffatt C (2010) Debugging decomposition data-comparative taphonomic studies and the influence of insects and carcass size on decomposition rate. J Forensic Sci 55:8–13

Smith CR, Baco AR (2003) Ecology of whale falls at the deep-sea floor. Oceanogr Mar Biol 41:311–354

Swift MJ, Heal OW, Anderson JM (1979) Decomposition in terrestrial ecosystems. Blackwell Scientific Publications, Oxford

Tomberlin JK, Benbow ME, Tarone AM, Mohr RM (2011) Basic research in evolution and ecology enhances forensics. Trends Ecol Evol 26:53–55

Tomberlin JK et al (2012) Interkingdom responses of flies to bacteria mediated by fly physiology and bacterial quorum sensing. Anim Behav 84:1449–1456

Towne EG (2000) Prairie vegetation and soil nutrient responses to ungulate carcasses. Oecologia 122:232–239

Turner MG, Romme WH, Gardner RH, Hargrove WW (1997) Effects of fire size and pattern on early succession in Yellowstone National Park. Ecol Monogr 67:411–433

VanLaerhoven SL, Anderson GS (1999) Insect succession on buried carrion in two biogeoclimatic zones of British Columbia. J Forensic Sci 44:32–43

von Berg K, Traugott M, Scheu S (2012) Scavenging and active predation in generalist predators: a mesocosm study employing DNA-based gut content analysis. Pedobiologia 55:1–5

Wallace JR (2015) Aquatic vertebrate carrion decomposition. In: Benbow ME, Tomberlin JK, Tarone AM (eds) Carrion ecology, evolution, and their applications. CRC Press, Boca Raton, pp 247–272

Walther GR et al (2002) Ecological responses to recent climate change. Nature 416:389–395

Wilmers CC, Stahler DR, Crabtree RL, Smith DW, Getz WM (2003) Resource dispersion and consumer dominance: scavenging at wolf- and hunter-killed carcasses in Greater Yellowstone, USA. Ecol Lett 6:996–1003

Wilson EE, Wolkovich EM (2011) Scavenging: how carnivores and carrion structure communities. Trends Ecol Evol 26:129–135

Woodcock BA, Watt AD, Leather SR (2002) Aggregation, habitat quality and coexistence: a case study on carrion fly communities in slug cadavers. J Anim Ecol 71:131–140

Yang LH (2004) Periodical cicadas as resource pulses in North American forests. Science 306:1565–1567

Yang LH (2006) Interactions between a detrital resource pulse and a detritivore community. Oecologia 147:522–532

Ecological Functions of Vertebrate Scavenging

James C. Beasley, Zachary H. Olson, Nuria Selva, and Travis L. DeVault

Contents

J. C. Beasley (✉)
Savannah River Ecology Lab, Warnell School of Forestry and Natural Resources, University of Georgia, Aiken, SC, USA
e-mail: beasley@srel.uga.edu

Z. H. Olson
Department of Psychology, University of New England, Biddeford, ME, USA
e-mail: zolson@une.edu

N. Selva
Institute of Nature Conservation, Polish Academy of Sciences, Kraków, Poland

T. L. DeVault
National Wildlife Research Center, Animal Plant Health Inspection Service, Wildlife Services, U.S. Department of Agriculture, Sandusky, OH, USA
e-mail: travis.l.devault@aphis.usda.gov

P. P. Olea et al. (eds.), *Carrion Ecology and Management*, Wildlife Research Monographs 2, https://doi.org/10.1007/978-3-030-16501-7_6

Introduction

The role of vertebrate scavenging in food web dynamics has historically been minimalized and portrayed as the activity of a select group of obligate scavengers (e.g., vultures), with a simplistic linkage between carrion and detritivores in food webs. Research in the last few decades, however, has revealed that the role of carrion in food web dynamics is severely underestimated, highly complex, and pervasive among ecosystems across the globe (DeVault et al. 2003; Selva and Fortuna 2007; Wilson and Wolkovich 2011). Such observations have led to a surge in research interest in scavenging ecology that continues to reveal new information on the ecological functions of scavenging and the complex role scavengers play in the cycling of nutrients within ecosystems.

Although the functional role of scavengers in ecosystems has been recognized by some ecologists for many years, the importance a single guild of scavengers can have on ecosystem processes was underappreciated by many prior to the collapse of *Gyps* vulture populations in Asia in the 1990s due to accidental poisoning with diclofenac, a non-steroidal anti-inflammatory drug that was administered to cattle (Oaks et al. 2004). This tragedy has served as a cornerstone example of the ecosystem services provided by scavengers, as well as the potential impacts of anthropogenic activities on ecosystem function (DeVault et al. 2016). Scavengers have historically been some of the most persecuted species across the globe (Ogada et al. 2012a). However, while direct persecution remains a threat to scavengers in some areas, ecosystem services provided by scavengers also are being eroded due to numerous other anthropogenic factors (e.g., climate change, trophic downgrading, habitat loss and fragmentation, contaminants). Many of these factors have far reaching impacts, from the deep sea to arctic ecosystems, and have the potential to disrupt nutrient cycling dynamics across all levels of biological organization.

As a rapidly growing sub-discipline within the field of ecology, substantial advancements continue to be made in our understanding of the role carrion and scavenging play in individual- to ecosystem-level processes (Benbow et al. 2019). In this chapter, we highlight the ecological functions of vertebrate scavenging in both terrestrial and aquatic ecosystems, including those functions that directly benefit humans (i.e. ecosystem services). Specifically, we have delineated specific sections devoted to nutrient cycling, biodiversity maintenance, and disease control, drawing upon examples and case studies from ecosystems across the globe. We also discuss the impact of anthropogenic activities on the availability of carrion within the context of ecosystem functions, as this topic is of growing importance among researchers studying scavenging dynamics. Although substantive progress has been made in our knowledge of scavenging dynamics over the last few decades, there are many aspects of scavenging ecology that remain unanswered and represent important areas for future research. Thus, we conclude the chapter with a brief section highlighting a few areas where additional research is particularly needed to continue advance our understanding of nutrient cycling and scavenging dynamics.

Nutrient Recycling

Food Web Complexity

Scavenging by vertebrates promotes food web complexity, provides routes by which communities may sequester resources at higher trophic levels, and may subsidize particular populations of vertebrates during otherwise food-limited times of the year, resulting in more stable food webs. Food webs are conceptual frameworks used to better understand ecosystems, but many food webs fail to account for trophic connections made through scavenging (Wilson and Wolkovich 2011). For example, early food webs were linear (Elton 1927), which precluded the recognition of many feeding relationships (Lindeman 1942). Later publication of curated food webs (Cohen 1978) and mathematical insights about connections in communities (May 1972) catalyzed a burst of scholarly activity devoted to determining the nature of the relationship between community complexity—that is, the number and strength of links in the food web—and community stability (Dunne 2006).

Yet, despite advancing the stability-complexity debate, these curated food webs (e.g., the cascade model of Cohen et al. 1990) were found to inaccurately characterize ecosystems because they specifically prohibited some categories of real feeding relationships that are commonly expressed as a function of scavenging. In particular, 'loops' were eliminated from a number of notable curated food webs (Pimm 1982; Lawton 1989; Cohen et al. 1990). Loops occur in real food webs because omnivores feed at more than one trophic level (top-down effects) and because consumers receive feedback from the trophic levels they consume (bottom-up effects; Dunne 2006). Thus, loops are an important facet of scavenging ecology, because scavenging facilitates feeding relationships that are typically limited by ecological barriers such as physical size or behaviors like aggression (e.g., see Olson et al. 2016; Turner et al. 2017; chapter "Vertebrate Scavenging Communities").

In counterpoint to the structural limitations of some curated webs, Polis (1991) describes incredibly detailed sub-webs from a desert habitat within the Coachella Valley of California, U.S. Another detailed food web was published by Martínez (1991), this time from a lake in Wisconsin, U.S. Notable differences between these and previous webs included the large number of species, high number of feeding relationships noted for each species, relatively long food chains making up the webs, and frequent observation of looping and omnivory (Polis 1991). Important implications for food web theory aside, Polis (1991) draws particular attention to the fact that connections between species become more numerous when researchers remove the various constraints they had placed on their search for connections. For example, the consideration of parasites as actors in food webs increased food web connectance by 93% (Lafferty et al. 2006). If, as Polis (1991) argues, there is no good reason to place limits on the feeding relationships that are entered into food webs, the acknowledgement of relationships that occur via scavenging of carrion will also increase complexity of food webs (Wilson and Wolkovich 2011). To take an example from the Coachella Valley food web, lagomorphs (*Sylvilagus* and *Lepus* spp.) are prey for perhaps 10–15 species throughout life (approximated from Polis 1991). Yet, McKinnerney (1978) noted 63 arthropod and four vertebrates engaged in feeding on experimentally-placed lagomorph carcasses in similar habitats, most of which would have been unaccounted-for links in the food web. The addition of carrion-related connections to food webs creates a large number of loops, both distant and tight. Wilson and Wolkovich (2011) estimated that in 23 well-studied food webs scavenging-related links were underestimated substantially at 1.4% of links. By considering connections likely to occur based on scavenging behavior, in those same food webs scavenging could have conceivably accounted for 22% of total links after the adjustment (Wilson and Wolkovich 2011). Clearly, feeding relationships that connect through animal carcasses increase food web complexity.

Scavenging and Food Web Stability

Although a complexity-stability theory for ecological food webs is not yet fully resolved, two themes have emerged from research on real food webs that support a potential role for scavenging to affect food web stability. Connectance is a measure of link saturation, and refers to the percentage of possible links that are realized in any given food web (May 1972). Food webs with higher connectance have been shown to exhibit greater stability (De Angelis 1975; Dunne et al. 2002). Further, the degree of a food web's nodes—that is, the number of predators and prey connected to each species (Wasserman and Faust 1994)—also influences food web stability, but in a more complicated way. Increased heterogeneity in degree distributions seems to foster a more stable food web (Neutel et al. 2002). Because scavenging results in sometimes unorthodox links between species (DeVault et al. 2003; chapter "Vertebrate Scavenging Communities"), scavenging increases connectance in food webs (Wilson and Wolkovich 2011; Schlichting et al. 2019). Moreover, the degree to which numerically dominant facultative scavengers participate in scavenging behavior varies widely. This variation highlights the fact that many scavenger relationships represent weak links—the very kind of links that influence degree heterogeneity in networks like food webs (McCann et al. 1998; Neutel et al. 2002; DeVault et al. 2003).

Another mechanism by which scavenging may affect food web stability stems from the inherent relationship between food web stability and the population dynamics of its constituent species (Tilman 1996). Few species are obligate scavengers (Ruxton and Houston 2004a), and as ubiquitous as scavenging behavior is at the species level among vertebrate taxa (DeVault et al. 2003), individual facultative scavengers often choose to engage in their preferred foraging method rather than scavenging. Scavenging is often described as a feeding relationship based on need, and as such the importance of scavenging-derived resources is thought to vary dramatically for many species. Gese et al. (1996) found that the ability to scavenge on ungulate carrion was central to coyote *Canis latrans* activity budgets during winter, particularly if deep snow made travel difficult. In that system, social dominance predicted access to carcasses, because dominant coyotes would sequester fresh carcasses and allow lower-ranked individuals access to scavenge only after feeding for several days (Gese et al. 1996). Grizzly bears *Ursus arctos horribilis* may also depend on carrion during winter, but seem to scavenge in order to replenish energy reserves after emerging from hibernation (Green et al. 1997). Alternatively, scavenging may provide a means for vertebrates to compensate for energetically-intensive activities such as migration. Bald eagles *Haliaeetus leucocephalus* follow this example where they have been shown to scavenge salmon *Oncorhynchus nerka* carcasses plentiful in a particular area along their migration route (Restani et al. 2000). In these situations, scavenging-derived nutrients represent important subsidies for some portion of the population or at some time of the year. Research in scavenging ecology is just beginning to elucidate the exact role that this unique feeding relationship has in maintaining ecosystem stability.

Scavenging Alters the Trophic Distribution of Carrion-Derived Nutrients

Studying the assemblage of species that feed on a vertebrate carcass provides insights into trophic connections that simply are not present when the animal is alive. If we were to count the number of necrophagy-based links that could be added to food webs based on our knowledge of scavenging by vertebrates (DeVault et al. 2003) and based on our knowledge of decomposition by invertebrates and microbes (VanLaerhoven 2010), the majority of new links—numerically, if not always in terms of overall energy flux—would connect the deceased vertebrate to invertebrates and microbial consumers (i.e., decomposers) rather than to other vertebrates (chapter "Carrion Decomposition"). The numerical importance of invertebrate and decomposer-derived links holds across terrestrial and marine ecosystems (DeVault et al. 2004; Beasley et al. 2012). However, the trend would not be without notable vertebrate exceptions. White-tailed deer *Odocoileus virginianus*, a large obligate herbivore common across much of North America, have been observed removing small mammal carcasses (Rooney and Waller 2003; DeVault et al. 2011). Because there are no published accounts of deer preying on live small mammals, the deer-small mammal link could be added to the food web only if we considered scavenging to be part of food web theory. Although this particular relationship is probably rare (1 in 234 mice based on DeVault et al. 2011), a body of work has emerged that demonstrates the near ubiquity of scavenging behavior across vertebrate taxa (DeVault et al. 2003).

Despite the numerical advantage enjoyed by invertebrate and microbial decomposers at carrion (Janzen 1977; DeVault et al. 2003), vertebrate scavengers are the focus of many investigations. The attention paid to vertebrate scavengers is undoubtedly enhanced by the fact that vertebrates are more easily recognizable than invertebrates or microbes, certainly at the species level, by means of direct (McKinnerney 1978; Hertel 1994; Cortés-Avizanda et al. 2012) or indirect observation (Jędrzejewska and Jędrzejewski 1998; DeVault et al. 2004; Selva et al. 2005;

Killengreen et al. 2012). Beyond that technical consideration, there are two main arguments for the importance of vertebrate scavenging on ecosystem processes. First, the most frequent scavengers in many systems are also predators (i.e., facultative scavengers; DeVault et al. 2003; Wilson and Wolkovich 2011), and predators have outsized influence on prey and on primary production via trophic cascades (Schmitz et al. 2000). Second, vertebrate scavenging is interesting because of its impact on the flow of carcass resources to the surrounding food web (Wikenros et al. 2013; Beasley et al. 2015). Vertebrate scavenging disproportionately impacts carcass resources, because even one vertebrate scavenger can deny carcass resources a direct path to the detrital pool, thereby maintaining carcass resources at higher trophic levels (Beasley et al. 2015). Simultaneously, vertebrate scavenging preempts or halts the acquisition of carrion resources by invertebrate and microbial decomposers by co-opting the carcass or by ingesting and killing invertebrate and microbial competitors along with the carcass (Janzen 1977; DeVault et al. 2003). The presence or absence of vertebrate scavenging, and the nature of the species doing the scavenging, therefore has important implications for the cycling of nutrients. The importance of vertebrates in determining the fate of individual carcasses is amplified by the fact that most carcasses in many ecosystems are scavenged by vertebrates (DeVault et al. 2003; Mateo-Tomás et al. 2015; Olson et al. 2016; Sebastián-González et al. 2016; chapter "Vertebrate Scavenging Communities").

Transportation of Carrion Within and Between Ecosystems

While decomposers mostly enhance/promote nutrient recycling into the soil surrounding the carcass, arthropods and vertebrate scavengers contribute to the transport of carrion-derived nutrients (Schlichting et al. 2019; chapter "Carrion Decomposition"). This is mainly done by consuming carrion and later producing excretions away from the carcass, but also by transporting parts of the carcasses. Although some arthropods can take away carcass portions, e.g. some carrion beetles *Necrophorus spp.* may roll and bury a carrion ball, vertebrate scavengers are mainly responsible for transport of carrion at longer distances. Many scavengers have been documented caching carrion in presumably less risky places and later returning to retrieve and consume the carcass (Magoun 1976; Selva 2004a). In Białowieża Forest, carrion retrieval was recorded in the jay *Garrulus glandarius*, raven *Corvus corax*, white-tailed eagle *Haliaeetus albicilla*, fox *Vulpes vulpes*, and pine marten *Martes martes*; wolves *Canis lupus* were often documented carrying carcass pieces few kilometers away from the kill. Caching meat is common among ravens and other corvids as well as many mammalian carnivores, and seems to be more widespread among scavengers at higher latitudes (Heinrich and Pepper 1998; Selva 2004a).

One of the most notable examples of carrion transport between ecosystems is the annual return of spawning salmon from the sea to the rivers, which represents a pulsed resource of marine nutrients into freshwater and terrestrial systems

Fig. 1 Salmon are a notable example of carrion transport between ecosystems, representing a pulsed resource of marine nutrients into freshwater and terrestrial systems. Photo credit: Adam Wajrak

(Cederholm et al. 1999; Gende et al. 2002). Salmon carrion "fertilizes" streams and provides high-quality food for fishes and invertebrates. Salmon, both alive and dead, are also consumed by a wide range of terrestrial vertebrates, including bears, wolves, coyotes, and numerous avian scavengers, and by a diverse community (>60 species) of terrestrial invertebrate decomposers (Hocking et al. 2009; Levi et al. 2015). Among salmon consumers, brown bears *Ursus arctos* are a keystone species for the transfer of salmon carrion and nutrients into adjacent forest (Fig. 1). After capturing salmon in the rivers, brown bears move to the adjacent land to feed on energy-rich portions, leaving for scavengers carcasses from the salmon they kill which are only partially consumed (Gende et al. 2004). Bear foraging and other scavengers' activities play a crucial role in making salmon-derived nutrients available to terrestrial vegetation and subsidizing the forest. Indeed, the salmon-bear interaction can provide up to 24% of riparian nitrogen (N) budgets (Helfield and Naiman 2006).

The transfer of marine carrion into adjacent lands is often mediated by scavengers. For example, seal (Phocidae) carrion may subsidize Arctic foxes *Vulpes lagopus* in Canada during low-density phases of the cyclic lemming *Dicrostonyx richardsoni* population (Roth 2003). Marine carrion reaching the coast also supported higher densities of coyotes in desert areas of California (Rose and Polis 1998). These allochthonous carrion inputs to scavengers exert notable effects on bottom-up regulation, top-down control and the complexity and structure of trophic interactions (Rose and Polis 1998). (see chapter "Carrion Availability in Space and Time" on Carrion Exchange at the Terrestrial-Aquatic Interface).

Aquatic vs. Terrestrial Systems

The fundamental processes of decomposition and scavenging are similar among ecosystems across the globe, with a diverse assemblage of microorganisms, invertebrates, plants, and vertebrates playing critical roles in the assimilation and redistribution of nutrients within the ecosystem. However, the composition and functional role of each of these groups can vary widely among ecosystems, as well as the resulting spatial and temporal diffusion of carcass nutrients, processes ultimately driven by fundamental differences in both abiotic conditions and emergent properties of air and water (Beasley et al. 2012).

In particular, the three-dimensional nature of aquatic ecosystems facilitates the movement of carrion by wave action, upwelling, and sinking, diffusing the nutrients within carrion sources across a much larger spatial scale than occurs in terrestrial ecosystems (Britton and Morton 1994). This diffusion of nutrients is most dramatically recognized in pelagic ocean habitats where disparate scavenging communities exist between the photic zone (where most nutrient production and sources of carrion occur) and organisms living in the benthic zone, where the existence of deep-sea scavengers is inextricably linked to productivity of nekton (i.e. raining organic matter) from the surface (Gooday et al. 1990; Britton and Morton 1994; Smith et al. 2008). Thus, the redistribution of nutrients sequestered in carcasses within aquatic ecosystems is complex and largely depends upon the speed of vertical dissemination of carrion through the water column. Carcass particulates can be even more mobile than whole carcasses, dispersing tens to hundreds of kilometers or more within aquatic ecosystems through wave action and diffusing aquatic carrion across vast spatial scales in three-dimensional space.

In contrast, within terrestrial ecosystems the processes of scavenging and decomposition are primarily localized around the origin of the carrion resource, creating a tight linkage between processes of nutrient production, scavenging, and decomposition in terrestrial ecosystems that is absent in many aquatic ecosystems (Wardle et al. 2004). Such localized concentrations of resources, even when scavenged, can create biogeochemical hotspots within terrestrial soils that alter the nutrient composition of forest soils, influencing the growth and species composition of local plant communities for years (Bump et al. 2009a, Hocking and Reynolds 2011; Barton et al. 2013; chapter "Carrion Decomposition"). Similar analogs also exist in benthic ecosystems where, once settled on the sea floor, large carrion falls (e.g., whales—Cetacea) create a localized flux of organically-enriched sediments surrounding carcasses that are colonized by dense assemblages (e.g., up to 40,000 per m^2) of crustaceans and annelid worms; although such enrichment of sediments does not appear to occur for carcasses <50 kg in size (Smith and Baco 2003).

However, even within terrestrial ecosystems food-web linkages between scavenging guilds can vary widely among ecosystems, primarily due to underlying differences in abiotic conditions (e.g., temperature, rainfall). Such differences are particularly pronounced when comparing arctic and tropical ecosystems, although a latitudinal gradient undoubtedly exists in the role of microbial, invertebrate, and vertebrates in the assimilation of carrion; gradients affecting guild membership also are pronounced between humid and arid environments (Parmenter and MacMahon 2009). Aquatic ecosystems (particularly benthic habitats), in contrast, are buffered to some degree by temporal changes in weather, potentially minimizing seasonal or even daily fluctuations in competitive interactions and nutrient assimilation between microbes, invertebrates and vertebrates that exists in terrestrial ecosystems (DeVault et al. 2003; Beasley et al. 2015). Such competitive interactions and consistency among guilds involved in recycling of carrion nutrients is further modulated by reduced temperatures in benthic ecosystems, which minimizes the role of bacteria in the redistribution of nutrients (Beasley et al. 2012). Thus, despite the similar functional role of the redistribution of carrion nutrients within food webs, the strength and consistency of individual linkages can vary widely among ecosystem types, complicating our understanding of food web dynamics and the impacts of anthropogenic activities upon such processes.

Biodiversity Maintenance

Obligate and Facultative Scavengers

Although most, if not all, carnivorous animals will eat carrion under certain conditions (DeVault et al. 2003; Fig. 2), some species rely heavily on carrion and thus are often referred to as scavengers rather than predators (chapter "Vertebrate Scavenging Communities"). Adaptations attributed to a scavenging lifestyle include rapid and efficient locomotion to locate an ephemeral and sometimes unpredictable food supply, well developed visual and/or olfactory senses, a large body size to withstand periods of food shortage, and the ability to withstand or detoxify products of microbial decomposition (Houston 1979; Ruxton and Houston 2004a, Shivik 2006). There appears to be a continuum of species with varying levels of adaptation for finding and consuming carrion (DeVault et al. 2003), and even within species some individuals scavenge more than others (Shivik and Clark 1999).

Despite this continuum, scavenging animals are generally classified as either obligate (species that scavenge exclusively) or facultative (predators that also occasionally scavenge, i.e., most carnivorous vertebrates) (chapter "Vertebrate Scavenging Communities"). Most researchers agree that the only terrestrial obligate scavengers are the Old and New World vultures (families Accipitridae and Cathartidae) (DeVault et al. 2003; Beasley et al. 2015). Indeed, even terrestrial species such as coyotes and spotted hyenas *Crocuta crocuta*, which are generally considered to be excellent scavengers, obtain most of their food by killing live prey (Gasaway et al. 1991; Linnell et al. 1995). In contrast, a great deal of discussion has been devoted to the question of whether aquatic obligate scavengers exist (e.g., Britton and Morton 1994; Kaiser and Moore 1999; Ruxton and Houston 2004b, Ruxton and Bailey 2005). Regardless of whether some aquatic species are able to

Fig. 2 Although vultures are globally recognized as important vertebrate scavengers, there is growing evidence that most vertebrates will opportunistically scavenge carrion when available. Facultative scavengers in Midwestern USA (clockwise) include golden eagle (*Aquila chrysaetos*), striped skunk (*Mephitis mephitis*), Virginia opossum (*Didelphis virginiana*), and coyotes (*Canis latrans*). Photo credit: Kelsey Turner and James Beasley

survive solely on carrion, it is clear that some are extremely well adapted for scavenging, having acute chemoreception, efficient locomotion, and being able to withstand long periods of food shortage (Tamburri and Barry 1999; Smith and Baco 2003). Additionally, the question of obligate scavenging has been argued with respect to long-extinct species (e.g., *Tyrannosaurus rex*; Ruxton and Houston 2003; Carbone et al. 2011).

Seemingly lost in discussions of "scavenger" vs. "predator" is the fact that even the clearest examples of obligate scavengers—vultures—occasionally kill and eat other animals, especially relatively helpless, newborn livestock (Lowney 1999; Avery and Cummings 2004; Margalida et al. 2014). It seems unreasonable that any carnivorous animal would choose not to kill and consume vulnerable prey, especially when carrion is scarce. Along these lines, we suggest that some well-adapted aquatic scavengers (e.g., hagfish; family Myxinidae) are likely just as "obligate" as many species of vultures (chapter "Vertebrate Scavenging Communities"), and that such classification exercises are generally unhelpful. Many terrestrial and marine organisms have developed mechanisms to survive largely, if not exclusively, on carrion, showing the importance of necrophagy to their livelihood.

Marine and Terrestrial Examples of Biodiversity

Carrion is a nutrient rich resource utilized by a diverse array of facultative and obligate scavengers across the globe and thus plays a critical role in shaping species diversity and nutrient cycling within ecosystems (Barton et al. 2013; chapters "Invertebrate Scavenging Communities", "Vertebrate Scavenging Communities", and "Carrion Decomposition"). Moreover, carrion can affect the composition of local soil (Melis et al. 2007; Bump et al. 2009a, b, Parmenter and MacMahon 2009), microbial (Yang 2004), arthropod (Melis et al. 2004), and plant (Towne 2000) communities and thus can have substantive effects on biodiversity and function within ecosystems. Such widespread use of carrion resources among levels of biological organization has led to intensive inter-kingdom competition for acquisition of nutrients sequestered within carrion resources. This competition has resulted in intensive selection pressure on many species that utilize carrion resources.

For example, some species of arthropods have evolved highly sensitive chemoreceptive abilities allowing them to locate vertebrate carrion resources at the time of, or soon after, death (Tomberlin et al. 2005). Oviposition by these flies is further influenced by complex interactions with microbes that dictates attraction and repulsion of conspecifics to a carcass and plays an important role in survivorship of deposited eggs (Lam et al. 2007; Zheng et al. 2013). Similarly, intensive intra- and interspecific competition exists among burying beetles for access to a carcass as larvae from a single clutch can consume an entire carcass (Müller et al. 1990). Among vertebrates, vultures are the most well-known exemplars of species with adaptations conducive of a scavenging lifestyle, with each species exhibiting unique adaptations to minimize niche overlap for a sparse and unpredictable food source. For example, although both black *Coragyps atratus* and turkey vultures *Cathartes aura* overlap extensively in distribution, both species have evolved disparate chemoreceptive abilities, social behavior, and flight ability (Kirk and Mossman 1998; Buckley 1999). Similarly, bearded vultures *Gypaetus barbatus* have evolved for specialized feeding on bone marrow and bones of animal carcasses to minimize interspecific competition with other old world vultures that primarily consume flesh (Margalida et al. 2009). However, adaptations that facilitate the detection (e.g., mobility, chemoreception) or reliance (e.g., fasting ability) on carrion are pervasive among terrestrial and aquatic vertebrate communities.

Regardless of ecosystem type, carrion is patchily distributed in both space and time, creating heterogeneity in community richness and composition. The spatial and temporal extent of such effects are largely influenced by the physical properties of carcasses (e.g., size), environmental conditions (e.g., temperature, humidity), and ecology (e.g., energy requirements, home range size, reliance on carrion resources) of scavenging guilds (DeVault et al. 2003; Moleón et al. 2015; Turner et al. 2017; chapter "Carrion Availability in Space and Time"). For example, a single small mammal carcass likely has a much more profound effect on the nutritive status of small- vs. large-bodied facultative scavengers (e.g., red fox vs. brown bear). Similarly, the influence of carrion on the survival and reproduction of facultative scavengers likely differs geographically as a function of ecosystem productivity, with the most profound effects of carrion resources occurring in the arctic or other landscapes with reduced vertebrate diversity (Fuglei et al. 2003). However, the role of carrion subsidies to survival and reproduction is largely unknown for most facultative scavengers.

Carcasses alter concentrations of inorganic N and phosphorous (P) in local soil, increasing the species richness and diversity of surrounding plant communities (Towne 2000; Parmenter and MacMahon 2009). Although such effects may be more pronounced for larger vertebrate carrion at a local scale, smaller vertebrates are often more abundant and thus likely equally significant in their contribution to nutrient cycling and dynamics at the landscape scale. For example, in portions of their invasive range coqui frogs *Eleutherodactylus coqui* can reach densities of >91,000 adult frogs/ha (Beard et al. 2009) and their carcasses are readily scavenged by vertebrates (Abernethy et al. 2016). Similarly, rodents, amphibians, invertebrates and other small organisms can achieve remarkably high densities in some ecosystems and thus represent large pools of potential carrion. Such influxes of nutrients through carrion (even scavenged carrion) create biogeochemical hotspots in local soils that alter plant community dynamics and thus aid in maintaining plant biodiversity by creating heterogeneity in resource availability throughout the landscape (Bump et al. 2009a). Shifts in abundance and community diversity at carcass sites also are observed in microbial and fungal communities where up to a 400% increase in microbial biomass has been observed in soil under a single rat *Rattus rattus* carcass (Bump et al. 2009b, Barton et al. 2013).

Aside from microbial communities, arthropods are generally the most abundant and species diverse group of organisms at a carrion source and their influence on decomposition processes can be profound (Barton et al. 2013; Pechal et al. 2014). Moreover, among biological communities that utilize carrion, niche separation and specialization across a multitude of carrion resources is probably most prevalent within the invertebrate community. Vertebrate carcasses can be further broken down into disparate components (e.g., flesh, skin, fluids, stomach content, hooves and horns), each of which can serve as a unique attractant to a diverse assemblage of invertebrate scavengers (Braack 1987; chapter "Invertebrate Scavenging Communities"). Of these, carcasses are dominated by soft tissues and thus such tissues support the greatest biomass and diversity of attendant arthropods. In particular, blow-fly larvae can achieve numbers in excess of 200,000 on a single carcass, dominating carcass colonization and biomass consumption of soft tissue, especially during warmer months (Payne 1965). Such competition for carrion resources has led to the evolution of a diverse community of species able to take advantage of other animal components such as rumen content (e.g., Scarabaeidae, Muscidae), skin (e.g., *Dermestes maculatus*), keratinized tissues (e.g., *Ceratophaga vastella*, trogid beetles), and remaining detritus following consumption of carcasses (e.g., ants) (Braack 1987). In addition, numerous invertebrate species (e.g., Histerid beetles) utilize carrion indirectly by preying on other invertebrates, particularly blowfly larvae at carcasses (see chapter "Invertebrate Scavenging Communities"). Such concentrations of invertebrates at carcasses is further exploited by parasitic species (e.g., parasitic wasps—Diapriidae, Pteromalidae), that target developing pupae (Braack 1987). Thus, carcass sites represent hotspots of invertebrate biodiversity that facilitate the maintenance and propagation of species diversity within invertebrate communities.

Such competition among invertebrates for resources sequestered within carrion is further enhanced through competition with vertebrates. In some ecosystems, vertebrates may assimilate as much as 90% of the available carrion and thus there is an intensive inter-kingdom "arms-race" for carrion resources that is driving the evolution of species reliant upon carrion (DeVault et al. 2003, 2011). Although only a few groups of vertebrate taxa have evolved as obligate scavengers (see above), nearly all vertebrates will utilize carrion opportunity and can be considered facultative scavengers (DeVault et al. 2003; Beasley et al. 2015). This propensity to scavenge underscores the importance of scavenging-derived nutrients in vertebrate communities and suggests scavenging behavior may have profound effects on food web stabilization (Wilson and Wolkovich 2011). Despite the growing awareness of the pervasiveness of scavenging among vertebrates across a wide range of taxa and ecosystems, the influence of carrion on survival and reproduction are largely unknown for many facultative scavengers, although such effects are potentially profound, particularly in resource-limited ecosystems.

The sequential colonization and use of carcasses by invertebrates has been well documented (Byrd and Castner 2010). Such patterns are sufficiently predictable and they represent a cornerstone of forensic investigations regarding time of death (chapter "Invertebrate Scavenging Communities"). Temporal heterogeneity in carcass use by vertebrates does not appear to be as predictable or prevalent, although there is recent evidence of some temporal partitioning in arrival times between sympatric black and turkey vultures, likely due to differences in their natural history (Beasley, unpublished data). Rather, spatial or temporal variation in carcass use by vertebrate species is likely more heavily influenced by local densities, abiotic factors, and community assembly (Selva and Fortuna 2007; Sebastián-González et al. 2013; Allen et al. 2014).

Succession among scavenging communities utilizing carrion is especially pronounced in deep sea ecosystems due to low temperatures, increased pressure, and reduced microbial activity (Smith and Baco 2003; chapter "Methods for Monitoring Carrion Decomposition in Aquatic Environments"). Although few studies have quantified community composition and succession at carcasses in abysmal ecosystems, observations at large cetacean carcasses suggest a single carcass can provide a nutrient-rich food source to hundreds of thousands of scavengers comprising nearly 200 species, lasting up to several years (Smith and Baco 2003). In contrast, large vertebrate carrion in terrestrial ecosystems can be assimilated by insects and microbes within days or months (Coe 1978; Selva et al. 2003; Carter et al. 2007). Thus, despite the spatio-temporal unpredictability of carrion in vast marine ecosystems, the persistence of carrion due to reduced microbial activity appears to be an important contributing factor to the evolution and maintenance of a diverse array of species in these ecosystems by broadening the range of niche space (Butman et al. 1995).

In particular, large cetacean carcasses clearly serve as hotspots of biological diversity within deep-sea ecosystems, the effects of which may last for decades (Smith and Baco 2003). Such carcasses have been hypothesized to serve a critical role as dispersal stepping stones for some deep sea vent and seep fauna and there is evidence some species may in fact be whale carcass specialists (Smith and Baco 2003). Whales can provide hotspots of activity even among large apex predators. For example, Fallows et al. (2013) observed up to 40 white sharks *Carcharodon carcharias* feeding on an individual whale carcass in a single day, with more than 5 individual sharks scavenging on the carcass in over 50% of their observational days. In addition to whale carcass specialists, there also is evidence to support the evolution of additional carrion specialists in marine ecosystems such as bone-eating *Osedax* worms (Rouse et al. 2004) and bone-eating *Rubyspira* snails (Johnson et al. 2010).

Creation of Localized Hotspots of Biodiversity
Because carrion is such an energy- and nutrient-rich resource, the availability of a carcass marks the beginning of an intense competition among microbes, invertebrates, and vertebrates to sequester carrion resources (DeVault et al. 2003). There are several attributes of carrion that might serve to maintain biodiversity in ecosystems. First, carrion is an ephemeral resource that is quickly attenuated (Coe 1978). The fast reduction time of dead animal matter is likely due to the fact that animal bodies represent one of the highest concentrations of N and P in ecosystems (Cornaby 1974; Swift et al. 1979; Carter et al. 2007). Rapid attenuation of carrion means that the resource is not ubiquitously available, at least at the level of an individual consumer (Ruxton and Houston 2004a). The individual consumer therefore perceives carrion as unpredictable in space and time (Wilmers et al. 2003a). Resource unpredictability is an important factor in biodiversity maintenance, because unpredictable resources are unlikely to be monopolized by a particular consumer (Cortés-Avizanda et al. 2012). Even the highly specialized, obligate scavenging vertebrates lose some of every carcass to combinations of microbes, invertebrates, and facultative vertebrate scavenger competitors (DeVault et al. 2003). Second, if carrion was a homogenous resource, the principle of competitive exclusion would suggest that it would be impossible for carrion to host the diverse field of consumers that are known to use it for food and reproduction (VanLaerhoven 2010). Rather than being used as homogenous resources, carcasses are partitioned by carrion obligate scavengers (Kruuk 1967); each species working to extract resources from their own realized niche (Hutchinson 1957). For example, past and present vulture assemblages are comprised of animals with different specialties in terms of their feeding strategy at carcasses (Hertel 1994; Houston 1979), and invertebrate species also partition carcasses by concentrating on particular areas or tissues (McKinnerney 1978; Hanski and Kuusela 1980). Carrion as a resource is defined as being rich in energy and nutrients, unpredictable in space and time, ephemeral where it occurs, and heterogeneous in composition, a list of properties that has primed carrion resources to be the focus of competitions that have shaped a number of species' evolutionary histories. Carrion itself provides

niche space through these competitive interactions that result in its hosting remarkable biodiversity, but carrion effects are not limited to the confines of the carcass.

Unless preserved after death, carcasses enter a linear and predictable sequence of stages leading to their reduction (Carter et al. 2007). Whether a carcass undergoes carrion decomposition on the one hand or carrion removal by vertebrates on the other represent a branching point for carrion nutrients in the ecosystem. In the absence of vertebrate scavenging, decomposition by microbes and invertebrates retains many carcass nutrients in situ. Carcass leachates created by (or released by) microbe and invertebrate action enrich the soil surrounding the carcass and aide in the formation of nutrient-abundant cadaver decomposition island (Carter et al. 2007). Carcass nutrients are then available for use by whole assemblages of soil microbes in addition to the nearby plant community (Melis et al. 2007; Bump et al. 2009a; see chapter "Carrion Decomposition" for more details on carrion decomposition).

Vertebrate scavengers can dramatically affect the outcome, this linear and predictable trajectory, of decomposition by removing (consuming) carrion resources. Carrion removal by vertebrates halts decomposition, halts the formation of a cadaver decomposition island (Carter et al. 2007), prevents the flush of nutrients to nearby soil and foliar communities, and generally disrupts the localized burst of biological activity that coincides with a decomposing carcass (Bump et al. 2009b; Beasley et al. 2015). Therefore, carrion can provide an intense resource pulse to nearby soil and foliar networks, causing a flush of growth in nearby plants and population growth in the soil microbiome (Wilson and Wolkovich 2011). Alternatively, carrion can be removed by vertebrate scavengers, mitigating further competitive interactions for carcass resources as the vertebrate monopolizes remaining resources and predates entire assemblages of microbial and invertebrate carcass colonizers (Polis 1991). In this way vertebrate scavenging strongly influences the biodiversity associated with carrion resources, and likely has influenced the evolutionary arms race that has produced so much biodiversity at carcass resources (DeVault et al. 2003).

Disease Control

Many authors have speculated on the role of scavenging in disease ecology (see Bellan et al. 2013 and references therein), but until recently few studies had considered how scavengers affect the spread of pathogens across landscapes (see chapter "The Role of Scavenging in Disease Dynamics"). Superficially, it would seem that scavengers could potentially proliferate diseases across large areas: animal carcasses can serve as hotspots for infectious materials, and scavengers often have large home ranges, foraging at widely-spaced carcasses (e.g., Houston 1979; DeVault et al. 2004). Many scavengers can visit a single large carcass before advanced decomposition (Jennelle et al. 2009). Also, at least some infectious materials can survive the digestive process of certain scavengers (VerCauteren et al. 2012). On the other hand, scavengers could reduce carcass-derived pathogens. Well adapted scavengers are often highly resistant to disease (Shivik 2006). For example, griffon vultures have highly acidic stomachs (pH of 1–2), which most pathogenic bacteria are unlikely to survive (Houston and Cooper 1975). Further, black and turkey vultures have extremely selective gastrointestinal tracts and most bacteria are killed during the digestive process (Roggenbuck et al. 2014).

Recent studies suggest that the composition of the scavenging community can strongly influence disease dynamics from carrion (Beasley et al. 2015). Ogada et al. (2012b) excluded African vultures from experimental carcasses and found that carcass persistence tripled, with resulting increased carcass use by mammals. Contacts between mammals at carcasses also increased in the absence of vultures, maximizing the probability of disease spread (Ogada et al. 2012b). Similarly, Hill et al. (2018) observed a >tenfold increase in the number of rabbit carcasses not fully scavenged among vulture-excluded carcasses. The dramatic decline of vultures in south Asia (caused by unintentional poisoning; Green et al. 2004; Oaks et al. 2004) also demonstrated how changes in the vertebrate scavenger community can influence disease. In the near-complete absence of vultures, human health costs from rabies increased dramatically (Markandya et al. 2008), apparently due to increased use of large carcasses by feral dogs and rats and subsequent population increases (Pain et al. 2003; Prakash et al. 2003).

It is clear that vertebrate scavengers can substantially influence disease dynamics. Under certain conditions, scavengers can facilitate the spread of some diseases, or alternatively, reduce carcass-derived pathogens across landscapes. Research suggests that the composition of the vertebrate scavenger community (e.g., one dominated by vultures vs. mammals) can mediate disease proliferation (e.g., Markandya et al. 2008; Ogada et al. 2012b). However, the role of vertebrate scavengers in disease dynamics is a fledgling area of study. Future research should examine the abiotic and biotic conditions that regulate the movement and propagation of carcass-derived pathogens within and across landscapes (see chapter "The Role of Scavenging in Disease Dynamics" for more information on the role of scavenging in disease dynamics).

Case Studies from Well-Studied Systems

Midwestern USA: Local Scavenger Guild Diversity and Its Effect on Carrion Removal

The midwestern United States is best described as a human-dominated landscape (Fig. 3a). Habitat fragmentation is severe with development sprawl, road networks, and mechanized agriculture serving to reduce overall habitat area and connectivity for wildlife species (Moore and Swihart 2005). Scavenging trials were conducted in northcentral Indiana, a state in the Midwestern US, to explore scavenging ecology in human-dominated landscapes. The original natural habitat across much of the Midwestern US and in Indiana was forest. However, fragmentation has reduced the

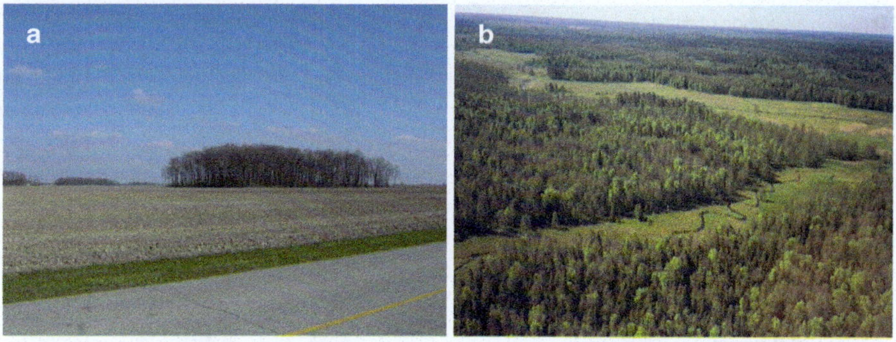

Fig. 3 (**a**) Example of a a highly fragmented agricultural ecosystem where vertebrate scavengers are often concentrated within remnant native habitat interspersed within a matrix of agriculture. Explore the area through Google Street View by clicking on https://goo.gl/maps/ZB5FW22iRUL2. (**b**) The Białowieża Primeval Forest along the Polish-Belarussian border, which is one of the best preserved European lowland temperate forests of its size. The Białowieża Primeval Forest (https://goo.gl/maps/Hb7DKihP8tk) is inhabited by a diverse animal community, including 5 species of ungulates and 40 species of raptors and carnivores. Photo credits: James Beasley (left) and Nuria Selva (right)

natural forest habitat into small patches referred to as woodlots. The woodlots in this study system acted essentially as habitat islands in a matrix of row crop agriculture and human development. Woodlots were small in general: 50% of forested habitat on the landscape exists in patches <2 ha in size, 75% exists in woodlots <5 ha in size, and only 1% of all forested habitat exists in woodlots >100 ha (Moore and Swihart 2005). From an ecological perspective, the landscape is inhabited by a truncated vertebrate fauna consisting mainly of those species that could adapt to a pervasive human presence. Top predators have been extirpated since at least the early 1900s (Whitaker and Mumford 2010), leaving a guild of mesopredators in their stead. Populations of some generalist mesopredators such as the Virginia opossum *Didelphis virginiana* and raccoon *Procyon lotor* have flourished for a number of reasons and have reached tremendous abundances (Beasley et al. 2011).

The pervasive ecological effects of such highly abundant mesopredator populations translated readily to affect scavenging ecology. When comparing the proportion of small mammal carcasses removed by vertebrates in the fragmented Midwestern study site to that in an intact, forested landscape in the southeastern US, the vertebrate scavenger community was much more efficient in the Midwest (Fig. 4a). Moreover, raccoon and opossum scavenging was responsible for removing nearly all of vertebrate-consumed small mammal carcasses in the Midwestern study site with a guild richness of just nine species (Fig. 4b; DeVault et al. 2011). In the intact southeastern US study site the vertebrate scavenger guild exhibited greater evenness (although raccoons and opossums still removed the majority of carcasses; DeVault et al. 2004) and a much greater scavenger guild richness (N = 17 species; Fig. 4b). Carrion removal in the fragmented Midwestern study system was effectively dominated by the highly abundant raccoon and opossum populations (DeVault et al. 2011).

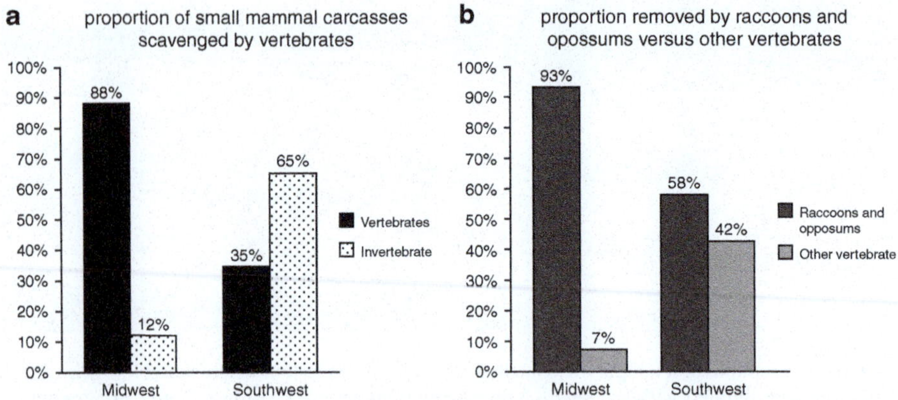

Fig. 4 (**a**) Proportion of small mammal carcasses removed by vertebrates and (**b**) the proportion removed specifically by raccoon (*Procyon lotor*) and Virginia opossum (*Didelphis virginiana*) relative to other vertebrates in the fragmented Midwestern USA study site compared to that in an intact, forested landscape in the southeastern USA. Credits: the chapter authors

Numerical and ecological dominance of mesopredators in the fragmented study system presented a unique opportunity to test the stability of ecosystem service provisioning under an altered scavenger fauna. To that end researchers removed the dominant scavenger, raccoons, from 13 woodlots and retained intact communities in 13 additional control woodlots (Olson et al. 2012), and then conducted scavenging trials using small mammals as carrion. Raccoons were allowed to recolonize removal woodlots naturally, but recolonization was still ongoing 3 years post-removal (Beasley et al. 2013). Given the behavioral plasticity of mesopredators (Prugh et al. 2009) and ecosystem-level efficiency of carcass removal in general (DeVault et al. 2003), it was hypothesized that any reduction in raccoon scavenging resulting from the experiment would be offset by a functional response in the next-most-dominant scavenger, the opossum. In other words, it was thought that opossums would be poised to take up any additional carcasses if carcass availability increased, and that invertebrates would claim any carcasses that the opossums did not remove. What was found instead was rather surprising.

Neither opossums nor invertebrates increased the proportion of carcasses they consumed in treatment (i.e., raccoon removal) woodlots relative to the proportion they consumed in control woodlots. This stability of carrion consumption by opossums and invertebrates occurred in juxtaposition to a >50% decline in raccoon scavenging in removal woodlots relative to control woodlots (i.e., the experimental effect). Interestingly, the number of carcasses that remained un-consumed at the ends of trials increased in those woodlots where raccoon scavenging was experimentally reduced relative to the control woodlots. Thus, an experimental alteration of the scavenging fauna resulted in a slight decrease in the efficiency with which carrion removal occurred on the landscape (Olson et al. 2012). The fact that the delivery of an important ecosystem service, carcass removal, could be altered by such a perturbation in a generalist-heavy ecosystem was not expected.

Given the importance of raccoons and opossums to the study system, and the large effect that fragmentation has on ecosystems (Fahrig 2003), additional work in the system focused on understanding how local scavenger guilds responded to raccoon and opossum carcasses in the context of landscape connectivity. Researchers deployed raccoon, opossum, and domestic rabbit carcasses into woodlots with high, intermediate, and low levels of habitat connectivity on the landscape (Olson et al. 2016). In this way the researchers could determine if carcass type or habitat connectivity was more influential in affecting facets of scavenging and carcass removal. In keeping with the efficiency of the scavenging guild in this Midwestern US study area, the vast majority (87%) of carcasses were scavenged at least in part by vertebrate scavengers. Invertebrates monopolized some (10%) carcasses during warmer weather and a small portion (3%) of carcasses went untouched through the experiment. Opossums occurred as scavengers most frequently in the study, making up 29% of all scavenger activity. In contrast to trials using small mammals as carrion, obligate scavenging turkey vultures accounted for the next-highest level of scavenger activity (17%) whereas raccoons accounted for 14% of scavenger activity. Vertebrates scavenging at frequencies ≥5% also included the red-tailed hawk (13%; *Buteo jamaicensis*), American crow (10%; *Corvus brachyrhynchos*), and coyote (5%), with overall guild richness of 15 species. The larger carcasses (2–6 kg) were not dominated by mesopredator scavenging to the extent that smaller carcasses (12–24 g) were in the same system.

Local scavenger guild richness was in line with that reported in other studies, between 2 and 3 scavengers on average, but was 24% lower for opossum carcasses than for raccoon carcasses with rabbits exhibiting no difference from other carcass types. At the same time, local scavenger guild richness was similar across categories of habitat connectivity. Similarly, beta diversity of local scavenger guilds was more affected by carcass type than habitat connectivity. Focusing on the process of carrion removal, the researchers found that each additional vertebrate scavenger that occurred at a carcass increased the proportion of the carcass that was consumed by 6% on average. From these findings, the researchers concluded that differences in carrion utilization exist among carcass types, these differences were not necessarily related to initial carcass mass, and that the reliability of carcass removal as an ecosystem service depends on the existence of a robust vertebrate and invertebrate community.

Białowieża Primeval Forest, Poland: Influence of Carrion Attributes on Scavenger Community Composition and Ecosystem Function

The Białowieża Primeval Forest (ca 1450 km^2, Polish-Belarussian border) is one of the best preserved European lowland temperate forests of its size (Fig. 3b). The forest, which is fairly flat (135–202 m a.s.l.) and sparsely populated, includes several distinct types: deciduous, mixed, coniferous and boggy associations. The most characteristic forest association is oak-lime-hornbeam (*Quercus robur*, *Tilia cordata*, *Carpinus betulus*). The climate is mostly continental, with cold (1 November–31 March) and warm seasons (1 April–31 October, Jędrzejewska and Jędrzejewski 1998). The forest is inhabited by a rich animal community. Ungulates are represented by five species: European bison *Bison bonasus*, moose *Alces alces*, red deer *Cervus elaphus*, roe deer *Capreolus capreolus*, and wild boar *Sus scrofa*, as well as forty species of raptors and carnivores, including wolves and Eurasian lynx *Lynx lynx*.

Scavenging is a crucial ecological process in Białowieża, as shown by studies conducted since the 1990s in the Polish part of the forest (Jędrzejewska and Jędrzejewski 1998; Selva 2004a). Ungulates constitute the bulk of the carrion supply; about 7–9% of the standing crop of ungulate biomass is removed by scavengers each year. Ungulate carrion is supplied as wolf and lynx kills, leftovers from hunting, and natural deaths caused by disease or malnutrition. This high "diversity" of ungulate carcasses includes different spatio-temporal availability and features (e.g. size, visibility, openness) of carrion resources. Whereas predator-kills are supplied more or less constantly at unpredictable locations, hunter-kills are more localized and natural deaths mostly appear as a pulse at the end of the winter. Carcasses also vary in how accessible and easy to find and consume they are. For instance, European bison dying from natural conditions can be available for several months in winter as they are often concealed, weigh a few hundred kg, and are hard to open and find (Selva et al. 2005). Small roe deer, perfectly camouflaged and guarded by lynx, also are not easily accessible to many scavengers. However, red deer killed by wolves are often partly consumed, covered in blood, and easy to find, take away, and consume, typically disappear quickly (Selva 2004a, b).

This "diversity" of carrion resources has facilitated different mechanisms of coexistence among the numerous species of scavengers and promoted a structured (nested) use of carrion resources (Selva 2004a, Selva and Fortuna 2007). Thirty-six species of birds and mammals were reported using carrion in Białowieża (Selva 2004a). Predator-kills were preferred by most scavengers, but predation risk, namely intraguild predation, clearly influenced the carcass choice by some species. Ravens, foxes, and pine martens selected wolf kills when scavenging, but less frequently used the guarded prey of lynx; lynx have often been reported to kill foxes and martens (Jędrzejewska and Jędrzejewski 1998; Palomares and Caro 1999). Raptors generally preferred to scavenge in open areas, and were frequent at hunters' kills. Ravens and foxes, present in 89 and 86% of the wolf-kills respectively, were documented to follow wolves on the ground and in the air to feed on their prey remains (Jędrzejewska and Jędrzejewski 1998). Dead ungulates were the least attractive to scavengers and show the lowest species richness. This may be related not only to difficulties in finding and opening carcasses, but also to the risk of disease and parasite transmission and the unpredictability of this carrion resource (Selva 2004a).

Apart from scavengers' preferences in relation to various types of carrion resources, other mechanisms of segregation structure the scavenging guild in Białowieża Forest (Selva 2004a). In particular, a clear temporal segregation among avian and mammalian scavengers, mostly active during the day and night respectively, exists. Scavengers also follow a sequence of arrival at carcasses, with ravens and foxes arriving first in most cases. Habitat heterogeneity also appears to promote scavenger coexistence (Selva 2004a). Pine martens and jays avoided carcasses in open areas, whereas raptors preferred them. All scavengers avoided carcasses located near villages. Weather conditions, namely temperature and snow cover, played a key role in mediating carcass use (Selva et al. 2005). In general, more species scavenged at lower temperatures and in deeper snow, reflecting the higher energetic requirements and lower availability of other food resources, like rodents, in winter. For instance, at cold temperatures and deep snow, pine marten reduce their activity and increases food intake, typically by staying close to and feeding on ungulate carcasses (Zalewski et al. 1995; Zalewski 2000) Therefore, carcass use by scavenger species varied seasonally with summer carcasses mostly consumed by invertebrates and decomposers. Among vertebrates, only the raccoon dog *Nyctereutes procyonoides* scavenged more frequently in summer than winter (Selva 2004a).

Ungulate carrion served as a crucial food resource for the vertebrate community in winter (Jędrzejewska and Jędrzejewski 1998; Selva 2004a). The decline of food resources and increased persistence of carrion during winter resulted in increased use of ungulate carcasses at the end of the cold season by medium-sized and small predators. For example, the proportion of rodents in pine marten diet decreased through the winter, while the consumption of ungulate carcasses increased, reaching a maximum in February (Jędrzejewski et al. 1993). The scavenger assemblage also was more nested in the cold season, with a more structured pattern of use observed at carcasses that were constantly supplied, either by large predators or by hunters (Selva and Fortuna 2007). Competition, particularly intraspecific, at ungulate carcasses was strong. Simultaneous feeding by different species at carcasses was uncommon and the monopoly of the carcass by a single species was prevalent (Selva 2004a). In general, dominant species arrived later than scavengers with a lower hierarchical rank. Facilitative interactions were also documented, particularly at intact ungulate carcasses. Wolves extensively used bison carcasses and were the only species able to open them and provide further access to other species Scavenging by wolves at bison carcasses always triggered scavenging by mesopredators (Selva et al. 2003).

Ungulate carcasses also had relevant effects on other components of the ecosystem, such as soil, vegetation, and alternative prey species (Melis et al. 2007; Cortés-Avizanda et al. 2009). Most carcass-derived nutrients are recycled via scavenging and therefore transferred by vertebrates to distant areas. Higher nutrient concentrations were found in the areas surrounding bison carcasses (up to 4 m), an effect which lasted up to 7 years after the death of the animal (Melis et al. 2007). However, the impact was not as dramatic as in other more homogeneous systems, such as tundra or prairie ecosystems, possibly due to the high scavenging activity and forest complexity of Białowieża Primeval Forest. Large carcasses also affected apparently distant trophic levels, such as herbivores, via facultative scavengers (Cortés-Avizanda et al. 2009) as these scavengers, which are also predators, aggregated at carcasses. For example, the probability of predator-prey encounters was significantly higher within the vicinity of carcasses (1-km radius). This was accompanied by a reduced presence of hares *Lepus europaeus* and other alternative prey, such as red squirrels *Sciurus vulgaris*, close to carcasses, probably as a consequence of direct killing and/or predator avoidance (Cortés-Avizanda et al. 2009).

Yellowstone National Park: Effect of Large Predators on Scavenger Communities Through Provisioning of Carrion

Yellowstone is well-known as the first National Park in North America. The landscape is dominated by large open valleys and shrub steppes, and coniferous forests at north slopes and higher elevations, which range from 1500 to 3400 m (Houston 1982). In winter, ungulates concentrate in large aggregations along the Yellowstone River drainage. Eight species of ungulates (elk *Cervus canadensis*, mule deer *Odocoileus hemionus*, white-tailed deer, moose, American bison *Bison bison*, bighorn sheep *Ovis canadensis*, pronghorn antelope *Antilocapra americana*, and the mountain goat *Oreamnos americanus*) coexist with five species of large carnivores (coyote, wolf, cougar *Puma concolor*, grizzly bear, and black bear *U. americanus*). Long cold snowy winters and short cool summers characterize the climate. Mean monthly temperatures range from −12 to +13 °C (Cook 1993). The Yellowstone ecosystem is one of the best studied ecosystems in the world and the reintroduction of wolves in 1995 has been a milestone in wildlife conservation and a natural experiment to investigate the role of large predators as carrion suppliers, and the induced trophic cascades and changes in scavenger communities.

An important change for scavengers brought by wolf restoration has been the alteration of the quantity and timing of carrion availability (Fig. 5; Wilmers et al. 2003a). Before wolf reintroduction, the mortality of elk, the most abundant ungulate, depended on the snow depth and, thus, carrion was plentiful during severe winters, but scarce during mild winters (Houston 1978). Now, wolves are the main cause of elk mortality in Yellowstone, and have changed the timing of ungulate carrion from a sudden pulse at the end of winter to a more constant resource throughout the year. Although they may lower the total amount of carrion available, wolves have reduced the temporal variation in the quantity of carrion and extended the period over which carrion is available (Wilmers et al. 2003a, Wilmers and Getz 2004). Moreover, wolf presence may buffer changes in carrion availability due to climate warming (Wilmers and Getz 2005). By mitigating the reduction in late-winter carrion due to shorter and milder winters, wolves support the adaptation of scavengers to climate change (Wilmers and Getz 2005).

Wolf-provided carrion may promote biodiversity and larger populations of scavenger species (Wilmers et al. 2003a). When compared to hunter kills, highly aggregated in time and space, wolf kills showed higher species diversity (Wilmers et al. 2003b). Bald eagles, golden eagles *Aquila chrysaetos*, ravens, magpies *Pica pica* and coyotes were common at wolf kills in Yellowstone whereas coyotes, and mam-

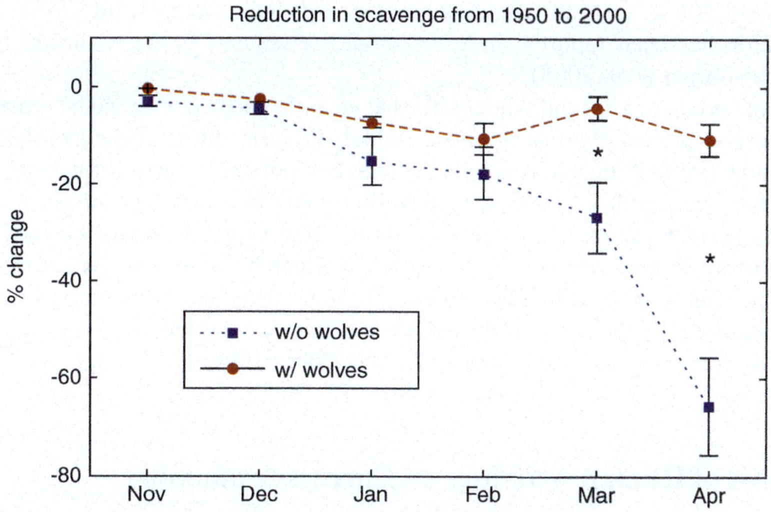

Fig. 5 Statistical model illustrating the extent to which the presence of wolves in Yellowstone National Park (https://goo.gl/maps/TuwcwD9Nd1P2) stabilize carrion availability and thus mitigate effects of climate change. "Reduction in winter carrion available to scavengers due to climate change 1950–2000: statistical model" is reproduced in its entirety from Wilmers and Getz (2005), and is licensed under CC BY 4.0 (https://creativecommons.org/licenses/by/4.0/legalcode)

mals in general, were absent at remains of hunter-kills in the surrounding landscape. Competitively dominant species, like coyotes, dominated scavenging at the more dispersed wolf kills whereas wide-ranging consumers, such as bald eagles and ravens, dominated consumption at highly aggregated hunter kills. However, the total number of individual scavengers at hunter kills exceeded those at wolf kills, likely due to social and information transfer mechanisms that eagles and ravens have developed to better track aggregated resources (Wilmers et al. 2003b). Indeed, ravens have quickly associated with wolves in order to scavenge first on the remains of their kills (Stahler et al. 2002).

The reintroduction of wolves clearly has affected the Yellowstone food web, and therefore, also the scavenger community (Peterson et al. 2014; Dobson 2014). For example, ungulate carrion is a key food for grizzly bears when they emerge from hibernation (Green et al. 1997) and opportunities to scavenge carcasses by grizzly bears have increased since the reestablishment of wolves (Wilmers et al. 2003a, b). Wolf-bear interactions, mostly at wolf kill sites, also are now more frequent, with a clear dominance by bears (Smith et al. 2003). Wolf-killed carrion also represents increased opportunities for intraguild predation. Although wolves clearly provide food for coyotes, ungulate carrion sites are risky and 7% of wolf-coyote interactions ended in coyote death (Merkle et al. 2009). Wolf aggression towards coyotes resulted in a strong decline in coyote densities and pack size in the years following wolf reintroduction (Smith et al. 2003). However, coyotes adapted to the new competitor at carcasses by modifying their behavior and maintaining spatial and temporal separation (Arjo and Pletscher 1999; Switalski 2003; Atwood and Gese 2008). In some ungulate species, a decrease in mortality from coyote predation has been reported in Yellowstone subsequent to the reintroduction of wolves (Berger et al. 2008).

Large predator kills and winter kills that are only partly consumed by scavengers are later exploited by insect larvae, which in turn, attract insectivorous birds (Houston 1978). Arthropods are also a main component of carrion food webs. The presence of ungulate carcasses in Yellowstone increases the local richness, abundance and occurrence of beetles (Sikes 1994). A total of 445 species of beetles were recorded at carcasses in Yellowstone, of which 57 were strongly associated to carrion. The family Silphidae ranked first in abundance, due to the species *Thanatophilus lapponicus*, followed by the family Staphylinidae, the richest in number of species, and the family Carabidae (Sikes 1994).

Impact of Human Activities on Carcass Availability

In addition to carrion provisioned through "natural" animal deaths (e.g., predator kills, starvation, exposure, diseases), modern humans have affected scavenging dynamics profoundly in both terrestrial and aquatic ecosystems by increasing the number, distribution, and temporal availability of carcasses to scavengers (see chapter "Human-Mediated Carrion: Effects on Ecological Processes"). In terrestrial ecosystems, human activities (e.g., collisions with vehicles, power lines and other structures, hunting) are responsible for billions of animal deaths worldwide, and most of these animals are left available for scavengers (Mateo-Tomás et al. 2015). In a review of cause-specific mortality of North American medium- and large-sized mammals, Collins and Kays (2011) found hunting accounted for about 35% of all mortality. Because hunters often leave inedible or undesirable portions of carcasses in the field, a substantial amount of carrion biomass can result from hunting mammals and birds that support a diverse assemblage of scavengers (Mateo-Tomás et al. 2015). Similarly, collisions with vehicles are responsible for an extensive number of vertebrate deaths in developed nations (Forman et al. 2003), with approximately 9% of all medium-and large-sized mammals in North America and 89–340 million birds killed annually by vehicle collisions (Collins and Kays 2011; Loss et al. 2014a). Other major human-related causes of vertebrate mortality in terrestrial ecosystems are less obvious than hunting and road mortality, but still significant. Conover et al. (2013) estimated thousands of ungulates die annually in the U.S. from colliding with or being entangled by fences. Nearly seven million birds in the U.S. and Canada also are killed each year by colliding with communication towers (Longcore et al. 2012), and 135–175 million by colliding with power lines (Conover et al. 2013). Further, in the U.S., 1.3–4.0 billion birds are killed each year by domestic cats (many of which are not eaten; Loss et al. 2013a), 365–988 million by collisions with buildings (Loss et al. 2014b), and 140,000–328,000 by wind turbines (Loss et al. 2013b).

Humans also have increased the amount of available carrion biomass in oceans and other aquatic ecosystems (Britton and Morton 1994). For example, many tons of fisheries discards ("by-catch") are returned to the water by commercial fishing boats each year globally. The ratio of discarded to harvested fish or other sea life can be as high as 10:1 (Hill and Wassenberg 1990) and most of the animal biomass returned to the water as by-catch is killed or severely injured during fishing operations (Broadhurst et al. 2006). Carrion from fisheries discards can subsidize marine food webs and well-adapted scavengers may increase in number due to the input of large quantities of fisheries discards (Ramsay et al. 1997; Catchpole et al. 2006). However, such carrion subsidies redistribute large quantities of biomass from the ocean floor and water column to the surface, likely altering ecosystem processes linked to nutrient cycling in marine ecosystems. Recreational fishing also can introduce carrion into aquatic ecosystems as mortality of released fish can be considerable (Bartholomew and Bohnsack 2005).

The causes of human-related vertebrate mortality summarized here are not exhaustive; other human activities (e.g., unintentional release of environmental toxins, intentional poisoning) commonly result in vertebrate mortality, and many of these animal carcasses become available to scavengers and decomposers. However, although it is clear human-derived carrion can have profound impacts on animal communities (e.g., Oro et al. 2013) the assimilation of animal carcasses produced as a result of human activities into ecosystems remains poorly characterized. (see chapter "Human-Mediated Carrion: Effects on Ecological Processes" for more information on effects of human-mediated carrion).

Conclusions and Future Perspectives

Despite a surge in scavenging research over the last few decades, scavenging dynamics and the ecological role of carrion in structuring individual to ecosystem level processes remains an understudied area of ecological research. At the most basic level there remains a paucity of data on scavenging dynamics in many common ecosystem types across the globe, especially island, arctic, tropical, and freshwater aquatic habitats, although many of these landscapes are becoming active areas of research (e.g., Abernethy et al. 2016). There also is a need for more integrative and manipulative research within well studied systems to better elucidate the effects of global change on scavenging dynamics, and the subsequent consequences to nutrient cycling and food web dynamics.

Disruptions to ecological communities due to climate change, overfishing, habitat fragmentation, etc. can alter competitive interactions among microbes, invertebrates, and vertebrates. The consequences of these shifts on scavenging dynamics are largely unknown, but likely profound. Similarly, there is a growing awareness of the potential role of carrion subsidies in the transport of nutrients among ecosystems (e.g., transport of biomass among aquatic ecosystems through disposal of fisheries discards, movement of nutrients sequestered within salmon between freshwater, pelagic, and terrestrial ecosystems), yet the impact of these subsidies on both the ecosystems from which they were removed and those in which they are input have not been adequately explored. There also has been extensive discussion of the role scavengers play in the control and dissemination of disease, although this remains an area of much needed research.

It is well established that temperature can have a confounding effect on investigations of carcass persistence and scavenging dynamics (DeVault et al. 2003; Farwig et al. 2014) and as such many researchers now routinely integrate temperature into the experimental design of many carrion fate studies. However, seemingly lost in many investigations of vertebrate scavenging ecology is the role of other potential factors on the composition of vertebrate scavenging communities. Recent research suggests carcass detection rates, persistence, and composition of scavenging communities can be highly influenced by the species and size of carrion (Olson et al. 2016; DeVault et al. 2017), as well as the local habitat type and accessibility of the carcass (Moleón et al. 2015; Turner et al. 2017). Thus, failure to integrate potential confounding factors, in addition to temperature, into experimental designs may bias inference of carrion and fate and scavenging dynamics.

Acknowledgements Contributions of James Beasley were partially supported by the U.S. Department of Energy under Award Number DE-EM0004391 to the University of Georgia Research Foundation.

References

Abernethy E, Turner K, Beasley JC et al (2016) Carcasses of invasive species are primarily utilized by invasive scavengers in an island ecosystem. Ecosphere 7:e01496

Allen ML, Elbroch LM, Wilmers CC, Wittmer HU (2014) Trophic facilitation or limitation? Comparative effects of pumas and black bears on the scavenger community. PLoS One 9:e102257

Arjo WM, Pletscher DH (1999) Behavioral responses of coyotes to wolf recolonization in northwestern Montana. Can J Zool 77:1919–1927

Atwood TC, Gese EM (2008) Coyotes and recolonizing wolves: social rank mediates risk-conditional behaviour at ungulate carcasses. Anim Behav 75:753–762

Avery ML, Cummings JL (2004) Livestock depredations by black vultures and golden eagles. Sheep Goat Res J 19:58–63

Bartholomew A, Bohnsack JA (2005) A review of catch-and-release angling mortality with implications for no-take reserves. Rev Fish Biol Fish 15:129–154

Barton PS, Cunningham SA, Lindenmayer DB, Manning AD (2013) The role of carrion in maintaining biodiversity and ecological processes in terrestrial ecosystems. Oecologia 171:761–772

Beard KH, Pitt WC, Price EA (2009) Biology and impacts of Pacific Island invasive species. 5. *Eleutherodactylus coqui*, the coqui frog (Anura: Leptodactylidae). Pac Sci 63:297–316

Beasley JC, Olson ZH, Dharmarajan G, Eagan TS II, Rhodes OE Jr (2011) Spatio-temporal variation the demographic attributes of a generalist mesopredator. Landsc Ecol 26:937–950

Beasley JC, Olson ZH, DeVault TL (2012) Carrion cycling in food webs: comparisons among terrestrial and marine ecosystems. Oikos 121:1021–1026

Beasley JC, Olson ZH, Beatty WS, Dharmarajan G, Rhodes OE Jr (2013) Effects of culling on mesopredator populations dynamics. PLoS One 8:e58982

Beasley JC, Olson ZH, DeVault TL (2015) Ecological role of vertebrate scavengers. In: Benbow ME, Tomberlin J, Tarone A (eds) Carrion ecology, evolution, and their application. CRC, Boca Raton, pp 107–128

Bellan SE, Turnbull PCB, Beyer W, Getz WM (2013) Effects of experimental exclusion of scavengers from carcasses of anthrax-infected herbivores on *Bacillus anthracis* sporulation, survival, and distribution. Appl Environ Microbiol 79:3756–3761
Benbow EM, Barton P, Ulyshen M, Beasley JC, DeVault TL, Strickland M, Tomberlin J, Jordan H, Pechal J (2019) Necrobiome framework for bridging decomposition ecology of autotrophically and heterotrophically derived organic matter. Ecol Monogr 89:e01331, pp. 1–29
Berger KM, Gese EM, Berger J (2008) Indirect effects and traditional trophic cascades: a test involving wolves, coyotes, and pronghorn. Ecology 89:818–828
Braack LEO (1987) Community dynamics of carrion-attendant arthropods in tropical African woodland. Oecologia 72:402–409
Britton JC, Morton B (1994) Marine carrion and scavengers. Oceanogr Mar Biol Annu Rev 32:369–434
Broadhurst MK, Suuronen P, Hulme A (2006) Estimating collateral mortality from towed fishing gear. Fish Fish 7:180–218
Buckley NJ (1999) Black vulture (*Coragyps atratus*). In: Poole A (ed) The birds of North America Online, Account 411. Cornell Lab of Ornithology, Ithaca, NY. http://bna.birds.cornell.edu.proxy-remote.galib.uga.edu/bna/species/411
Bump JK, Webster CR, Vucetich JA, Peterson RO, Shields JM, Powers MD (2009a) Ungulate carcasses perforate ecological filters and create biogeochemical hotspots in forest herbaceous layers allowing trees a competitive advantage. Ecosystems 12:996–1007
Bump JK, Peterson RO, Vucetich JA (2009b) Wolves modulate soil nutrient heterogeneity and foliar nitrogen by configuring the distribution of ungulate carcasses. Ecology 90:3159–3167
Butman CA, Carlton JT, Palumbi SR (1995) Whaling effects on deep-sea biodiversity. Conserv Biol 9:462–464
Byrd JH, Castner JL (2010) Forensic entomology: the utility of arthropods in legal investigations. CRC, Boca Raton
Carbone C, Turvey ST, Bielby J (2011) Intra-guild competition and its implications for one of the biggest terrestrial predators, *Tyrannosaurus rex*. Proc R Soc B 278:2682–2690
Carter DO, Yellowlees D, Tibbett M (2007) Cadaver decomposition in terrestrial ecosystems. Naturwissenschaften 94:12–24
Catchpole TL, Frid CLJ, Gray TS (2006) Importance of discards from the English *Nephrops norvegicus* fishery in the North Sea to marine scavengers. Mar Ecol Prog Ser 313:215–226
Cederholm CJ, Kunze MD, Murota T, Sibatani A (1999) Pacific salmon carcasses: essential contributions of nutrients and energy for aquatic and terrestrial ecosystems. Fisheries 24:6–15
Coe MJ (1978) The decomposition of elephant carcasses in the Tsavo (East) National Park, Kenya. J Arid Environ 1:71–86
Cohen JE (1978) Food webs and niche space. Princeton University Press, Princeton, NJ
Cohen JE, Briand F, Newman CM (1990) Community food webs: data and theory. Springer, New York
Collins C, Kays R (2011) Causes of mortality in North American populations of large and medium-sized mammals. Anim Conserv 14:474–483
Conover MR, Dinkins JB, Haney MJ (2013) Impacts of weather and accidents on wildlife. In: Wildlife management and conservation: contemporary principles and practices. Johns Hopkins University Press, Baltimore, MD, pp 144–155
Cook RS (1993) Ecological issues on reintroducing wolves into Yellowstone National Park. US Department of the Interior, National Park Service Scientific Monograph
Cornaby BW (1974) Carrion reduction by animals in contrasting tropical habitats. Biotropica 6:51–63
Cortés-Avizanda A, Selva N, Carrete M, Donázar JA (2009) Effects of carrion resources on herbivore spatial distribution are mediated by facultative scavengers. Basic Appl Ecol 10: 265–272
Cortés-Avizanda A, Jovani R, Carrete M, Donázar JA (2012) Resource unpredictability promotes species diversity and coexistence in an avian scavenger guild: a field experiment. Ecology 93:2570–2579

De Angelis DL (1975) Stability and connectance in food web models. Ecology 56:238–243

DeVault TL, Rhodes OE Jr, Shivik JA (2003) Scavenging by vertebrates: behavioral, ecological, and evolutionary perspectives on an important energy transfer pathway in terrestrial ecosystems. Oikos 102:225–234

DeVault TL, Reinhart BD, Brisbin IL Jr, Rhodes OE Jr (2004) Home ranges of sympatric black and turkey vultures in South Carolina. Condor 106:706–711

DeVault TL, Olson ZH, Beasley JC, Rhodes OE Jr (2011) Mesopredators dominate competition for carrion in an agricultural landscape. Basic Appl Ecol 12:268–274

DeVault TL, Beasley JC, Olson ZH, Moleón M, Carrete M, Margalida A, Sánchez-Zapata JA (2016) Ecosystem services provided by avian scavengers. In: Şekercioğlu CH, Wenny DG, Whelan CJ (eds) Why do birds matter? Avian ecological function and ecosystem services. University of Chicago Press, pp 235–270

DeVault TL, Seamans TW, Linnell KE, Sparks DW, Beasley JC (2017) Scavenger removal of bird carcasses at simulated wind turbines: Does carcass type matter? Ecosphere 8(11):e01994

Dobson AP (2014) Yellowstone wolves and the forces that structure natural systems. PLoS Biol 12:e1002025

Dunne JA (2006) The network structure of food webs. In: Pascual M, Dunne JA (eds) Ecological networks: linking structure to dynamics in food webs. Oxford University Press, Oxford, pp 27–86

Dunne JA, Williams RJ, Martínez ND (2002) Food-web structure and network theory: the role of connectance and size. Proc Natl Acad Sci U S A 99:12917–12922

Elton C (1927) Animal ecology. Sidgwick and Jackson, London

Fahrig L (2003) Effects of habitat fragmentation on biodiversity. Annu Rev Ecol Evol Syst 34:487–515

Fallows C, Gallagher AJ, Hammerschlag N (2013) White sharks (*Carcharodon carcharias*) scavenging on whales and its potential role in further shaping the ecology of an apex predator. PLoS One 8:e60797

Farwig N, Brandl R, Siemann S, Wiener F, Müller J (2014) Decomposition rate of carrion is dependent on composition not abundance of the assemblages of insect scavengers. Oecologia 175:1291–1300

Forman RTT, Sperling D, Bissonette JA, Clevenger AP, Cutshall CD, Dale VH, Fahrig L, France RL, Goldman CR, Haanue K, Jones J, Swanson F, Turrentine T, Winter TC (2003) Road ecology: science and solutions. Island Press, Washington

Fuglei E, Øritsland NA, Prestrud P (2003) Local variation in arctic fox abundance on Svalbard, Norway. Polar Biol 26:93–98

Gasaway WC, Mossestad KT, Stander PE (1991) Food acquisition by spotted hyenas in Etosha National Park, Namibia: predation versus scavenging. African J Ecol 29:64–75

Gende SM, Edwards RT, Willson MF, Wipfli MS (2002) Pacific salmon in aquatic and terrestrial ecosystems. Bioscience 52:917–928

Gende SM, Quinn TP, Hilborn R, Hendry AP, Dickerson B (2004) Brown bears selectively kill salmon with higher energy content but only in habitats that facilitate choice. Oikos 104:518–528

Gese EM, Ruff RL, Crabtree RL (1996) Foraging ecology of coyotes (*Canis latrans*): the influence of extrinsic factors and a dominance hierarchy. Can J Zool 74:769–783

Gooday AJ, Turley CM, Allen JA (1990) Responses by benthic organisms to inputs of organic material to the ocean floor: a review. Philos Trans R Soc Lond Ser B Biol Sci 331:119–138

Green GI, Mattson DJ, Peek JM (1997) Spring feeding on ungulate carcasses by grizzly bears in Yellowstone National Park. J Wildl Manag 61:1040–1055

Green RE, Newton I, Shultz S, Cunningham AA, Gilbert M, Pain DJ, Prakash V (2004) Diclofenac poisoning as a cause of vulture population declines across the Indian subcontinent. J Appl Ecol 41:793–800

Hanski I, Kuusela S (1980) The structure of carrion fly communities: differences in breeding seasons. Ann Zool Fenn 17:185–190

Heinrich B, Pepper JW (1998) Influence of competitors on caching behavior in the common raven, Corvus corax. Anim Behav 56:1083–1090

Helfield JM, Naiman RJ (2006) Keystone interactions: salmon and bear in riparian forests of Alaska. Ecosystems 9:167–180

Hertel F (1994) Diversity in body size and feeding morphology within past and present vulture assemblages. Ecology 75:1074–1084

Hill BJ, Wassenberg TJ (1990) Fate of discards from prawn trawlers in Torres strait. Aust J Marine Freshwater Res 41:53–64

Hill JE, DeVault TL, Beasley JC, Rhodes OE Jr, Belant JL (2018) Effects of vulture exclusion on carrion consumption by facultative scavengers. Ecol Evol 8(5):2518–2526

Hocking MD, Reynolds JD (2011) Impacts of salmon on riparian plant diversity. Science 331:1609–1612

Hocking MD, Ring RA, Reimchen TE (2009) The ecology of terrestrial invertebrates on Pacific salmon carcasses. Ecol Res 24:1091–1100

Houston DB (1978) Elk as winter-spring food for carnivores in northern Yellowstone National Park. J Appl Ecol 15:653–661

Houston DC (1979) The adaptations of scavengers. In: Sinclair ARE, Griffiths NM (eds) Serengeti, dynamics of an ecosystem. University of Chicago Press, Chicago, pp 263–286

Houston DB (1982) The northern Yellowstone elk: ecology and management. Macmillan, New York

Houston D, Cooper J (1975) The digestive tract of the whiteback griffon vulture and its role in disease transmission among wild ungulates. J Wildl Dis 11:306–313

Hutchinson G (1957) Concluding remarks. Cold Spring Harb Symp Quant Biol 22:415–427

Janzen D (1977) Why fruits rot, seeds mold, and meat spoils. Am Nat 111:691–713

Jędrzejewska B, Jędrzejewski W (1998) Predation in vertebrate communities. The Białowieża Primeval Forest as a case study. Springer, Berlin

Jędrzejewski W, Zalewski A, Jędrzejewska B (1993) Foraging by pine marten *Martes martes* in relation to food resources in Białowieża National Park, Poland. Acta Theriol 38:405–426

Jennelle C, Samuel MD, Nolden CA, Berkley EA (2009) Deer carcass decomposition and potential scavenger exposure to chronic wasting disease. J Wildl Manag 73:655–662

Johnson SB, Warén A, Lee RW, Kano Y, Kaim A, Davis A, Strong EE, Vrjenhoek RC (2010) Rubyspira, new genus and two new species of bone-eating deep-sea snails with ancient habits. Biol Bull 219:166–177

Kaiser MJ, Moore PG (1999) Obligate marine scavengers: do they exist? J Nat Hist 33:475–481

Killengreen ST, Strømseng E, Yoccoz NG, Ims RA (2012) How ecological neighbourhoods influence the structure of the scavenger guild in low arctic tundra. Divers Distrib 18:563–574

Kirk DA, Mossman MJ (1998) Turkey vulture (*Cathartes aura*). Account 339 in Poole A (ed) The Birds of North America Online. Cornell Lab of Ornithology, Ithaca, NY. http://bna.birds.cornell.edu.proxy-remote.galib.uga.edu/bna/species/339

Kruuk H (1967) Competition for food between vultures in East Africa. Ardea 55:171–193

Lafferty KD, Dobson AP, Kuris AM (2006) Parasites dominate food web links. Proc Natl Acad Sci U S A 103:11211–11216

Lam K, Babor D, Duthie B, Babor EM, Moore M, Gries G (2007) Proliferating bacterial symbionts on house fly eggs affect oviposition behavior of adult flies. Ani Behav 74:81–92

Lawton J (1989) Food webs. In: Cherrett J (ed) Ecological concepts. Blackwell Scientific, Oxford, pp 43–78

Levi T, Wheat RE, Allen JM, Wilmers CC (2015) Differential use of salmon by vertebrate consumers: implications for conservation. PeerJ 3:e1157. https://doi.org/10.7717/peerj.1157

Lindeman RL (1942) The trophic-dynamic aspect of ecology. Ecology 23:399–417

Linnell JD, Ronny A, Reidar A (1995) Who killed bambi? The role of predation in the neonatal mortality of temperate ungulates. Wildl Biol 1:209–223

Longcore T, Rich C, Mineau P, MacDonald B, Bert DG, Sullivan LM, Mutrie M, Gauthreaux SA Jr, Avery ML, Crawford RL, Manville AM II, Travis ER, Drake D (2012) An estimate of avian mortality at communication towers in the United States and Canada. PLoS One 7:e34025

Loss SR, Will T, Marra PP (2013a) The impact of free-ranging domestic cats on wildlife of the United States. Nat Commun 4:1396

Loss SR, Will T, Marra PP (2013b) Estimates of bird collision mortality at wind facilities in the contiguous United States. Biol Conserv 168:201–209
Loss SR, Will T, Marra PP (2014a) Estimation of bird-vehicle collision mortality on U.S. roads. J Wildl Manag 78:763–771
Loss SR, Will T, Loss SS, Marra PP (2014b) Bird-building collisions in the United States: estimates of annual mortality and species vulnerability. Condor 116:8–23
Lowney MS (1999) Damage by black and turkey vultures in Virginia, 1990-1996. Wildl Soc Bull 27:715–719
Magoun AJ (1976) Summer scavenging activity in northeastern Alaska. MS thesis, University of Alaska, Fairbanks
Margalida A, Bertran J, Heredia R (2009) Diet and food preferences of the endangered Bearded Vulture *Gypaetus barbatus*: a basis for their conservation. Ibis 151:235–243
Margalida A, Campión D, Donázar JA (2014) Vultures *vs.* livestock: conservation relationships in an emerging conflict between humans and wildlife. Oryx 48:172–176
Markandya A, Taylor T, Longo A, Murty MN, Murty S, Dhavala K (2008) Counting the cost of vulture decline—an appraisal of the human health and other benefits of vultures in India. Ecol Econ 67:194–204
Martínez ND (1991) Artifacts or attributes? Effects of resolution on the Little Rock Lake food web. Ecol Monogr 61:367–392
Mateo-Tomás P, Olea PP, Moleón M, Vicente J, Botella F, Selva N, Viñuela J, Sánchez-Zapata JA (2015) From regional to global patterns in vertebrate scavenger communities subsidized by big game hunting. Divers Distrib 21:913–924
May RM (1972) Will a large complex system be stable? Nature 238:413–414
McCann K, Hastings A, Huxel GR (1998) Weak trophic interactions and the balance of nature. Nature 395:794–798
McKinnerney M (1978) Carrion communities in the northern Chihuahuan Desert. Southwest Nat 23:563–576
Melis C, Teurlings I, Linnell JDC, Andersen R, Bordoni A (2004) Influence of a deer carcass on Coleopteran diversity in a Scandinavian boreal forest: a preliminary study. Eur J Wildl Res 50:146–149
Melis C, Selva N, Teurlings I, Skarpe C, Linnell JDC, Andersen R (2007) Soil and vegetation nutrient response to bison carcasses in Białowieża Primeval Forest, Poland. Ecol Res 22:807–813
Merkle JA, Stahler DA, Smith DW (2009) Interference competition between gray wolves and coyotes in Yellowstone National Park. Can J Zool 87:56–63
Moleón M, Sánchez-Zapata JA, Sebastián-González E, Owen-Smith N (2015) Carcass size shapes the structure and functioning of an African scavenging assemblage. Oikos 124:1391–1403
Moore JE, Swihart RK (2005) Modeling patch occupancy by forest rodents: incorporating detectability and spatial autocorrelation with hierarchically structured data. J Wildl Manag 69:933–949
Müller JK, Eggert AK, Dressel J (1990) Intraspecific brood parasitism in the burying beetle, *Nicrophorus vespilloides* (Coleoptera: Silphidae). Anim Behav 40:491–499
Neutel AM, Heesterbeek JA, de Ruiter PC (2002) Stability in real food webs: weak links in long loops. Science 296:1120–1123
Oaks JL, Gilbert M, Virani MZ, Watson RT, Meteyer CU, Rideout BU, Shivaprasad HL, Ahmed S, Chaudhry MJI, Arshad M, Mahmood S, Ali A, Khan AA (2004) Diclofenac residues as the cause of vulture population decline in Pakistan. Nature 427:630–633
Ogada DL, Keesing F, Virani MZ (2012a) Dropping dead: causes and consequences of vulture population declines worldwide. Ann N Y Acad Sci 1249:57–71
Ogada DL, Torchin ME, Kinnaird MF, Ezenwa VO (2012b) Effects of vulture declines on facultative scavengers and potential implications for mammalian disease transmission. Conserv Biol 26:453–460
Olson ZH, Beasley JC, DeVault TL, Rhodes OE Jr (2012) Scavenger community response to the removal of a dominant scavenger. Oikos 121:77–84

Olson ZH, Beasley JC, Rhodes OE Jr (2016) Carcass type affects local scavenger guilds more than habitat connectivity. PLoS One 11:e0147798

Oro D, Genovart M, Tavecchia G, Fowler MS, Martínez-Abrain A (2013) Ecological and evolutionary implications of food subsidies from humans. Ecol Lett 16:1501–1514

Pain DJ, Cunningham AA, Donald PF, Duckworth JW, Houston DC, Katzner T, Parry-Jones J, Poole C, Prakash V, Round P, Timmins R (2003) Causes and effects of temporospatial declines of *Gyps* vultures in Asia. Conserv Biol 17:661–671

Palomares F, Caro TM (1999) Interspecific killing among mammalian carnivores. Am Nat 153:492–508

Parmenter R, MacMahon J (2009) Carrion decomposition and nutrient cycling in a semiarid shrub-steppe ecosystem. Ecol Monogr 79:637–661

Payne JA (1965) A summer carrion study of the baby pig Sus scrofa Linnaeus. Ecology 46: 592–602

Pechal JL, Benbow EM, Crippen TL, Tarone AM, Tomberlin JK (2014) Delayed insect access alters carrion decomposition and necrophagous insect community assembly. Ecosphere 5(4):45

Peterson RO, Vucetich JA, Bump JM, Smith DW (2014) Trophic cascades in a multicausal world: Isle Royale and Yellowstone. Annu Rev Ecol Evol Syst 45:325–345

Pimm SL (1982) Food webs. Chapman and Hall, New York

Polis GA (1991) Complex trophic interactions in deserts: an empirical critique of food-web theory. Am Nat 138:123–155

Prakash V, Pain DJ, Cunningham AA, Donald PF, Prakash N, Verma A, Gargi R, Sivakumar S, Rahmani AR (2003) Catastrophic collapse of Indian white-backed *Gyps bengalensis* and long-billed *Gyps indicus* vulture population. Biol Conserv 109:381–390

Prugh LR, Stoner CJ, Epps DW, Bean WR, Ripple WJ, Laliberte AS, Brashares JS (2009) The rise of the mesopredator. Bioscience 59:779–791

Ramsay K, Kaiser MJ, Moore PG, Hughes RN (1997) Consumption of fisheries discards by benthic scavengers: utilization of energy subsidies in different marine habitats. J Anim Ecol 66:884–896

Restani M, Harmata AR, Madden EM (2000) Numerical and functional responses of migrant bald eagles exploiting a seasonally concentrated food source. Condor 102:561–568

Roggenbuck M, Schnell IB, Blom N, Baelum J, Bertelsen MF, Pontén TS, Sørensen SJ, Gilbert MTP, Graves GR, Hansen LH (2014) The microbiome of New World vultures. Nat Commun 5:5498

Rooney TP, Waller DM (2003) Direct and indirect effects of white-tailed deer in forest ecosystems. For Ecol Manag 181:165–176

Rose MD, Polis GA (1998) The distribution and abundance of coyotes: the effects of allochthonous food subsidies from the sea. Ecology 79:998–1007

Roth JD (2003) Variability in marine resources affects arctic fox population dynamics. J Anim Ecol 72:668–676

Rouse GW, Goffredi SK, Vrijenhoek RC (2004) *Osedax*: Bone-eating marine worms with dwarf males. Science 305:668–671

Ruxton GD, Bailey DM (2005) Searching speeds and the energetic feasibility of an obligate whale-scavenging fish. Deep Sea Res Part 1 Oceanogr Res Pap 52:1536–1541

Ruxton GD, Houston DC (2003) Could Tyrannosaurus rex have been a scavenger rather than a predator? An energetics approach. Proc R Soc Lond B Biol Sci 270:731–733

Ruxton GD, Houston DC (2004a) Obligate vertebrate scavengers must be large soaring fliers. J Theor Biol 228:431–436

Ruxton GD, Houston DC (2004b) Energetic feasibility of an obligate marine scavenger. Mar Ecol Prog Ser 266:59–63

Schlichting PE, Love CN, Webster SC, Beasley JC (2019) Efficiency and composition of vertebrate scavengers at the land-water interface in the Chernobyl Exclusion Zone. Food Webs 18:e00107

Schmitz OJ, Hambäck PA, Beckerman AP (2000) Trophic cascades in terrestrial systems: a review of the effects of top carnivore removals on plants. Am Nat 155:141–153

Sebastián-González M, Sánchez-Zapata JA, Donazar JA, Selva N, Cortéz-Avizanda A, Hiraldo F, Blázquez M, Botella F, Moleón M (2013) Interactive effects of obligate scavengers and scavenger community richness on lagomorph carcass consumption patterns. Ibis 155:881–885
Sebastián-González E, Moleón M, Gibert JP, Botella F, Mateo-Tomás P, Olea PP, Guimarães PR Jr, Sánchez-Zapata JA (2016) Nested species-rich networks of scavenging vertebrates support high levels of interspecific competition. Ecology 97:95–105
Selva N (2004a) The role of scavenging in the predator community of Białowieża Primeval Forest (E Poland). PhD thesis, University of Sevilla, Spain
Selva N (2004b) Life after death – scavenging on ungulate carcasses. In: Jędrzejewska B, Wójcik JM (eds) Essays on mammals of Białowieża Forest. Mammal Research Institute, PAS, Białowieża, pp 59–68
Selva N, Fortuna MA (2007) The nested structure of a scavenger community. Proc R Soc Lond B Biol Sci 274:1101–1108
Selva N, Jędrzejewska B, Jędrzejewski W, Wajrak A (2003) Scavenging on European bison carcasses in Bialowieza Primeval Forest (eastern Poland). Ecoscience 10:303–311
Selva N, Jędrzejewska B, Jędrzejewski W, Wajrak A (2005) Factors affecting carcass use by a guild of scavengers in European temperate woodland. Can J Zool 83:1590–1601
Shivik JA (2006) Are vultures birds, and do snakes have venom, because of macro- and microscavenger conflict? Bioscience 56:819–823
Shivik JA, Clark L (1999) Ontogenetic shifts in carrion attractiveness to brown tree snakes (*Boiga irregularis*). J Herpetol 33:334–336
Sikes DS (1994) Influences of ungulate carcasses on coleopteran communities in Yellowstone National Park, USA. MS thesis, Montana State University, Montana
Smith CR, Baco AR (2003) Ecology of whale falls at the deep-sea floor. Oceanogr Mar Biol Annu Rev 41:311–354
Smith DW, Peterson RO, Houston DB (2003) Yellowstone after Wolves. Bioscience 53:330–340
Smith CR, De Leo FC, Bernardino AF, Sweetman AK, Arbizu PM (2008) Abyssal food limitation, ecosystem structure and climate change. Trends Ecol Evol 23:518–528
Stahler D, Heinrich B, Smith D (2002) Common ravens, *Corvus corax*, preferentially associate with grey wolves, Canis lupus, as a foraging strategy in winter. Anim Behav 64:283–290
Swift MJ, Heal OW, Anderson JM (1979) Decomposition in terrestrial ecosystems. Blackwell, Oxford
Switalski TA (2003) Coyote foraging ecology and vigilance in response to gray wolf reintroduction in Yellowstone National Park. Can J Zool 81:985–993
Tamburri MN, Barry JP (1999) Adaptations for scavenging by three diverse bathyla species, *Eptatretus stouti*, *Neptunea amianta* and *Orchomene obtusus*. Deep Sea Res 46(Pt 1):2079–2093
Tilman D (1996) Biodiversity: population versus ecosystem stability. Ecology 77:350–363
Tomberlin JK, Sheppard DC, Joyce JA (2005) Black soldier fly (Diptera: Stratiomyidae) colonization of pig carrion in South Georgia. J Forensic Sci 50:JFS2003391-2
Towne EG (2000) Prairie vegetation and soil nutrient responses to ungulate carcasses. Oecologia 122:232–239
Turner KL, Abernethy EF, Conner LM, Rhodes OE Jr, Beasley JC (2017) Abiotic and biotic factors modulate carrion fate and scavenging community dynamics. Ecology 98:2413–2424
VanLaerhoven S (2010) Ecological theory and its application in forensic entomology. In: Byrd J, Castner J (eds) Forensic entomology: the utility of arthropods in legal investigations. CRC Press, Boca Raton, pp 493–518
VerCauteren KC, Pilon JL, Nash PB, Phillips GE, Fischer JW (2012) Prion remains infectious after passage through digestive system of American crows (*Corvus brachyrhynchos*). PLoS One 7:e45774
Wardle DA, Bardgett RD, Klironomos JN, Setala H, van der Putten WH, Wall DH (2004) Ecological linkages between aboveground and belowground biota. Science 304:1629–1633
Wasserman S, Faust K (1994) Social network analysis: Methods and applications, vol 8. Cambridge University Press, Cambridge

Whitaker JO Jr, Mumford RE (2010) Mammals of Indiana. Revised an. Indiana University Press, Bloomington

Wikenros C, Sand H, Ahlqvist P, Liberg O (2013) Biomass flow and scavengers use of carcasses after re-colonization of an apex predator. PLoS One 8(10):e77373

Wilmers CC, Getz WM (2004) Simulating the effects of wolf-elk population dynamics on resource flow to scavengers. Ecol Model 177:193–208

Wilmers CC, Getz WM (2005) Gray wolves as climate change buffers in Yellowstone. PLoS Biol 3:e92

Wilmers CC, Crabtree RL, Smith DW, Murphy KM, Getz WM (2003a) Trophic facilitation by introduced top predators: grey wolf subsidies to scavengers in Yellowstone National Park. J Anim Ecol 72:909–916

Wilmers CC, Stahler DR, Crabtree RL, Smith DW, Getz WM (2003b) Resource dispersion and consumer dominance: scavenging at wolf- and hunter-killed carcasses in Greater Yellowstone, USA. Ecol Lett 6:996–1003

Wilson EE, Wolkovich EM (2011) Scavenging: how carnivores and carrion structure communities. Trends Ecol Evol 26:129–135

Yang LH (2004) Periodical cicadas as resource pulses in North American forests. Science 306:1565–1567

Zalewski A (2000) Factors affecting the duration of activity by pine martens (Martes martes) in the Białowieża National Park, Poland. J Zool 251:439–447

Zalewski A, Jędrzejewski W, Jędrzejewska B (1995) Pine marten home ranges, numbers and predation on vertebrates in a deciduous forest (Białowieża National Park, Poland). Ann Zool Fenn 32:131–144

Zheng L, Crippen TL, Holmes L, Singh B, Pimsler ML, Benbow ME, Tarone AM, Dowd S, Yu Z, Vanlaerhoven SL, Wood TK, Tomberlin JK (2013) Bacteria mediate oviposition by the black soldier fly, *Hermetia illucens* (L.), (Diptera: Stratiomyidae). Sci Reports 3:2563

Part II
Human and Carrion: The Impact of Humans on Carrion Ecology and Management

The Role of Scavenging in Disease Dynamics

Joaquín Vicente and Kurt VerCauteren

Contents

Introduction

Scavenging is a widespread strategy among vertebrates with an important contribution to food webs and ecosystem function (see chapter "Carrion Decomposition"). Among the important ecosystem functions and services supported by scavengers, the control of disease spreading from decomposing carcasses is frequently highlighted as a major argument for scavenger conservation. Particularly, obligate scavengers (vultures) seem to reduce the risk of disease in the community, including

J. Vicente (✉)
SaBio - Instituto de Investigación en Recursos Cinegéticos, IREC (UCLM-CSIC), Ciudad Real, Spain
e-mail: Joaquin.Vicente@uclm.es

K. VerCauteren
National Wildlife Research Center, USDA/APHIS/Wildlife Services, Fort Collins, CO, USA
e-mail: kurt.c.vercauteren@aphis.usda.gov

P. P. Olea et al. (eds.), *Carrion Ecology and Management*, Wildlife Research Monographs 2, https://doi.org/10.1007/978-3-030-16501-7_7

humans (e. g. Markandya et al. 2008). Scavenging can potentially reduce the spread of disease among wildlife because many pathogenic organisms in carcasses cannot survive passage through the highly acidic digestive system of vultures (Houston and Cooper 1975). A reduction in the numbers of vultures and/or the effects of anthropogenic factors can lead to increasing consumption by facultative scavengers, i.e. those apecies not relying on scavenging to survive but utilizing carrion when it is available (Markandya et al. 2008; Ogada et al. 2012; Mateo-Tomás et al. 2015; Inger et al. 2016). The role these changes in scavenging dynamics have on ecological and sanitary issues is a matter of debate. Moreover, important knowledge gaps exist relative to the role of different scavenging species in disease dynamics at carcasses, especially under the current scenario of increasing anthropization of natural ecosystems.

In this chapter, we address aspects of the transmission of diseases by or facilitated by scavenging, and under which conditions, reviewing and synthesizing past research. We define future research needs in this field aimed to address knowledge gaps on the issue. We must keep in mind that in many ecosystems, diseases provide important mechanisms to regulate animal populations, and that death from disease produces from occasional short-term to large long-term pulses of carrion into ecosystems (see chapter "Carrion Availability in Space and Time" and this chapter). Scavenged material can potentially favor transmission and spread of infectious diseases or contaminants to wild animals, the environment and the human-related compartment; or contrary, scavenging can limit pathogen spread. Finally, we address human-related factors that influence pathogen spread by scavenging and its management.

It is still necessary to improve the basic knowledge of the epidemiology of wildlife diseases in order to understand the role of scavenging in disease dynamics. Studying wildlife disease is challenging, particularly the study of aspects regarding pathogen transmission and the mechanisms of perpetuation in the host community and the environment. There are many concurrent factors of very different nature, especially when diseases are shared among different hosts, some of which can act as facilitators or vectors of pathogen spread. Also, the role varying environments play poses unique challenges. For instance, ferrets (*Mustela furo*) and wild boar (*Sus scrofa*) are considered to act as a host complex (via reciprocal scavenging) relative to the transmission of tuberculosis (TB) in New Zealand, and the combined effects of these species on pathogen transmission are proposed to be density dependent (Barron et al. 2015). We must also consider host-pathogen adaptations and life history strategies. Different mechanisms, modes of transmission and strategies for pathogens to persist are possible, and they can operate differently according to local circumstances. A given pathogen, host and scavenging community assemblage, under different conditions, may present variable contributions to pathogen spread by scavenging, from promotion to prevention. Most pathogens that are transmitted through predation or scavenging do not have to leave the body of the host for transmission to occur, but are passively ingested, such as mycobacteria species causing tuberculosis. Some other pathogens may leave the carcass and persist in the environment, or are spread by vectors, which may be facilitated by previous carcass opening by scavengers; which includes pathogens highly resistant in the environment, such as anthrax – an acute fatal multi-species disease caused by the bacterium *Bacillus anthracis*. In the last instance, understanding all the epidemiological and ecological processes is key to managing disease and scavenging in the wild for a given purpose; for instance, reducing pathogen spread in human intervened systems where the number of vultures has been reduced, and preventing endangered facultative scavengers from becoming infected via scavenging.

We identify some important aspects that are central in current and future research regarding the role of scavenging in disease dynamics. (1) The use of animal remains by scavengers and their exposure to disease. Little is still known about this issue, in spite of it being a first step required to understand the exposure of scavenger communities to disease, and to subsequently guide specific research on the potential role of scavenging in pathogen spread. This aspect is also key to identify different epidemiological contexts and patterns of consumption. (2) The relevance of scavenging for pathogens to spread and persist. Scavenging is likely limiting pathogen spread in many systems. Scavengers have also been implicated as potential vectors for a number of diseases, and facilitators of disease spread once they open the carcass, contaminating the vicinity and enabling the action of vectors. (3) Human-related factors resulting in increased risk of pathogen transmission through scavenging. Disease emergence and re-emergence in the wild, impacting natural populations, tends to occur more in ecosystems altered by human activities, affecting host-pathogen-environment relationships. This phenomena may lead to increased amounts of contaminated material, leaving extremely resistant agents in the environment and may also affect scavenger community-carrion dynamics and contact rates among susceptible scavenger hosts and infected material. (4) Management of scavenging to reduce disease risks. Preventing the entrance and/or controlling levels of disease in environments is the first approach. Once disease is present, re-directing scavenging activity and restoring scavenger communities might provide useful strategies to reduce disease risks.

The Use of Animal Remains and the Exposure of Scavengers to Disease

Exposure to disease mediated by scavenging results as a consequence of both a given pattern of carrion use by scavengers and the presence of a disease in the system. While consumption poses the highest risk, the use of carrion in relation to pathogen exposure does not necessarily involve scavenging, but other risky behaviors such as exploring the environment (direct exposure to pathogens in the environment or vectors) or consuming invertebrates. For instance, wild boar explore in the soil next to and under the carcasses of conspecifics, sniffing and poking the carcass (Probst et al. 2016). Regarding carrion use by scavengers, several studies have been conducted using a variety of carrion (e.g. from entire carcasses to gut piles) to monitor scavenger activity (e. g. DeVault et al. 2003; Wilmers et al. 2003a, b, Roen and

Yahner 2005; Selva et al. 2005; Mateo-Tomás et al. 2017; Carrasco-Garcia et al. 2018). These studies determine rates of activity and consumption by scavengers, and the decomposition of animal remains in specific situations. Every study addresses a particular situation, representing different scavenger assemblages, localizations and types of carrion material (carcasses, gut piles, abortions, etc.). A review by Mateo-Tomás et al. (2015) concerning ungulate carrion availability on a global scale (and references therein) found that although scavenger communities consuming remains of hunter-killed animals vary among regions, a common pattern emerges in which birds and mammals dominate consumption. They found scavenging birds to consume twice as much as mammals, but that more mammal than bird species took part in scavenging (Mateo-Tomás et al. 2015). Facultative scavengers (i.e. mainly generalists and apex predators) dominate consumption everywhere, especially where the presence of obligate scavengers (vultures) and apex predators (e.g. wolves *Canis lupus*, hyenas, and eagles) is low. Suids (i.e. wild boar) are also frequent scavengers where present. Often species of other taxonomic groups, such as reptiles and invertebrates, function as scavengers (see chapters "Invertebrate Scavenging Communities" and "Vertebrate Scavenging Communities"). For example, blowflies have been suspected to play an important role in anthrax transmission (Bellan et al. 2013). Determining the relative contribution of specialized as compared to facultative scavengers would provide relevant insights into their risk of exposure to disease and the relevance of scavenging in disease dynamics. Vultures consume carcasses quickly, and their absence is usually associated with significantly slower decomposition rates (Mateo-Tomás et al. 2017), leading to higher abundances of facultative scavengers that spend a longer time in close to carcasses (Ogada et al. 2012). These studies also differentiate between which species or individuals are true scavengers and which act merely as visitors, or exceptionally, as scavengers due to very particular rare behavior displayed by particular individuals. For instance, it has even been reported that some deer species may consume animal tissues, including flesh, digestive content and bone of dead animals (e.g., Cook et al. 2004; Vicente et al. 2011).

Researchers must also pay attention to the specific tissues that scavengers consume, relating this potentially infection risk. For instance, large carrion birds do not usually consume bovine central nervous tissue, but do consume their intestines and, occasionally, ovine central nervous system tissue (or at least the spinal medulla), which may determine their exposition to prions (European Commission 2002). Also, selective particularities for different scavengers can be determined. For instance, there is anecdotal evidence that feral pigs may avoid feeding on ferret carcasses, whereas they readily consume brushtail possum (*Trichosurus vulpecula*) carcasses in New Zealand (Byrom 2004). Determining the temporal patterns of scavengers consuming carcasses, as well as the degree of persistence of carcasses and/or remains is very informative because it may facilitate the understanding of disease dynamics for animals and vectors that visit carcasses and/or consume carrion. The persistence of carcasses and/or remains also determine the relative contribution of microbial and invertebrates in the removal of dead material from the environment. Studying the biotic and abiotic factors that affect rates of carcass consumption and decomposition in the environment is essential to explain variations in the relative contribution of scavengers under different scenarios (Jennelle et al. 2009; Vicente et al. 2011; Turner et al. 2017) and subsequent exposure to disease. For instance, the role of facultative scavengers, wild boar and red fox (*Vulpes vulpes*), greatly varies between habitats (open vs vegetation covered) and scavenging community (presence of vultures vs absence; Vicente et al. 2011).

The typical procedure to perform these studies consists in randomly placing carcasses and/or remains throughout the landscape, or performing the study *in situ* where remains are produced (naturally through, for example, predation, disease, accident, or hunting). The activity of scavengers in recent research is usually monitored by using remotely activated cameras, and then it is quantified (e.g. Mateo-Tomás et al. 2015, 2017). The use of infrared lights minimizes disturbance to scavengers, particularly mammals, which can be affected by direct observations. It is recommended to check *in situ* the status of the carrion to describe the amount of material that is left, the parts that have been removed, and any activity that is difficult to record with cameras (e.g. invertebrate activity). This has to be carried out such that normal activity of scavengers is not disrupted. The use of cameras can reveal the representative scavenger communities (and visitors) of carrion (carcasses or gut piles). Nonetheless, photographic evidence that some scavengers consume some specific carcass materials is difficult to assess since different species may overlap in a period of time, and pictures or videos do not always clarify this. At this stage, scavenging can only be related with increased exposure to infectious materials, but this does not necessarily involve transmission of pathogens.

As aforementioned, it is necessary to combine descriptive data on scavenging activity with the collection, analysis and interpretation of wildlife and/or livestock disease data. Most of the available information of disease in many animal communities is based on punctual "surveys" of disease, which are not specifically designed to analyze health risks. Therefore, this information is not sufficient. By contrast, disease "surveillance" and "monitoring" programs usually refer to ongoing processes and provide opportunities to analyze risk related to scavenging activity. These programs first provide information on the causes of wildlife loss other than predation, such as starvation, disease, trauma (e.g. collisions with vehicles, fences), injuries sustained while fighting, other wounds, and pesticide exposure, among others. Surveillance is specifically designed to analyze and manage any associated health risks (Artois et al. 2009), and data are used to identify the areas to be targeted for control, and to anticipate management interventions to reduce disease risks. The actions taken can be passive (or reactive, obtaining information on the disease agent in sick or dead animals) or active (actively searching for the disease in animals through sampling). Further, it is important to detect spatial and temporal trends as well as the presence/absence of the disease. Monitoring addresses the estimation of epidemiological parameters related to a defined disease, such as prevalence (infected relative to population), incidence (new infected relative to susceptible population) and the basic reproductive number R_0 (the number of secondary cases which a first infection case would produce in a population free of pathogens; it explains whether a disease will die out or whether it may become epidemic if $R_0 > 1$). Monitoring is always based on an active approach (searching for diseased animals, sampling and testing). Data and samples gathered can then be used for detecting emerging diseases, spatiotemporal variation in patterns of disease spread and persistence, and finally, for assessing what the relationships with variations in scavenging activity maybe (i.e. the factors determining risks of pathogen transmission by scavenging). A nice example is brucellosis (*Brucella abortus*) in the Greater Yellowstone Area in western Wyoming, USA, where it is endemic in elk (*Cervus elaphus nelsoni*) using winter feed grounds. This is so, presumably, because of increased animal density, duration of attendance, and subsequent contact with aborted fetuses. Maichak et al. (2009), rather than only measuring levels of disease in the population, or just describing scavenging on aborted fetuses, addressed both aspects complementary.

Sentinel surveillance can be used for the detection of pathogens, especially those that are newly because sentinel species or individuals are often more at risk of infection than other animals (e.g. reservoir of disease or obligate scavengers). A sentinel wildlife species acquires infection primarily or exclusively from a sympatric wildlife maintenance host, or from the environment where the infected host was located (Anderson et al. 2015). A sentinel species normally is not itself a major part of the pathogen but is predominantly a spillover host, it is easily infected and remains alive in an infected state for months or years, and thus available for detection. Sentinel species may also have much larger ranging areas than maintenance hosts. The use of wildlife sentinels may be a particularly valuable approach to surveillance for emerging zoonotic infections, many of which have their origins in wild hosts. Certain taxa may be relatively more efficient at concentrating some pathogens, for instance predators at the top of food chains or scavengers that may be exposed to infectious carcasses (Smith 1994; Leighton et al. 1995). For example, coyotes (*Canis latrans*), wild boar and ferrets carrying *Mycobacterium bovis* that causes tuberculosis in wildlife may be used as sentinels during disease surveillance (VerCauteren et al. 2008; Byrom et al. 2015).

The Relevance of Scavenging for Pathogens to Spread and Persist

Although for many pathogens horizontal transmission via scavenging (or facilitated by) of contaminated material is possible, the consequences for disease dynamics are unknown in most cases. Studying wildlife diseases is challenging, and it is especially difficult to disentangle the relative contribution of scavenging-related factors to the spread and maintenance of diseases in ecosystems. Experiments may help to assess the possibility of cross-species pathogen transmission by scavenging, and empirical observations in the field can support those assessments on the potential risk for transmission. Nonetheless, the relevance of scavenging for pathogens to persist must be specifically assessed for each system and epidemiological context.

Scavenging is probably limiting disease spread in most cases, even when scavengers become infected. The concept of disease reservoir is relevant to understand disease maintenance and management in relation to scavenging. Briefly, an infected animal population can be classed as either a maintenance or spillover host, depending on the dynamics of the infection. In a maintenance (true reservoir) host, infection can persist by intraspecies transmission alone, and may also be the source of infection for other species. In a spillover host, infection will not persist indefinitely unless there is re-infection from another species or the environment. For instance, surveys of carnivores and omnivores from Michigan have confirmed *M. bovis* (one of the causative agents of animal TB) infection in coyotes, bobcats (*Felis rufus*), red foxes, black bears (*Ursus americanus*), opossums (*Didelphis virginiana*), raccoon (*Procyon lotor*) and domestic cats (*Felis silvestris catus*; Bruning-Fann et al. 2001). Molecular techniques have evidenced that deer and other wildlife are infected with a common strain of *M. bovis,* and the species above-mentioned likely become infected through scavenging of deer or domestic cow carcasses. However, infection with limited lesion development in the scavenger species suggests they are spillover hosts and not important in the maintenance of the epizootic in deer or transmission to other susceptible hosts. A similar situation regards the red fox in Europe (Delahay et al. 2006; Millan et al. 2008). TB cases in red fox only suggest opportunities for infection with the MTC when it is present in the natural environment, but this species is considered a spillover host. TB may affect species of interest for conservation, such as the endangered Iberian lynx (*Lynx pardinus*) that inhabits the Iberian Peninsula of Spain, although its epidemiological role for TB maintenance in the host community is insignificant (Briones et al. 2000; Pérez et al. 2001). The consumption of infected ungulate carrion in the Iberian Peninsula may contribute to the spread and persistence of TB in wildlife with regard to omnivorous species like wild boar, rather than avian scavengers (e.g. *Gyps* vultures; see below) that effectively remove infectious sources (Vicente et al. 2011). Regarding *Brucella* in elk, Maichak et al. (2009) monitored pseudo-aborted elk fetuses, placentas, and fluids in winter. Since many scavengers in the study area were migratory, canids and sedentary facultative scavenger birds accounted for most of the scavenging. Among mammals, coyotes, wolves and, presumably, red fox can be infected with brucellosis likely from ingestion of contaminated tissues or fluids, although they are spillover hosts. Interestingly, they found that scavengers reduced the proportion of elk exposed to aborted material, illustrating that scavengers native to the area likely function as biological control agents of disease. Cannibalism after Aujeszky's disease outbreaks (a viral disease of swine and wild boar that can affect most mammals except humans and primates) in wild boar has been related to pathogen transmission (Gortazar et al. 2002), although the relevance of scavenging is punctual and short-term. Anecdotally, scavenging Aujeszky's virus infected material is a risk for carnivores, such as the Florida panther (*Felis concolor corii*), the European brown bear (*Ursus arctos*; Glass et al. 1994; Zanin et al. 1997) and the endangered Iberian lynx (Masot et al. 2016), although without relevance for disease maintenance.

The number of contacts among conspecific mammals at a carcass can also potentially facilitate pathogen transmission. Ogada et al. (2012) determined a nearly threefold increase in the mean number of contacts among mammalian scavengers at carcasses without vultures, highlighting the role of vultures in carcass decomposition and level of contact among mammalian scavengers. Nonetheless, while aggregation and intraspecific contact at carcasses is frequently reported as a likely source of horizontal pathogen transmission for several scavengers (e.g. carnivores; Markandya et al. 2008; Ogada et al. 2012; Newsome et al. 2014), the role of most scavengers in the transmission dynamics of pathogens from carcasses remains widely unknown and even misunderstood. A special example is the potential risk of anthrax spread associated with scavenging activity (Bellan et al. 2013). Anthrax epidemiology is driven by environmental conditions and its preferred hosts vary amongst habitats and ecosystems. After an animal dies from anthrax, when the carcass is opened by scavengers sporulation of bacteria takes place. Spores can be disseminated indirectly by: blowflies, which contaminate browse in the vicinity; vultures and mammalian scavengers, which contaminate water supplies (by bathing and defecating in it or the soil nearby); water run-off contaminating the grazing in the vicinity; and directly by scavengers eating from the carcass or chewing on old bones (herbivores with osteophagia; Hutson et al. 2013). Though scavenger species are relatively resistant to this pathogen, the spores are very resistant in the environment. Sporulation of bacteria occurs primarily after scavengers have opened an infected fresh carcass and the tissues have been exposed to the air, and the number of spores produced is therefore dependent on an early opening of an anthrax-infected carcass. Vultures may disseminate the infection, yet they may curtail its spread by locating carcasses shortly after death and minimizing contamination by rapidly consuming smaller and thin-skinned carcasses before most of the vegetative forms have had time to sporulate (Bellan et al. 2013).

Experimental research has determined susceptibility of scavenger species to particular infections and the likelihood of pathogen passage through the gastrointestinal tract. The digestive system of obligate scavengers such as vultures present conditions of extreme acidity, around pH 1–1.2 in the stomach (Houston and Cooper 1975; Farner 1967). These conditions in the digestive tract of carrion birds probably provide helpful factors both to digest carrion and as a form of protection against biological agents. Nonetheless, their effect on prions has not been fully confirmed, since prion proteins are more sensitive to alkalinity than to acidity. Transmissible spongiform encephalopathies (TSEs) are a group of diseases caused by prions that affect the brain and nervous system of humans and animals. These include bovine spongiform encephalopathy (BSE), Scrapie and chronic wasting disease (CWD). CWD is a geographically expanding prion TSE of North American cervids (Nichols et al. 2015) first described in the United States in 1967. Prions can remain viable in the environment for many years (Seidel et al. 2007). During decomposition, ungulate carcasses release nutrients into surrounding soils, stimulate subsequent plant biomass production that attracts herbivores, and serve as a potential source of infectious material (Towne 2000; Miller et al. 2004). Avian and mammalian scavengers that consume infected materials could spread prions through deposition of feces in the environment (Houston and Cooper 1975) or by transport of infectious carrion during food-catching or young-provisioning. For instance, American crows (*Corvus brachyrhynchos*) have the potential to translocate infectious CWD prions in their feces to disease-free areas, which suggests that avian scavengers may play a role in the transmission and translocation of prion diseases (VerCauteren et al. 2012). It has been reported that the ability of CWD-infected brain material to pass through the gastrointestinal tract of coyotes following oral ingestion, and be infectious, demonstrating that mammalian scavengers could contribute to the translocation and contamination of CWD in the environment (Nichols et al. 2015). Currently, there is very low incidence of BSE in Europe; for instance, only 5 cases were diagnosed in cattle during 2016 (out of 1,352,585 animals tested) in the EU (8EFSA 2017). Only one of these cases was classified as classical BSE. However, five cases of CWD were reported in Norway: three in wild reindeer (*Rangifer tarandus*) and two in moose (*Alces alces*). This situation, in addition to the fact that no role for BSE spread by obligate scavengers has been demonstrated in Europe, led to modifying EU legislation that allowed for the conservation of scavenging birds to allow them to assist with carcass removal (Margalida et al. 2012, Margalida and Colomer 2012).

The modelling of disease dynamics can be useful to study the contribution of scavenging to disease spread and/or persistence, although it is difficult to parameterize scavenging rates between hosts. It has rarely been estimated between-host transmission parameter from scavenging estimates for a particular disease, particularly in the case of TB (Barron et al. 2015). The rates reported were likely not very accurate (as it assumed all scavenging of an infected carcass resulted in disease transmission and assumed that carcasses are contacted or scavenged within the period that *M. bovis* bacteria remain viable). Nonetheless, in multi-host systems, these rates provide comparable data among species that can be useful to model disease and the contribution of scavenging. Barron et al. (2015) concluded that TB could persist in the wild boar/ferret complex in the absence of possum hosts since scavenging would provide a source of spillback infection for possums once their numbers recovered even many years after control. Nonetheless, transmission rate estimation is complex and model outputs are sensitive to variations of several parameters, generating uncertainties. Estimation of these transmission parameters is required to enable managers to assess if multi-host disease dynamics are important for their disease control programs.

Figure 1 illustrates the risk of infectious disease spread, or contrary, disease control, through scavenging on mammals. This is a simplified scenario and the diversity of pathogens, epidemiological contexts, routes of transmission and scavenging assemblages make every case singular. Since the range of hosts a pathogen infects (host specificity) depends on shared phylogenetic history (Clark and Clegg 2017), pathogen circulation through scavenging is more probable from mammal to mammal. For instance, when ungulate carcasses are available, facultative ungulate scavengers, such as wild boar, are more susceptible to pathogens present in the scavenged material (especially when cannibalism occurs) than are birds.

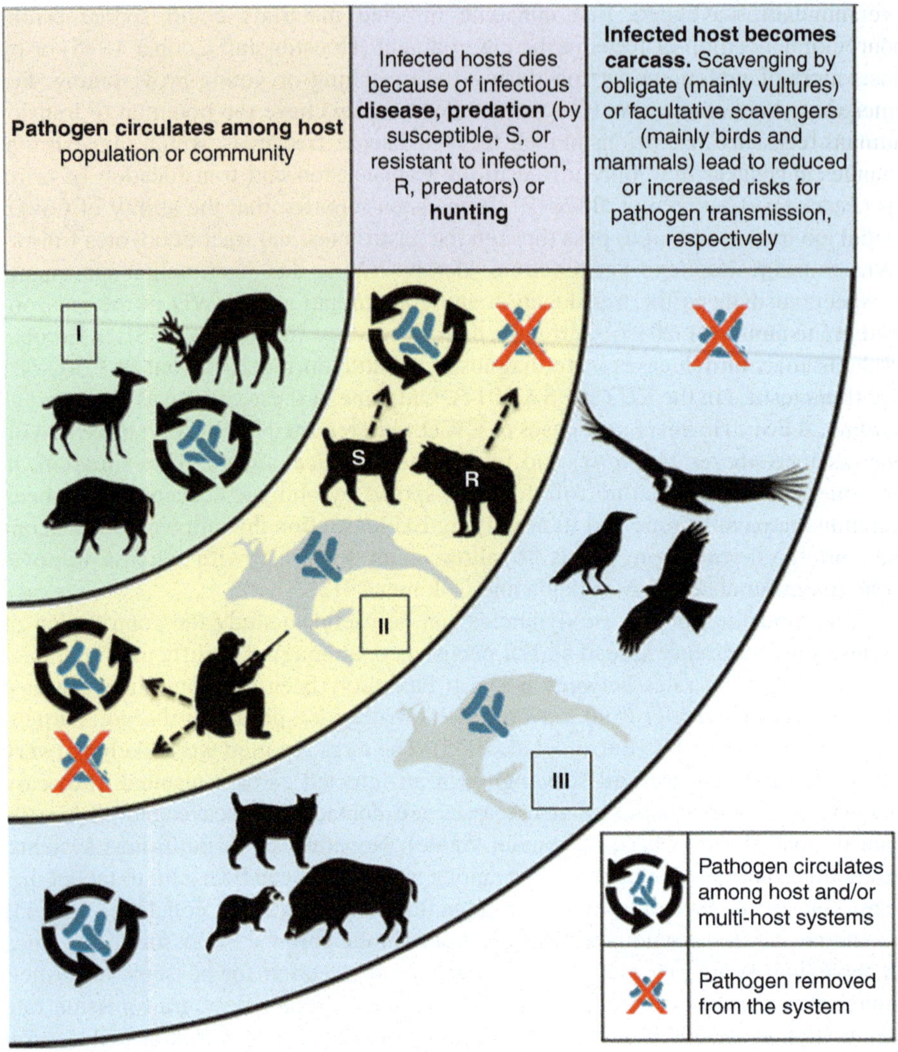

Fig. 1 Simplified scenario on the risk of infectious disease spread, or contrary, disease spread prevention through scavenging on mammal carrion. Credit: the chapter authors

Human Related Factors Resulting in Increased Risk for Disease Transmission Through Scavenging

Human activities may lead to increasing amounts of contaminated carrion, leaving extremely resistant agents in the environment. This, together with human impacts on scavenger communities (e.g. threats on top predators like vultures or other scavengers with key roles in ecosystems; Mateo-Tomás et al. 2017; or big game overabundance consequences, and particularly, facultative scavengers, such as wild boar; Gortazar et al. 2006) may increase contact rates among susceptible scavenger hosts and infected material. These situations are likely particular cases of disease emergence and re-emergence tending to occur more in ecosystems altered by humans (e.g. Daszak et al. 2000). Alterations to host-parasite relationships result in changes in host survival and contact rates among susceptible and infected individuals (Pedersen et al. 2007). This issue could acquire special relevance in the light of global environmental changes that can alter the patterns and prevalence of disease (Wilson and Wolkovich 2011).

Species considered big game vary with geographical areas and generally includes medium to large size ungulates and predators. Such a diverse group of species plays essential roles in the ecological dynamics of natural or semi-natural systems (e.g. Apollonio et al. 2010). Harvests that collect only part of animals, such as big game hunting, leave the remaining biomass for scavengers (Mateo-Tomás et al. 2015). For instance, there are some 20 species of big game within Europe (Cervidae, Bovidae, Ovidae and Suidae) adding up to 15 million animals and representing a standing biomass of more than 0.75 billion kg (Apollonio et al. 2010). In practice, there may be slight differences in the final presentation of the dressed carcass due to local cultural practices, type of trophy and final carcass use. In such ways, human activities could directly increase carrion supply and hence the prevalence of scavenging. Big game management by humans may contribute to the establishment and subsequent maintenance of pathogens that affects scavengers (especially facultative), the rest of the animal community and the environment (Vicente et al. 2011). High rates of pathogen circulation driven by human management are often associated to overabundance of big game; and in these cases, some infected species often play a role as facultative scavenger (e. g. wild boar, Gortazar et al. 2006). In addition, the way big game waste is disposed of may allow for facultative disease-susceptible scavengers to access infected material (Moreno-Opo et al. 2012; Cano-Terriza et al. 2018). For instance, in Europe, as in many other parts of the world, the changes occurring across the last 40 years have had a pronounced effect on the environment, creating a dynamic situation where pathogens or new hosts emerge o re-emerge. In particular, there have been important changes in big game population densities and/or host behavior (management favoring aggregation, Acevedo et al. 2007), which affect disease prevalence and, in some cases, may allow disease agents to boost their virulence and widen their host range (Ferroglio et al. 2011). In south and central Spain, big game is frequently managed in order to promote high densities on hunting estates, and also reach considerably high densities in protected areas (Acevedo et al. 2007). This risk is increased by the sanitary consequences of ungulate overabundance caused by intensive management, which favor the transmission of disease and its persistence in ungulates (e.g. by increasing aggregation, and therefore contacts, at feeding and water points, Gortazar et al. 2006; Acevedo et al. 2007). These systems often generate a huge amount of hunting remains highly aggregated in both space (the remains are often disposed of at the same dumps) and time (e.g. big game hunting occurs mostly in autumn and winter; Vicente et al. 2011). TB averages 45% (locally, even higher, Gortazar et al. 2008) of prevalence based on macroscopic lesions in wild boar in south and central Spain (Martín-Hernando et al. 2007). The importance of the intraspecific transmission of TB may be especially relevant in certain areas because of the absence of vultures, and the fact that wild boar, the main carrier of the disease in Spain (Naranjo et al. 2008), is the principal consumer of ungulate carrion provided by hunting. The consumption of infected ungulate carrion and waste may subsequently contribute to the spread and persistence of TB in wildlife with regard to carnivorous and omnivorous species rather than avian scavengers, which effectively remove infectious sources (see above and Fig. 1 in chapter "Ecological Functions of Vertebrate Scavenging"). In this context, controversy has arisen concerning vulture conservation, since European sanitary policies (e.g. Decision 2000/418/EC) encouraged the destruction of domestic animal carcasses (although EU Regulation 142/2011 allows now some exceptions; see this chapter), rather than being left in the open (Donázar et al. 2009a, b), and this scenario could also apply to the management of big game carcasses and hunting remains (Margalida and Moleón 2016).

Also, collateral effects of human activities may occur on diseases that are not (mainly) transmitted by scavenging. These cases usually relate to human intervention altering the host community, for instance, reducing the number of specialist scavengers, or favoring the abundance of generalist scavengers that participate in the spread and/or perpetuation of diseases, not necessarily transmitted by scavenging. Small predators, which are well-known disease reservoirs (e.g. rats *Rattus* spp., feral dogs *Canis familiaris*) appear to increase in abundance when vultures are absent in India (Pain et al. 2003). Diclofenac (a non-steroidal anti-inflammatory drug) residues in the tissues of treated dead cattle have been found to be highly toxic to vultures, resulting in up to 99% mortality in some species in India (also in Pakistan and Nepal, Markandya et al. 2008). The decline in vultures has favored an increase in packs of rabies-carrying feral dogs scavenging cattle remains. As a consequence, the number of cases of rabies in people due to dog bites has since increased (Oaks et al. 2004; Sharp 2006; Markandya et al. 2008). Similarly, spotted hyenas (*Crocuta crocuta*) and jackals (*Canis aureus*, *C. mesomelas* and *C. adustus*) host a number of pathogens that infect a wide range of other wild and domestic species (Alexander et al. 1994; Harrison et al. 2004). Two major diseases that affect African carnivores, rabies and canine distemper, spread through direct contact between infected and susceptible individuals, and carnivores often interact closely at carcasses (Mills 1993; Roelke-Parker et al. 1996; Butler et al. 2004; Ogada et al. 2012). Also, numbers of free-ranging dogs have reached unprecedented numbers in many areas of Africa. Dogs outcompete vultures on wildlife reserve boundaries owing to their high densities, nocturnal and diurnal activity, physical dominance and greater tolerance to human disturbance (Butler and du Toit 2002). The cascade effects, and its potential effects for disease spread as vultures decline, are complex. For example, changes in abundance of other avian scavengers (tawny eagles *Aquila rapax*) were most pronounced in the absence of vultures (Ogada et al. 2012).

Finally, in natural systems scavengers are part of a food chain that may be initiated by the action of predation. Changes in predation rates or predator-prey dynamics are among the factors that may affect patterns of disease emergence, reemergence, and persistence (Choo et al. 2003; Packer et al. 2003; Holt and Roy 2007). Ecosystems altered by removal of natural predators by humans may respond differently to endemic or novel pathogens than intact systems (Heesterbeek and Roberts 1995; Gross and Miller 2001; Packer et al. 2003). A mathematical model evaluated how dynamics of prion disease in deer populations may respond to nonrandom removal resulting from selective predation by wolves compared with effects of random removal through harvest or culling under different scenarios. The conclusion was that as CWD distribution and wolf range overlap in the future, wolf predation may suppress disease emergence or limit prevalence (Wild et al. 2011).

Management of Scavenging to Reduce Disease Risks

Preventing the entrance and/or controlling levels of disease in a system are the first strategy. Attempts to control disease in wildlife populations have been based on a variety of methods. These include setting up barriers, improving hygienic measures, culling, habitat management and feeding bans, vector control, treatments, and vaccination (Wobeser 2002; Artois et al. 2009; Karesh et al. 2005; Gortazar et al. 2006). Leaving entrails or carcasses *in situ* is an option when game animals are hunted in their natural habitat. Also, when livestock are not suspected of being infected with diseases transmissible to humans or other animals or affected with any potential hazard and good hunting practices are observed. This is recommended to protect natural ecological processes. In the case of big game remains should follow a veterinarian inspection (at least of the offal), if there is evidence that relevant pathogens circulate in the region. In this case, it becomes essential disease surveillance and monitoring of big game diseases (see above), and in general of wildlife in the area.

Once disease is present, re-directing scavenging activity and restoring predator and scavenger communities might provide useful strategies to reduce disease risks, realizing that eradication of disease may be impossible in the presence of wildlife disease reservoirs. For example, the observation of scavenging in TB multi-host systems from New Zealand may provide a mechanism for the potential control strategy of the disease in different contexts (Barron et al. 2015). Ferrets and pigs were predicted to act as a host complex via reciprocal scavenging, and it was suggested that moderately intensive control of either species was able to reduce their combined abundance below the threshold for disease persistence. Nonetheless, in most cases the selection and implementation of appropriate management is complex. Following, we provide examples and list different management approaches of scavenging that could help to reduce disease risks, although without exception, these types of approaches warrant further evaluation. In addition to surveillance of scavenging activity (e.g. using camera traps), monitoring disease in the community and how it responds to changes in scavenging activity and management becomes important, but to date has been rarely addressed.

Restoration of Large Predators

Wolves, for example, may be able to recover functional predator-prey-scavenger food webs (Beschta and Ripple 2009). The restoration of natural predator-prey dynamics in ecosystems previously altered by removal of natural predators may contribute to counteract disease emergence (Heesterbeek and Roberts 1995; Gross and Miller 2001; Choo et al. 2003; Packer et al. 2003; Holt and Roy 2007; Wild et al. 2011). This provides an opportunity to observe and evaluate the effects that predation (selective or not) may have on the prevalence of diseases. For instance, wolves may be a promising approach for control of CWD and therefore risks associated to scavenging (Wild et al. 2011).

Elimination of Hunting of Scavengers

This has been suggested to maintain coyote populations on or adjacent to elk feed grounds in the Greater Yellowstone Area (GYA), because it may increase scavenging rates and subsequently decreases investigation rates (contact with infected material) by elk and thus risk for *Brucella* infection from abortions associated with elk feeding grounds (Cook et al. 2004; Maichak et al. 2009). Elk fetuses (elk abortions may transmit brucellosis) were placed in the GYA to determine which predators visited and how long the carrion lasted on the landscape. Inside Yellowstone, where scavengers such as coyotes are not hunted, the fetuses disappeared more quickly (most before 20 days, Cook et al. 2004). Scavengers have not been implicated in transmission of brucellosis from wildlife to livestock. Thus, the authors suspect that scavengers reduce brucellosis transmission by limiting the time an infectious fetus remains in the environment.

Destruction of Big Game and Domestic Animal Carcasses

We previously exemplified that, in most cases, no research has concluded what the relevance of different species of scavengers, carrion and the environment are to maintain and transmit pathogens. Nonetheless, drastic measures are often preventively taken to prevent an alleged disease spread. For instance, the destruction of big game and domestic animal carcasses has been encouraged by some European policies, rather than allowing carcasses or offal from being left in the open (Donázar et al. 2009a, b). This still arises controversy concerning vulture conservation (see above). As suggested for CWD in high-risk regions from North America (Jennelle et al. 2009), recent laws on TB endemic areas in south and central Spain have considered the removal of hunted animals (including offal) to limit potential TB deposition near a kill site that mammals, and particularly wild boar, can access. It is therefore desirable that management and legislation be applied in a way that would allow for the selective access of vultures to the abandoned carcasses and offal piles that appear during hunting seasons. Although sanitary authorities should consider the removal of infected hunted animals and viscera to limit potential pathogen contamination where facultative scavengers can access, the conservation of obligate scavengers and other birds requires of selective disposal that guarantees their food supply. Burial or burning carcasses (the last is considered often the optimal carcass decontamination measure) is not always logistically feasible. Other options include spraying carcasses with considerable quantities of 5 to 10% formalin or covering them with locally available obstructive materials (e.g., bushes, branches) to deter some scavengers, such as birds, but probably not mammals. In parts of Africa with fuel shortages and poor infrastructures, anthrax contaminated carcasses have been covered in thorn scrub and left. In Canada, the helicopter dump 400 gallons of 5% formaldehyde over carcasses very effectively discouraged all scavenging until the burning team could reach the site; which presumably limited the ground contamination (Hugh-Jones and de Vos 2002). Neither of these methods will deter hyenas, which were seen to overturn a drum with 10% formaldehyde and eat formalin-saturated animal organs. In most cases, it is unlikely that all carcasses can be located (e. g. natural deaths, injured animals during hunting activities), even following intensive searches, therefore measures can only be taken on what can be found. The effectiveness of removing carcasses is not always clear, as when employed during botulism outbreaks in wild birds do not appear to have reduced avian mortality (Wobeser 2007). Nonetheless, in this particular example pathogen transmission is not mediated by carcass consumption, and probably the rapid effect of scavengers would limit environmental contamination. One final remark, which is discussed elsewhere in this chapter, is that the root cause often is related to big game overabundance contexts (Gortazar et al. 2006) favoring both high prevalence of pathogens in the population (as a density dependent mechanism) and elevated production of carrion.

Restoration of the Effects of Overabundance

We above highlighted the relationship between overabundance and disease, and therefore control of the situation will lead to limiting the associated risks. The assessment of overabundance and the available management tools have been discussed recently (Gortazar et al. 2006). A multidisciplinary approach is needed to diagnose if a given wildlife population is overabundant, which includes signs such as adverse effects on the soil, vegetation or fauna, poor body condition scores, low reproductive performance or increased parasite burdens and infectious disease prevalence. A close monitoring of wildlife densities and diseases, the establishment of reference values for all signs of overabundance, and the mapping of the disease and density hotspots will be needed to design adequate management for each particular situation.

Excluding Mammalian and Avian Scavengers from Natural Carrions

We found just one study that experimentally evaluated the effect of excluding mammalian and avian scavengers from natural carrion on the risk for disease persistence and transmission, particularly anthrax (Bellan et al. 2013). When the only alternative is to leave the carcass unmoved and to protect it adequately from scavengers and humans, stopping scavengers can significantly reduce spore production, but this can be difficult. Nonetheless, Bellan et al. (2013) found no significant effect of scavengers on soil spore density in Etosha National Park (ENP), Namibia. They concluded that scavenging by vertebrates is not a critical factor in the life cycle of *Bacillus anthracis* and that anthrax control measures relying on deterrence or exclusion of vertebrate scavengers to prevent sporulation are unlikely to be effective. Consumption of carcass material early after host death may reduce the quantity of vegetative cells available to sporulate.

Excluding Mammalian and Avian Scavengers from Vulture Restaurants

Vulture restaurants, where access to facultative mammalian scavengers is impeded, can be used as a conservation tool for threatened vultures, as these places can be supplied with a source of supplementary carrion free from risky pathogens, toxicants and veterinary drugs. Nonetheless, not all are advantages, and there may be several negative impacts of supplementary feeding sites on ecosystem and scavenging community dynamics from an ecological and conservation perspective (Donázar et al. 2009a, b; Moreno-Opo et al. 2015). The provision of carrion at specific locations close to breeding sites, in accordance with requirements of sanitary regulations for the management of animal by-products not intended for human consumption has partly mitigated the negative effects of widespread and compulsory removal of livestock carcasses from the wild. Because of the apparent contradictions between the application of sanitary and conservation strategies, managers and policy makers must solve a problem of availability of food for one of the most threatened wildlife groups (vultures), at the same time that make it compatible with food security policies. Wild ungulate waste may mitigate these effects. For instance, a sufficient amount of carrion from big game is already generated in large areas (in Mediterranean areas from Spain may reach about 40% of obligate scavenger diet, and nearly 100% during hunting season, Margalida et al. 2012; see also Mateo-Tomás and Olea 2010 in northern areas of the Iberian peninsula), and the question is how ensuring the supply of big game waste management with not sanitary risks for other groups of animals and humans. Nonetheless, when not protected against mammals, there is evidence that vulture restaurants increase the local abundance of mammalian carnivores in South Africa (i.e. brown hyaena *Hyaena brunnea* and black-backed jackal; Yarnell et al. 2015), which require that the wider impacts of providing supplementary carrion for conserving threatened species are fully investigated. In fact, vultures may serve as sanitary filters because they clean up the environment by eating carrion. One possible solution is the use of mammal-proof enclosures or other effective barriers (high game fences), having a minimum area for allowing large birds landing and taking off. It has been demonstrated that some models of electrified fences were effective for preventing facultative mammal scavengers from gaining access into an enclosure for a period of time sufficient for avian scavengers to access and remove potentially infectious hunting remains (Moreno-Opo et al. 2012). Nonetheless, it is difficult to ensure that rodents and other mammals that easily burrow or climb cannot gain access to infected carcasses. We note that restricting carrion access to birds also may affect the species that benefit from carrion (e.g. endangered species such as the brown bear and wolf in Spain); which often exert a natural effect of controlling diseases. In addition there may occur other collateral effects, such as an increase of attacks by predators on cattle or hunting when carrion is not available anymore. A condition previous to the exclusion of mammalian and avian scavengers from natural carrions in vulture restaurants should be a rigorous assessment of how they depend on carrion and what their ecosystem role is (e. g. for large predators, usually very threatened). Understanding the functionality of the assemblage of scavenging species is essential, as they "work together", benefitting from each other; for instance, wolves opening the carcass may make innards more available to crows.

A recent observational study in south-central Spain (Cano-Terriza et al. 2018) evaluated the management of big game remains aimed at reducing the access of facultative scavengers, using wild boar-proof vulture restaurants or eliminating the carrion (burial or removal by waste management firms authorized by the administration). In the study region, among other factors, intensive management of wild ungulates for hunting has led to an increase in their populations, implying a higher generation of hunting waste, which in turn can favor the transmission of infectious diseases, including TB. The above-mentioned study assessed the usefulness of disposal/elimination of hunting remains as a TB control measure in wild boar, comparing two adjacent regions where the management of these remains according to legislation varied. The *Mycobacterium tuberculosis* complex (MTC) seroprevalence detected in wild boar was significantly higher before the disposal of big game hunting by-products compared to the second period where control measures were applied (after this control measure a 25% reduction was achieved). By contrast, no significant differences in seroprevalence against MTC between periods were found in wild boar from the adjacent control area where carrion management was not implemented. This study suggests that the management of hunting waste in specific sites may yield some outcome of limit TB prevalence in wild boar. However, further studies are needed to assess the effect of scavenging management on the health status of wildlife species.

Conclusions and Future Perspectives

There is an increasing concern on the exposure of scavenger communities to disease, their effects on scavengers, and the epidemiological role that scavenging plays in disease dynamics. This chapter evidences that, with some rare exceptions, research up to date does not reliably conclude what the role of different species of scavengers in the maintenance and transmission of diseases and/or the role of carrion are. Disease transmission associated to scavenging is extremely difficult to study in the field because these phenomena are not straightforward to observe and parameterize. Inferences about the pathogen maintenance by scavengers, carrion and the environment have been made mainly using sources of evidence that may be considered circumstantial. To now, it is only by the accumulation of additional pieces of correlational evidence that the acceptation of such inferences has been generally adopted. There are some examples where scavengers, including specialist and facultative, have been regarded as potential vectors for a number of diseases, in which they ingest or are exposed to infected material and later shed the disease agent in new areas. Scavenging of infectious remains poses a confounding duality between the removal of infectious materials by consumption and potential dispersion of infectious material in the environment. We illustrate in this chapter that we can benefit enormously in learning from previous experiences. Current evidences support the need for further research regarding scavenger ecology within the context of wildlife and human health, and also for wildlife management and conservation. First, research must determine the possible cascade of pathogens through food webs. Research should also assess which disposal regime is most beneficial to obligate scavengers, thus guaranteeing their food supply, while reducing the exposure of susceptible animals to potentially infectious material. Experimental research should determine susceptibility of scavenger species to infections and the likelihood pathogen passage through the gastrointestinal tract. Management of diseases and subsequent reduction in the generation of hazardous carcasses and remains usually require changes in human activities and close collaborations among wildlife ecologists, veterinarians and public health professionals. We should work towards a more informed management of scavenging. In accordance, the regulations concerning carrion provisioning need to be flexible, reconciling sanitary and environmental objectives.

Acknowledgements We thank Pedro P. Olea and Patricia Mateo-Tomás their valuable guidance and comments on the chapter.

References

Acevedo P, Vicente J, Höfle U, Cassinello J, Ruiz-Fons F, Gortazar C (2007) Estimation of European wild boar relative abundance and aggregation: a novel method in epidemiological risk assessment. Epidemiol Infect 135:519–527

Alexander KA, Kat PW, Wayne RK, Fuller TK (1994) Serologic survey of selected canine pathogens among free-ranging jackals in Kenya. J Wildl Dis 30:486–491

Anderson DP, Ramsey DSL, de Lisle GW, Bosson M, Cross ML, Nugent G (2015) Development of integrated surveillance systems for the management of tuberculosis in New Zealand wildlife. N Z Vet J 63:89–97
Apollonio M, Andersen R, Putman R (2010) Introduction. In: Apollonio M, Andersen R, Putman R (eds) European ungulates and their management in the 21st century. Cambridge University Press, Cambridge
Artois M, Bengis R, Delahay RJ, Duchêne M, Duff JP, Ferroglio E et al (2009) Wildlife disease surveillance and monitoring. In: Delahay RJ, Smith GC, Hutchings MR (eds) Management of disease in wild mammals. Springer, Tokyo, pp 187–213
Barron MC, Tompkins DM, Ramsey DSL, Bosson MAJ (2015) The role of multiple wildlife hosts in the persistence and spread of bovine tuberculosis in New Zealand. N Z Vet J 63:68–76
Bellan SE, Turnbull PCB, Beyer W, Getz WM (2013) Effects of experimental exclusion of scavengers from carcasses of anthrax-infected herbivores on Bacillus anthracis sporulation, survival, and distribution. Appl Environ Microbiol 79:3756–3761
Beschta RL, Ripple WJ (2009) Large predators and trophic cascades in terrestrial ecosystems of the western United States. Biol Conserv 142:2401–2414
Briones V, de Juan L, Sánchez C et al (2000) Bovine tuberculosis and the endangered Iberian lynx. Emerg Infect Dis 6(2):189–191
Bruning-Fann CS, Schmitt SM, Fitzgerald SD et al (2001) Bovine tuberculosis in free-ranging carnivores from Michigan. J Wildl Dis 37:58–64
Butler JRA, du Toit JT (2002) Diet of free-ranging domestic dogs (Canis familiaris) in rural Zimbabwe: implications for wild scavengers on the periphery of wildlife reserves. Anim Conserv 5:29–37
Butler JRA, du Toit JT, Bingham J (2004) Free-ranging domestic dogs (Canis familiaris) as predators and prey in rural Zimbabwe: threats of competition and disease to large wild carnivores. Biol Conserv 115:369–378
Byrom A (2004) Spread of Tb by ferrets in the northern South Island high country. Landcare Research Contract Report: LC0304/146
Byrom AE, Caley P, Paterson BM, Nugent G (2015) Feral ferrets (*Mustela furo*) as hosts and sentinels of tuberculosis in New Zealand. N Z Vet J 63:42–53
Cano-Terriza D, Risalde MA, Jiménez-Ruiz S et al (2018) Management of hunting waste as control measure for tuberculosis in wild ungulates in south-central Spain. Transbound Emerg Dis 65(5):1190–1196
Carrasco-Garcia R, Barroso P, Perez-Olivares J, Montoro V, Vicente J (2018) Consumption of big game remains by scavengers: a potential risk as regards disease transmission in central Spain. Front Vet Sci 5:4
Clark NJ, Clegg SM (2017) Integrating phylogenetic and ecological distances reveals new insights into parasite host specificity. Mol Ecol 26:3074–3086
Choo K, Williams PD, Day T (2003) Host mortality, predation, and the evolution of parasite virulence. Ecol Lett 6:310–315
Cook WE, Williams ES, Dubay SA (2004) Disappearance of bovine fetuses in northwestern Wyoming. Wildl Soc Bull 32:254–259
Daszak P, Cunningham AA, Hyatt AD (2000) Emerging infectious diseases of wildlife—Threats to biodiversity and human health. Science 287:443–449
Delahay RJ, Smith GC, Barlow AM, Walker N, Harris A, Clifton-Hadley RS, Cheeseman CL (2006) Bovine tuberculosis infection in wild mammals in the south-west region of England: a survey of prevalence and a semi-quantitative assessment of the relative risks to cattle. Vet J 164:90–105
DeVault TL, Rhodes OE, Shivik JA (2003) Scavenging by vertebrates: behavioural, ecological, and evolutionary perspectives on an important energy transfer pathway in terrestrial ecosystems. Oikos 102:225–234
Donázar JA, Cortés-Avizanda A, Carreta M (2009a) The role of trophic resource predictability in the structure of scavenger guild. Vultures, feeding stations and sanitary legislation, a conflict and its consequences from the perspective of conservation biology. Munibe 29:344–352

Donázar JA, Margalida A, Carrete M, Sánchez-Zapata JA (2009b) Too sanitary for vultures. Science 326:664

EFSA (European Food Safety Authority) (2017) Scientific report on the European Union summary report on surveillance for the presence of transmissible spongiform encephalopathies (TSE) in 2016. EFSA J 15(5069):68

European Commission – Health & Consumer Protection Directorate-General (2002) Scientific Steering Committee. Opinion on: necrophagous birds as possible transmitters of TSE/BSE. Adopted by the scientific steering committee at its meeting of 7–8 November 2002. https://ec.europa.eu/food/sites/food/files/safety/docs/sci-com_ssc_out295_en.pdf

Farner DS (1967) The hydrogen ion concentration in avian digestive tracts. Poultry Sci 21:445

Ferroglio E, Gortazar C, Vicente J (2011) Wild ungulate diseases and the risk for livestock and public health. In: Putman R, Apollonio M, Andersen R (eds) Ungulate management in Europe: problems and practices. Cambridge University Press, Cambridge

Glass CM, Mclean RG, Katz JB et al (1994) Isolation of pseudorabies (Aujeszky's disease) virus from a Florida panther. J Wildl Dis 30:180–184

Gortazar C, Acevedo P, Ruiz-Fons F, Vicente J (2006) Disease risks and overabundance of game species. Eur J Wildl Res 52:81–87

Gortazar C, Torres MJ, Vicente J et al (2008) Bovine tuberculosis in Doñana Biosphere Reserve: the role of wild ungulates as disease reservoirs in the last Iberian lynx strongholds. PLoS One 3:1–8

Gortazar C, Vicente J, Fierro Y, Leon L, Cubero MJ, González M (2002) Natural Aujeszky's disease in a Spanish wild boar population. Ann N Y Acad Sci 969:210–212

Gross JE, Miller MW (2001) Chronic wasting disease in mule deer: disease dynamics and control. J Wildl Manag 65:205–215

Harrison TM, Mazet JK, Holekamp KE, Dubovi E, Engh AL, Nelson K, Van Horn RC, Munson L (2004) Antibodies to canine and feline viruses in spotted hyena (*Crocuta crocuta*) in the Masai Mara National Reserve. J Wildl Dis 40:1–10

Heesterbeek JAP, Roberts MG (1995) Mathematical models for microparasites of wildlife. In: Grenfell BT, Dobson AP (eds) Ecology of infectious diseases in natural populations. Cambridge University Press, Cambridge, pp 90–122

Holt RD, Roy M (2007) Predation can increase the prevalence of infectious disease. Am Nat 169:690–699

Houston DC, Cooper JE (1975) The digestive tract of the whitebacked griffon vulture and its role in disease transmission among wild ungulates. J Wild Dis 11:306–313

Hugh-Jones ME, De Vos V (2002) Anthrax and wildlife. Rev Sci Tech 21:359–383

Hutson JM, Burke C, Haynes G (2013) Osteophagia and bone modifications by Giraffe and other large ungulates. J Archaeol Sci 40:4139–4149

Inger R, Cox DTC, Per E, Norton BA, Gaston KJ (2016) Ecological role of vertebrate scavengers in urban ecosystems in the UK. Ecol Evol 6:7015–7023. https://doi.org/10.1002/ece3.2414

Jennelle CS, Samuel MD, Nolden CA, Berkley EA (2009) Deer carcass decomposition and potential scavenger exposure to chronic wasting disease. J Wildl Manag 73:655–662

Karesh WB, Cook RA, Bennett EL, Newcomb J (2005) Wildlife trade and global disease emergence. Emerg Infect Dis 11:1000–1002

Leighton FA, Artois M, Capucci L, Gavierwiden D, Morisse JP (1995) Antibody-response to rabbit viral hemorrhagic disease virus in red foxes (*Vulpes vulpes*) consuming livers of infected rabbits (*Oryctolagus cuniculus*). J Wildl Dis 31:541–544

Maichak EJ, Scurlock BM, Rogerson JD, Meadows LL, Barbknecht AE, Edwards WH, Cross PC (2009) Effects of management, behavior, and scavenging on risk of brucellosis transmission in elk of western Wyoming. J Wildl Dis 45:398–410

Margalida A, Colomer MA (2012) Modelling the effects of sanitary policies on European vulture conservation. Sci Rep 2:753

Margalida A, Carrete M, Sánchez-Zapata JA, Donázar JA (2012) Good news for European vultures. Science 335:284–284

Margalida A, Moleón M (2016) Toward a carrion-free ecosystem. Front Ecol Environ 14:182–183

Markandya A, Taylor T, Longo A, Murty MN, Murty S, Dhavala K (2008) Counting the cost of vulture decline: an appraisal of the human health and other benefits of vultures in India. Ecol Econ 67:194–204

Martín-Hernando MP, Höfle U, Vicente J et al (2007) Lesions associated with *Mycobacterium tuberculosis* complex infection in the European wild boar. Tuberculosis 87:360–367

Masot AJ, Gil M, Risco D, Jiménez OM, Núñez JI, Redondo E (2016) Pseudorabies virus infection (Aujeszky's disease) in an Iberian lynx (Lynx pardinus) in Spain: a case report. BMC Vet Res 13:6

Mateo-Tomás P, Olea PP (2010) When hunting benefits raptors: a case study of game species and vultures. Eur J Wildl Res 56:519–528

Mateo-Tomás P, Olea PP, Moleón M, Vicente J, Botella F, Selva N, Viñuela J, Sánchez-Zapata JA (2015) From regional to global patterns in vertebrate scavenger communities subsidized by big game hunting. Divers Distrib 21:913–924

Mateo-Tomás P, Olea PP, Moleón M, Selva N, Sánchez-Zapata JA (2017) Both rare and common species support ecosystem services in scavenger communities. Glob Ecol Biogeogr 26:1459–1470

Millan J, Jimenez MA, Viota M, Candel MG, Peña L, León-Vizcaino L (2008) Disseminated bovine tuberculosis in a wild red fox (*Vulpes vulpes*) in southern Spain. J Wildl Dis 44:701–706

Miller MW, Williams ES, Hobbs NT, Wolfe LL (2004) Environmental sources of prion transmission in mule deer. Emerg Infect Dis 10:1003–1006

Mills MGL (1993) Social systems and behaviour of the African wild dog *Lycaon pictus* and the spotted hyaena *Crocuta crocuta* with special reference to rabies. Onderstepoort J Vet Res 60:405–409

Moreno-Opo R, Trujillano A, Margalida A (2015) Optimization of supplementary feeding programs for European vultures depends on environmental and management factors. Ecosphere 6(7):127

Moreno-Opo R, Margalida A, García F, Arredondo A, Rodríguez C, González LM (2012) Linking sanitary and ecological requirements in the management of avian scavengers: effectiveness of fencing against mammals in supplementary feeding sites. Biodivers Conserv 21:673–1685

Naranjo V, Gortazar C, Vicente J, de la Fuente J (2008) Evidence of the role of European wild boar as a reservoir of Mycobacterium tuberculosis complex. Vet Res 127:1–9

Newsome TM, Dellinger JA, Pavey CR et al (2014) The ecological effects of providing resource subsidies to predators. Glob Ecol Biogeogr 24:1–11

Nichols TA, Fischer JW, Spraker TR, Kong Q, VerCauteren KC (2015) CWD prions remain infectious after passage through the digestive system of coyotes (*Canis latrans*). Prion 9:367–375

Oaks JL, Gilbert M, Virani MZ et al (2004) Diclofenac residues as the cause of vulture population decline in Pakistan. Nature 427:630–633

Ogada DL, Torchin ME, Kinnaird MF, Ezenwa VO (2012) Effects of vulture declines on facultative scavengers and potential implications for mammalian disease transmission. Conserv Biol 26:453–460

Packer C, Holt RD, Hudson PJ, Lafferty KD, Dobson AP (2003) Keeping the herds healthy and alert: implications of predator control for infectious disease. Ecol Lett 6:797–802

Pain DJ et al (2003) Causes and effects of temporospatial declines of Gyps vultures in Asia. Conserv Biol 17:661–671

Pedersen AB, Jones KE, Nunn CL, Altizer S (2007) Infectious diseases and extinction risk in wild mammals. Conserv Biol 21:1269–1279

Pérez J, Calzada J, León-Vizcaino L, Cubero MJ, Velarde J, Mozos E (2001) Tuberculosis in an Iberian lynx (*Lynx pardina*). Vet Rec 148:414–415

Probst C, Globig A, Knoll B, Conraths FK, Depner K (2016) Behaviour of free ranging wild boar towards their dead fellows: potential implications for the transmission of African swine fever. R Soc Open Sci 4(5):170054

Roelke-Parker ME et al (1996) A canine distemper outbreak in Serengeti lions (Panthera leo). Nature 379:441–445

Roen KT, Yahner RH (2005) Behavioral responses of avian scavengers in different habitats. Northeast Nat 12:103–112

Seidel B, Thomzig A, Buschmann A, Groschup MH, Peters R, Beekes M, Terytze K (2007) Scrapie agent (strain 263k) can transmit disease via the oral route after persistence in soil over years. PLoS One 2:E435.10.1371

Selva N, Jedrzejewska B, Jedrzejewski W, Wajrak A (2005) Factors affecting carcass use by a guild of scavengers in European temperate woodland. Can J Zool 83:1590–1601

Sharp D (2006) Meloxicam to prevent rabies? Lancet 367:887–888

Smith CR (1994) Wild carnivores as plague indicators in California: a cooperative interagency disease surveillance program. In: Halverson WS, Crabb AC (eds) Proceedings of 16th vertebrates pest conference. University of California, Davies, pp 192–199

Towne EG (2000) Prairie vegetation and soil nutrient responses to ungulate carcasses. Oecologia 122:232–239

Turner KL, Abernethy EF, Conner LM, Rhodes OE Jr, Beasley JC (2017) Abiotic and biotic factors modulate carrion fate and vertebrate scavenging communities. Ecology 98:2413–2424

VerCauteren KC, Atwood TC, DeLiberto TJ, Smith HJ, Stevenson JS, Thomsen BV, Gidlewski T, Payeur J (2008) Sentinel-based surveillance of coyotes to detect bovine tuberculosis, Michigan. Emerg Infect Dis 14:1862–1869

VerCauteren KC, Pilon JL, Nash PB, Phillips GE, Fischer JW (2012) Prion remains infectious after passage through digestive system of American crows (*Corvus brachyrhynchos*). PLoS One 7(10):e45774

Vicente J, Carrasco R, Acevedo P, Montoro V, Gortazar C (2011) Big game waste production: sanitary and ecological implications. In: Kumar S (ed) Integrated waste management, vol 2. Intech, Rijeka

Wild MA, Hobbs NT, Graham MS, Miller MW (2011) The role of predation in disease control: a comparison of selective and nonselective removal on prion disease dynamics in deer. J Wild Dis 47:78–93

Wilmers CC, Crabtree RL, Smith DW, Murphy KM, Getz WM (2003a) Trophic facilitation by introduced top predators: grey wolf subsidies to scavengers in Yellowstone National Park. J Anim Ecol 72:909–916

Wilmers CC, Stahler DR, Crabtree RL, Smith DW, Getz WM (2003b) Resource dispersion and consumer dominance: scavenging at wolf- and hunter-killed carcasses in Greater Yellowstone, USA. Ecol Lett 6:996–1003

Wilson EE, Wolkovich EM (2011) Scavenging: how carnivores and scavenging structure communities. Trends Ecol Evol 26:129–135

Wobeser GA (2007) Disease in wild animals: investigation and management. Springer, Berlin

Wobeser GA (2002) Disease management strategies for wildlife. Rev Sci Tech 21:159–178

Yarnell RY, Phipps WL, Dell S, MacTavish LM, Scott DM (2015) Evidence that vulture restaurants increase the local abundance of mammalian carnivores in South Africa. Afr J Ecol 53:287–294

Zanin E, Capua I, Casaccia C, Zuin A, Moresco A (1997) Isolation and characterization of Aujeszky's disease virus in captive brown bears from Italy. J Wildl Dis 33:632–634

Human-Mediated Carrion: Effects on Ecological Processes

Rubén Moreno-Opo and Antoni Margalida

Contents

Introduction

Carrion occurrence in the wild is determined by factors originally related to natural processes of death, depredation and availability of different organic unexcreted remains as a result of births, plumage molting or detachment of other tissues. Within

R. Moreno-Opo (✉)
Ministry for the Ecological Transition, Madrid, Spain

Evolution and Conservation Biology Research Group, University Complutense of Madrid, Madrid, Spain

A. Margalida
Department of Animal Science, Faculty of Life Sciences and Engineering, University of Lleida, Lleida, Spain

Institute for Game and Wildlife Research, IREC (CSIC-UCLM-JCCM), Ciudad Real, Spain
e-mail: a.margalida@csic.es

P. P. Olea et al. (eds.), *Carrion Ecology and Management*, Wildlife Research Monographs 2, https://doi.org/10.1007/978-3-030-16501-7_8

this dynamic of appearance determined solely by ecological pressures, carrion constitutes a pulsed, ephemeral and random resource (DeVault et al. 2003; Selva and Fortuna 2007). It emerges in an unpredictable way in time and space, based on epidemiological (prevalence of wildlife diseases), population (according to the relative abundance of animals), ecological (predator-prey relationships) and meteorological factors (Wilson and Wolkovich 2011; Oro et al. 2013; chapter "Carrion Availability in Space and Time"). Humans have altered this scenario throughout the Anthropocene, understood as the period within the Quaternary during which anthropic activities began to impact the environment significantly and transform the global functioning of ecosystems (Darimont et al. 2015; Sarrazin and Lecomte 2016). Therefore, human activities regarding the dynamics of carrion occurrence should be considered from a changing perspective in time, with three differentiated periods: (1) when humans exhibited scavenging behavior and, later, hunting activity to obtain foods of animal origin, (2) the onset and development of agriculture and livestock (approximately 8000 years ago), and (3) from the beginning of the Industrial Revolution (S. XVIII). These phases were not equally developed globally, and were more advanced over time in the most industrialized countries, especially on the European continent.

During the first phase, humans were not only facilitators and modulators of carrion availability in ecosystems, but were also carrion consumers (Domínguez-Rodrigo 2001; Moleón et al. 2014). As animals were a limited resource and their acquisition through hunting was energetically costly, all organic remains obtained from prey were exploited (Domínguez-Rodrigo and Pickering 2003), and thus anthropogenic carrion was very scarce in the wild. With the onset of animal husbandry, humans began to raise a greater amount of biomass of animal origin, for food, pets and as working animals for agriculture. In this period, different factors increased the availability of carrion subsidized by humans. Namely, some body parts of animals intended for consumption were now discarded due to their low nutritional, commercial and cultural value. These remains (for instance, bones, viscera or heads) were disposed of in the surroundings of slaughterhouses and butcheries. On the other hand, carcasses of draft animals or pets were often moved to dumps close to urban centers, to limit their presence in villages due to potential health risks and smells provoked by the process of rotting. Thus, in most regions and continents, an anthropogenic carcass management system was created in which different scavenger animals, both domestic and wild, took advantage of this resource in the vicinity of urban environments. This type of carrion provisioning in fixed locations around villages or towns constituted a traditional and sustainable system of elimination of organic remains not usable by humans, which supplements the feeding of different raptors and corvids (see review in Donázar et al. 2009). This system also enhanced mutualism between humans and vultures, favored by the commensalism of the latter of the resources provided by the former and the disposal of biomass not intended for human consumption (Moleón et al. 2014). This general pattern varies geographically depending on cultures and religions, as well as on the perception of human societies about the role of wild scavengers (Mundy et al. 1992; Donázar 1993; Allen 2016; Stara et al. 2016; Morales-Reyes et al. 2018).

With the Industrial Revolution, there was an increase in the consumption of food of animal origin, associated with the continued growth of human populations and an increased capacity for transport of people and other items in motorize vehicles (Lucas 2003). Domestic livestock grew in number to meet the demand of the growing population, and thus intensive farms were established and large areas of land were exploited for livestock grazing. The increase in cattle substantially improved carrion availability due to the wasting of many parts of the carcasses for commercial reasons, such that a greater amount of food being transferred to wild scavengers. However, this pattern was not homogeneous geographically: while in some areas carrion availability for avian scavengers was supported (for instance, around the Mediterranean basin and on the Asian continent), in other regions different byproducts from the carcasses were used for other purposes (cosmetics, domestic animal feeding, fats, etc.) and the rest of the unusable remains were destroyed to avoid potential health risks (Donázar et al. 1997; Margalida et al. 2010; Jayathilakan et al. 2012). Throughout the twentieth century, a trend towards the industrialization of carrion management occurred. As a result, the establishment of sites for disposing of carcasses and other remains from farms ("vulture restaurants") and the increase in the extensive cattle herds generated greater carrion availability, which favored the feeding of avian scavengers (Donázar et al. 1997) and, in part, the recovery of large carnivores (Chapron et al. 2014). However, wild ungulate populations decreased globally in the twentieth century due to hunting and poaching, reducing their availability for scavengers (Blázquez and Sánchez-Zapata 2009; Mateo-Tomás and Olea 2010; Margalida et al. 2011a, b; Ripple et al. 2015). This substitution of carrion sources in the natural environment influenced the population dynamics of scavengers and, consequently, their conservation status (Lambertucci et al. 2009b; Margalida and Colomer 2012), based on the feeding behavior and the prey selection of each species (Moreno-Opo et al. 2015a, 2016). This pattern has mainly been observed in Europe and North America. In Africa, Asia and South America, vulture restaurants for economic or commercial reasons have been less common, and are authorized only for specific conservation actions (Mundy et al. 1992). On the other hand, other practices have managed to generate a greater amount of anthropogenic organic waste over the twentieth century, such as fishing and its discards worldwide (Votier et al. 2004; Heath et al. 2014) or sport hunting in regions such as Europe, North America, and southern and eastern Africa (Mateo-Tomás et al. 2015). All of these circumstances have led to a paradigm shift in the availability and management of human-mediated carrion: from scarce amounts around urban environments to a complex system of food subsidies for wildlife (Oro et al. 2013). This transformation has brought about changes in the dynamics of resource use by fauna, and both positive and negative effects, which are dealt with in this chapter.

Types of Human-Mediated Carrion

Commercial activity has increased in the last few decades, resulting in a supply of key food subsidies within the trophic webs of ecosystems. Carrion, along with the remains of agricultural crops, is the main nutrient provided by anthropic activities like food production, recreation, sanitation and environmental management (Oro et al. 2013). The different carrion types provided by humans are described below.

Livestock

Livestock husbandry is an activity occurring throughout the world whose management typology determines the supply of carrion to the natural environment. The abundance of existing livestock also influences carrion provisions and is generally higher in temperate regions where primary productivity and other environmental conditions (i.e. temperature, rainfall, zoonoses) are more favorable (Flanders and Gillespie 2015).

Two main livestock management systems determine the way carrion is produced: *intensive* and *extensive* management. Although there are intermediate systems, *intensive* farming is understood as the type of management in which animals inhabit closed enclosures, are fed with human-processed products (fodder or harvested grass) and in which higher levels of care are provided (medication, temperature control, etc.) to increase yield. Livestock carcasses are mainly destined for transformation plants for the industrial use of their by-products and/or for their destruction (Herrero et al. 2013). However, depending on the livestock species, in some cases the carcasses may be destined for fenced feeding sites for avian scavengers. This is one of the practices that has led to an increase in vulture populations in Spain, thanks to the abundant inputs of pork and chicken carcasses associated with the growth of this economic activity since the end of the twentieth century (Donázar and Fernández 1990; Parra and Tellería 2004). To be sustainable, this system requires large populations of wild scavengers for the consumption of the total biomass provided, and thus it is not applicable to regions without an abundant guild of carrion-eating birds. Another model of carrion provisioning within this intensive system is the supply of unusable remains for human consumption from slaughterhouses and butcheries. These remains are usually destined for supplementary feeding points as a part of specific conservation programs (see below) and have played an important role in the conservation of threatened species, such as the bearded vulture *Gypaetus barbatus* (Sesé et al. 2005; Oro et al. 2008).

The *extensive* management of livestock can be described as the use of natural pastures by different livestock species throughout most of the year, such that there is a presence of animals in open areas, concentrated in herds or scattered in the field, which allows a higher level of interaction with wildlife (Bignal and McCracken 2000; Blench 2000). As these livestock inhabit wild areas, are far from infrastructures and are frequently not subject to programs of removal and destruction of carcasses, the likelihood that carrion will be available for scavengers increases (Gavashelishvili and McGrady 2006; Olea and Mateo-Tomás 2009; Margalida et al. 2018). Given the abundance of extensive livestock, its presence in almost all regions and the difficulty of human-mediated collection and disposal of carcasses, extensive livestock is currently the main food resource for obligate scavengers worldwide, replacing wildlife (Chamberlain et al. 2005; Costillo et al. 2007b; Lambertucci et al. 2009a, b).

The occurrence of livestock carrion varies depending on the region, the type of carcass management, the legislation on by-product treatment and the local culture. A greater or lesser supply is determined by different non-mutually exclusive factors:

the livestock transhumant movements performed, application of sanitary policies, implementation of conservation programs for scavenger species, degree of use of carcasses for human activities (feeding, industrial uses), geographic accessibility to carcasses for their removal or availability of financial resources for processing the carcasses (Margalida et al. 2018).

The quality of the carrion, in terms of the prevalence of pathogens or substances that can generate a negative effect in the environment, varies depending on the applied management. In general, carrion from extensive livestock has a lower risk of disseminating diseases to other animals and lower levels of veterinary drugs (Casas-Díaz et al. 2016), except in cases of high prevalence of transmissible diseases in the herds due to disruptions of population levels of their vectors and reservoirs (Gortazar et al. 2015). In this sense, the scavenger vertebrates and particularly obligate scavengers, tolerate the intake of carcasses of animals that died from most transmissible diseases without suffering negative effects on their own health status, thanks to evolutionary adaptations to the ingestion of this food and certain native pathogens (Houston and Cooper 1975; Bellan et al. 2013). On the other hand, the remains of intensive livestock hold a greater likelihood of adverse effects for the animals consuming them. The more intensive management of livestock, the higher the likelihood of containing residues of different substances (Casas-Díaz et al. 2016). This poses a risk to the vital fitness of the organisms feeding on them, for example due to antiparasitic, antibiotic or hormonal products (Margalida et al. 2014a), or may even lead to severe intoxication, as occurs with certain anti-inflammatory drugs (Green et al. 2004; Zorrilla et al. 2015).

Hunting

Hunting has evolved with humans' ability to use tools and technology. Originally, it supported the subsistence of human populations, with little carrion remains left in the environment as most parts of the animals hunted were used for food, clothing, or a variety of structures and tools (Mithen 1999; Moleón et al. 2014). However, it has shifted conceptually since the twentieth century in the temperate regions and, especially, in the most industrialized countries (Darimont et al. 2017). Currently, hunting is mainly a recreational practice, and has increased in commercial importance and unleashed new patterns of wildlife management. Thus, several game species, as well as non-game species that interfere with hunting, have seen their population abundances modulated. Many have been favored and their populations have increased in the last 50–60 years, while others have been negatively affected by conflicts with this practice (Lindsey et al. 2007).

In relation to food subsidies, hunting has served as a growing source of scavenging resources in places where this activity has increased. The manner in which it is performed determines the availability of carrion: if the practice involves a large number of animals simultaneously shot, carrion appearance increases and triggers important effects on feeding patterns of wild scavengers. For example, in the

Mediterranean region hunting events or population controls usually involve the dropping of large numbers of ungulates (red deer *Cervus elaphus*, wild boar *Sus scrofa*, fallow deer *Dama dama*, etc.) on the same day. This implies significant amounts of carrion consumed and eliminated by vultures, constituting a prime source of food during autumn and winter (Mateo-Tomás and Olea 2010; Mateo-Tomás et al. 2015). In this type of hunting, the shot animals are collected and moved to specific points where they are sanitarily inspected and where the pieces of commercial interest (i.e. for human consumption) are selected. The discarded by-products are usually transported to specific points far from populated areas and serve as a food source for wild fauna (Vicente et al. 2011; Moreno-Opo et al. 2012).

The most common practice worldwide is the selective hunting of an individual in a contained area by hunter. This modality is directed, in general, at a greater spectrum of species and produces a smaller volume of carrion in a given event. In this case, the viscera and legs, which will not be consumed by humans, or in some cases the whole body with the exception the trophy parts, are most often left for scavengers to feed upon. At a global scale, this is the most common activity contributing to carrion disposal in the wild, due to its wide geographical occurrence in regions such as North America, Europe, South and East Africa and Asia, and the temporary continuity of the activity, which allows for the appearance of carrion in small quantities dispersed across a territory. This favors not only obligate but also an enriched variety of facultative scavengers (Wilmers et al. 2003; Blázquez et al. 2009).

Carrion from hunting activity may be hazardous to species that consume it, including humans, if the ammunition contains toxic materials (Mateo-Tomás et al. 2016). In this sense, lead ammunition provokes negative and even lethal effects in different bird species frequently consuming shot carcasses (Gangoso et al. 2009; Hernández and Margalida, Kelly et al. 2014). Despite the fact that lead intoxication may be partially due to natural sources (i.e. soil composition) and that, in general, the population trends of avian scavengers are not conditioned by lead poisoning, negative effects of lead ammunition use have been shown. Moreover, there are safe chemical alternatives that could be used to avoid risks to animal health (Thomas 2013). On the other hand, in places where the management of ungulate game populations is intensive and high population densities are reached, feeding on hunted carrion can affect the sanitary status of scavengers due to potentially higher prevalences of epizootics such as bovine tuberculosis or sarcoptic mange (Vicente et al. 2011). In these cases, authorities usually propose measures to limit carrion ingestion by wildlife (Margalida and Moleón 2016).

Fisheries

Marine fishing has grown exponentially over the last century, with some 81.5 million tons of animals extracted per year around the world (FAO 2016). Technological improvements in the fishing fleet and the increase in the human population have generated a growing demand for marine food products, which has led to overexploitation of resources and the modification of ecological relationships in the seas (Coleman and Williams 2002). As a result, fishing exploitation produces huge amounts of carrion returned to the sea, consisting of waste from marine animal catches without commercial interest, non-target fishing species or individuals not meeting the minimum size requirements (Kelleher 2005). These are the so-called fishing discards, whose exact quantity is unknown given the absence of official records due to their release into the sea from the vessels before disembarking. Nevertheless, discards are estimated to reach 7.5% of all fishing catches and about 7.5 million tons annually (Davies et al. 2009; Oro et al. 2013).

Fishing discards are the main source of anthropogenic-mediated carrion in the marine environment. They are associated with certain fishing modalities, mainly those yielding a greater volume of catches of non-target species (bycatch), such as bottom and pelagic trawl nets as well as drift gillnets (Kelleher 2005). Therefore, greater carrion availability occurs in seasons and areas where these fishing gears are used, especially where vessels with a high capture and storage capacity operate (e.g. deep sea fishing). Discards are returned to the sea in all the fisheries of the world, with the aim of avoiding their inclusion as catches in the controls in harbors made by competent authorities. Disposal of discards at sea occurs after the capture is complete, usually during the transit back to the harbors.

In places with abundant fishing discards, there are often concentrations of animals feeding on these resources. Thus, the continued exploitation of these remains by seabirds has been demonstrated, as this is an essential part of the diet for species such as Audouin's gull *Larus audouinii* and the northern gannet *Morus bassanus* (Bodey et al. 2014; García-Tarrasón et al. 2015). Discards are also consumed in the pelagic and benthic environment (mainly by fishes and invertebrates), although their ecological relevance and patterns of use are not well known (Bozzano and Sarda 2002; Heath et al. 2014).

The management of discards has changed to allow its use as a food resource for people or for other purposes, thus avoiding commercial waste, at least within the European countries, New Zealand and several North American regions (Condie et al. 2014; Catchpole et al. 2017). For this reason, the obligation to transfer all the catches made at sea to harbors, including discards, has been regulated. In this way, they can be included within the authorized quotas and the use of more selective fishing gear is subsequently promoted (Sigurðardóttir et al. 2015). This should reduce the amount of carrion in the sea in the next few years and may trigger ecological effects on facultative scavengers whose feeding behavior has adapted to the exploitation of these resources (Heath et al. 2014; Sardà et al. 2015).

Rubbish Dumps

Rubbish dumps are more abundant, more geographically widespread and more predictable in time and space, compared to other sources of anthropogenic subsidies (Donázar 1992; Oro et al. 2013). At landfills, carrion serves as an important portion of the usable resources for scavengers. These sites collect waste from houses and small businesses that generate a continuous and large amount of animal by-products. In fact, several scavenger populations have increased their dependence on these resources resulting in behavioral adaptations that led to an increase in numbers and even distribution ranges (e.g. storks, gulls, raptors, corvids, rats, etc., Newsome et al. 2015; Plaza and Lambertucci 2017). The study of scavenging patterns and the effects generated at the demographic, behavioral and ecosystem levels has been very extensive (e.g. Sarrazin et al. 1994; Oro et al. 2008; Plaza and Lambertucci 2017; Tauler-Ametller et al. 2017). This has triggered the primary anthropogenic change in the energy balance of scavenger species, due to the relatively recent appearance and rapid increase in garbage dumps on our planet and at such a wide scale (Hobbs et al. 2009; Plaza and Lambertucci 2017).

The way in which carrion is presented in landfills varies. There are large garbage dumps with continuous and large inputs that receive waste from several towns and/or large cities. There are also dumps associated with small villages that receive less biomass. Regardless of the size, if carrion is accessible, scavengers congregate to feed, especially facultative and opportunistic species (Oro et al. 2013). The health risks associated with food ingestion in landfills are usually high due to the abundance of pathogens, especially bacteriological, in the available by-products (e. g. salmonella, Benskin et al. 2009). In addition, the ingestion of undigestible misidentified items (plastics mainly, Peris 2003), the great competition and high predation rates in the environment (Newsome et al. 2015; Plaza and Lambertucci 2017) and even the mortality of individuals due to crushing from garbage transported by vehicles are all risk factors associated. Nonetheless, these risks may be demographically compensated for by the high availability of food that allows increased survival rates and productivity (Donázar 1992).

Supplementary Feeding Programs

Supplementary feeding is a management tool for animal populations in which humans provide food to certain individuals and under specific conditions with the purpose of solving a conservation problem (Fielding et al. 2014; Cortés-Avizanda et al. 2016). In general, this method aims to reduce the risk of extinction of an endangered population in which the lack of food or factors associated with its acquisition (food quality, inter- and intraspecific competition, habitat alteration or human disturbances) is limiting (González et al. 2006; Cortés-Avizanda et al. 2010; Moreno-Opo et al. 2015a, b). In general, supplementary feeding programs are proposed to increase the values of parameters such as survival, generally of pre-adult life-cycle phases (Oro et al. 2008; Margalida 2010; García-Heras et al. 2013), productivity or breeding success (González et al. 2006; Margalida et al. 2017a).

Supplementary feeding programs are heterogeneous and depend on the purpose and the target species. As they address the main limiting factors in the life-cycle of animals, such as obtaining trophic resources, this approach must be implemented in a precise manner balancing its positive and negative effects (Donázar et al. 2009; Cortés-Avizanda et al. 2016). In the case of carrion, supplementary feeding mainly consists of the supply of remains in locations where the target species/population is present. Thus, feeding initiatives are developed for avian scavengers to facilitate carrion consumption in supplementary feeding sites ("vulture restaurants", Mundy et al. 1992; Donázar et al. 2009). This satisfies the demand of a high number of wild animals, although the most competitive species and individuals are favored and better adapted (Cortés-Avizanda et al. 2010; Moreno-Opo et al. 2015a). A variant of this model is the provision of certain carcass parts, in specific amounts, times and periods, and in particular habitats, which increases the probability of consumption by targeted individuals of specialized species in poorer conservation status (Moreno-Opo et al. 2015a, b). For the case of territorial species, such as large eagles or large carnivores, the programs are usually intended for breeding animals and their brood and require a greater effort of continued provision during a planned period (González et al. 2006). This practice usually leads to more positive results of population stability and a greater safety for animals, as well as an increase in the values of reproductive parameters (Davis et al. 2005; Robb et al. 2008).

This anthropogenic-mediated carrion source has ecological pros and cons. It is a very powerful conservation tool, which yields quick results at the population level, increasing population numbers (Robb et al. 2008). In addition, it can ensure a higher quality of the carrion supplied, concentrating the animals at safe points with respect to other feeding areas where illegal practices such as intentional poisoning can occur (Oro et al. 2008). It is also often used for purposes other than conservation, such as nature tourism, wildlife photography or hunting (Dubois and Fraser 2013; Fielding et al. 2014; Massé et al. 2014). However, in some cases, the programs are executed without a specific design, without assessing the ecological effects and without considering the prey selection patterns of the target species (Cortés-Avizanda et al. 2010; Moreno-Opo et al. 2015a).

Other Types of Human-Mediated Carrion

Human activity can result in the output of carrion for wildlife in several other ways. These other activities have little global repercussion in general although they may have important effects on populations. *Accidents* with wildlife caused by human traffic on roads or railways cause the death of hundreds of million animals annually (Rytwinski et al. 2016), considering only vertebrates. This huge number of kills generates an important biomass available as carrion (Lambertucci et al. 2009a, b). A large portion of the carcasses killed by vehicles are removed by maintenance services from the roads, although a portion remains available for scavenging both on the infrastructure itself and in nearby areas (Teixeira et al. 2013; see chapter "Carrion Availability in Space and Time"). The presence of this carrion along linear infrastructures can create an ecological trap, since the scavengers that come to feed on roads or railways can also be hit by cars and trains. There are cases, mainly on railways, of accidents involving a large number of vultures feeding on the carcasses of ungulates previously hit by the train (https://www.4vultures.org/2017/04/09/31-vultures-dead-after-being-hit-by-a-train-in-india/).

Electrocution and collision in power lines result in the death of about 10 to 41 millions of birds in Canada (Rioux et al. 2013) and 12 to 64 millions in the USA (Loss et al. 2014b). In addition, casualties due to electrocution generally have an aggregated distribution depending on the technical characteristics of the pylons, the landscape traits and the availability of food (Guil et al. 2011). These deaths also generate carrion-like biomass exploited by facultative scavengers that prospect along power lines, especially near those with higher mortality rates. In fact, it is estimated that only 30–84% of the carcasses of electrocuted birds are found in studies or surveys assessing mortality rates and wild scavengers are the main cause of their disappearance (Ponce et al. 2010; Schutgens et al. 2014). Other types of accidents with human infrastructures also generate carrion, such as the collision of migrating birds with buildings or other glazed infrastructures (Loss et al. 2014a). This favors the feeding of both domestic animals (cats) and opportunistic wild scavengers (corvids, rats, etc). Finally, drownings in stagnant water (wells, reservoirs, ponds, etc.), collisions in wind farms or the deaths of insects around artificial sources of light provide carrion as well, although in small amounts.

Humans cause *incidents* with wildlife that trigger the occurrence of carrion through activities such as shooting, trapping and poisoning. The legality or illegality of the action determines the subsequent effects on wild scavengers. A highlighted case is the supply of poisoned carrion for the purpose of killing vultures by ivory traffickers in certain regions of Africa to prevent the location of poached elephants and rhinos by the authorities, as large flocks of flying vultures around a carcass are observable from long distances (Ogada et al. 2016a). In addition to these purposeful incidents, other accidental ones can introduce poisoned carrion into trophic chains. Exploitation of carcasses from cattle treated with the toxic anti-inflammatory diclofenac in South Asia resulted in the collapse of vulture populations there (Green et al. 2004; Oaks et al. 2004). Poisoning scavengers through the offer of carrion is an illegal practice commonly occurring worldwide, mainly when these animals are perceived as competitors with hunting and agriculture interests (Hernández and Margalida 2008, 2009b; Margalida 2012; Mateo-Tomás et al. 2012; Berny et al. 2015) (see chapter "What Makes Carrion Unsafe for Scavengers? Considerations for Appropriate Regulatory Policies and Sound Management Practices" for more information on contaminated carrion).

Spatiotemporal Patterns of Appearance of Anthropogenic Carrion

Understanding how carrion of anthropic origin appears in the landscape is essential. Both biomass abundance and, above all, the location and times at which the carrion occurs determine the *predictability* of the resource (Deygout et al. 2010;

Cortés-Avizanda et al. 2012). This concept is relevant given that most species tend to optimize their search for food by reducing their associated energy costs, through saving foraging time and reducing competition (Robb et al. 2008; Kane et al. 2017). Therefore, an adaptive pressure towards predictable resources may act at individual or population levels, which a priori generates benefits in their fitness balanced with the difficulty of obtaining naturally limited trophic resources (Oro et al. 2013; Cortés-Avizanda et al. 2016). In addition, regardless of the gregarious (colonial) or territorial character, habitat quality determines food availability for scavengers so that areas with a greater abundance of resources are considered more favorable (Carrete and Donázar 2005; Kendall 2013). To assess the dynamics and effects of this resource in the ecology and behavior of wild scavengers, it is necessary to know the spatiotemporal patterns of occurrence of the different carrion subsidies.

In relation to livestock husbandry, extensive management resembles the natural supply conditions generated by wild ungulates. Carrion availability increases where livestock are concentrated, corresponding to the most favorable areas of palatable pasture (Fryxell 1991). This greater concentration of domestic animals attracts scavengers as the probability of finding carcasses is higher. In this sense, the geographic component is complemented with time aspects given that the management of the herds and their location depend on the season (Fernández-Giménez and Fillat 2012). Thus, there are migratory grazing practices, such as transhumance, that are common in temperate regions in which the extensively managed livestock is moved depending on the season (summers vs. winters), always in search of areas with an abundance of pastures (Ruiz and Ruiz 1986; Olea and Mateo-Tomás 2009; Margalida et al. 2018). In addition to animal abundance, the life cycle of the livestock is important. During the birthing season the supply is greater, both from the deaths of mothers and offspring and from the placentas remaining in the field (Breck et al. 2011; Margalida et al. 2014c). Meteorological factors are also influential and adverse events, such as storms or cold or heat waves, also increase livestock mortality (Hahn 1981; Nardone et al. 2010). Epizootics are important as well, and an increase in their prevalences in herds increases mortality rates. In brief, the scavenger species, and especially vultures, will frequent territories with a greater amount of livestock during periods in which the likelihood of finding carrion is higher; that is, the time of birthing (usually spring and, sometimes, autumn in sheep/goats) and in summer in areas of high mountain pastures (Nardone et al. 2010). With regard to intensively managed livestock, the way they are handled determines the predictability of carrion: in cases where the carcasses end up as carrion for wild scavengers, these become important places of attraction, generally of birds. Thus, farms with a large amount of livestock will produce a greater number of carcasses, so that the associated deposit points (vulture restaurants) will host higher numbers of scavengers on a regular basis (Cortés-Avizanda et al. 2010; Deygout et al. 2010). In these cases, food sources are predictable and the patterns of food searching and habitat selection are altered to a greater extent, thus resulting in demographic, behavioral and health consequences for necrophagous species (Robb et al. 2008; García-Heras et al. 2013; Oro et al. 2013).

Hunting for sporting purposes occurs in certain seasons according to existing regulations. In most regions, it is practiced outside the breeding season of the game species, maintaining the resource in a sustainable manner. However, selective hunting takes place almost year-round in many regions, since it targets certain individuals with a special morphological value (Di Minin et al. 2016). Thus, carrion from hunting can be available during most of the year. On the contrary, hunting in which more than one individual is pursued is usually performed during the winter, as is the case in Mediterranean environments (Delibes-Mateos et al. 2009). At this time, and especially in regions with large hunting estates, carrion becomes predictable providing a percentage of the diet that can reach 30% of the total biomass ingested annually in the case of the cinereous vulture *Aegypius monachus* (Costillo et al. 2007b). In fact, vultures forage nomadically around the zones where ungulate hunting is practiced. Small game hunting yields smaller amounts of biomass in the form of carrion per piece hunted, although its practice is much more widespread. In the cases in which the fallen animals are eviscerated in the field, the scavengers may take advantage of these resources, especially the facultative species (mesocarnivores, corvids and raptors other than vultures, Carrasco-García et al. 2018).

The availability of fishing discards is linked to the regulations on exploitation quotas and seasons established both throughout the year and according to the fishing grounds. Therefore, carrion provisions at sea are subject to fishing activity depending on the gear that generates such discards. For this reason, the areas and seasons overlapping with fisheries using trawls for benthic species and those intended for pelagic species with gillnets offer a greater availability of carrion (Garthe et al. 1996; Shester and Micheli 2011). These gears also capture the highest quantities of fish due to their massive and less selective nature, subsequently resulting in large amounts of discards (Kelleher 2005). In the case of deep-sea fishing, the disposal is located far from the coast, while in inshore fishing the discard waste coincides to a large extent with the return times and routes of the vessels to ports (Kelleher 2005).

In relation to other anthropogenic sources of carrion, supplementary feeding programs provide this food resource based on proposed objectives of conservation, tourism or hunting. Aiming at improving the conservation status of threatened species, carrion is provisioned at times when there is a greater energy demand, usually during breeding, or when a natural supply of food is scarcer (Robb et al. 2008; Fielding et al. 2014; Cortés-Avizanda et al. 2016). There is also some predictability with carrion occurrence from deaths caused by human infrastructures. For example, many accidents with wildlife may occur on roads crossing habitats rich in biodiversity such as wetlands, forests or wildlife corridors (Gunson et al. 2011). The rates of vehicular impacts with wildlife increase at times of greater movement of animals, either due to post-juvenile dispersion, mating season or migration (Neumann et al. 2012; Morelle et al. 2013). Hence, black kites *Milvus migrans* and red kites *Milvus milvus* have been observed prospecting roads at dawn with high traffic levels, to take advantage of the carcasses of small mammals, herpetes or macroinvertebrates killed during the night (personal observation). A similar situation takes place at power lines where electrocuted birds are concentrated; these are selected as foraging sites by facultative scavengers, usually mammals (Ponce et al. 2010; Schutgens et al. 2014).

Effects on Feeding Patterns and Viability of Scavenger Populations

Exploitation by Vertebrate Scavengers

Changes in the occurrence patterns of human-mediated carrion have triggered effects on the feeding patterns of scavenger species, varying according to the necrophagous character and the traits of the resources provided. In the case of obligate scavengers, namely the vultures of the New and Old World (Ruxton and Houston 2004), the species less specialized in the exploitation of certain parts of carcasses (Moreno-Opo et al. 2015a, 2016), have been favored by the increase in human-mediated carrion. Thus, the food sources of species that are more generalist in their prey selection, such as vultures of the genus *Gyps*, are augmented. As a result, there has been a population increase (Parra and Tellería 2004), in cases where there were no other limiting factors, such as the availability of suitable breeding sites or non-natural mortality (Oaks et al. 2004; Virani et al. 2011; Ogada et al. 2016b). Species selecting carrion parts that are scarcer have not shown this positive population response despite the net increase in food subsidies. This is the case of the Egyptian vulture *Neophron percnopterus*, hooded vulture *Necrosyrtes monachus* and the bearded vulture *Gypaetus barbatus*, which feed on small parts, offal and/or bones (Brown and Plug 1990; Margalida et al. 2007, 2009). Consequently, the effects of the increase of anthropogenic trophic resources have varied depending on the prey selection patterns derived from morphological and behavioral adaptations to the exploitation of certain parts of the carcasses, which generate niche partitioning (Hertel 1994; Moreno-Opo et al. 2016). However, some species of vultures exhibit plasticity in their diet allowing the exploitation of new or more abundant resources (Costillo et al. 2007a, b; Donázar et al. 2010). This adaptability of the diet is compromised by the evolutionary mechanisms of carrion exploitation, foraging behavior and social relationships (Margalida 2008; Moreno-Opo et al. 2015a, 2016), such that not all species have been able to adapt their feeding patterns to the resources offered by humans. At the European level, species such as the cinereous and the griffon vulture *Gyps fulvus* have shown a greater plasticity in their diet, varying their diet composition as a result of changes in the abundances of the different types of carrion. Other species with a greater heterogeneity in their diet, such as the Egyptian or bearded vulture have not shown this functional response (Donázar et al. 2010; Cortés-Avizanda et al. 2010; Moreno-Opo et al. 2015a, b).

In the case of facultative scavengers, which take advantage of carrion in a variable but never exclusive manner and that are generalists in their selection of existing carrion types, the increase in human subsidies has produced effects at the population, behavioral and community levels (Wilson and Wolkovich 2011; Oro et al. 2013). Given their opportunistic nature and plasticity (Hulme-Beaman et al. 2016), there has been an increase in food availability and feeding opportunities due to the decreasing competition for limiting resources such as live prey (DeVault et al. 2003; Wilson and Wolkovich 2011). Therefore, a general population increase of facultative scavengers in areas with a greater availability of human subsidies has been observed (Oro et al. 2013). In the same way, other species like insectivorous birds and reptiles include carrion and its associated resources (e.g., necrophagous arthropods concentrated around carcasses) as an important source of food (Barton et al. 2013b; Moreno-Opo and Margalida 2013). At the behavioral and community level, the wide trophic valence of some opportunistic species has allowed greater exploitation of a predictable resource. This plasticity has occurred in the case of terrestrial mammals, such as carnivores (felids, mustelids, bears and canids), ungulates (wild boar and of several deer species) and even rodents, favored mainly by the increase in the remains of hunted wild ungulates and domestic refuse (Gompper and Vanak 2008; Oro et al. 2013). In birds, the number of favored species is higher, especially among those exploiting carrion in landfills, like storks, raptors such as harriers, kites, buzzards, eagles, gulls and corvids, and in the marine environment, like seagulls and petrels (Oro et al. 2013; Plaza and Lambertucci 2017). Knowledge about other animal groups such as reptiles, fishes and invertebrates or species dwelling in pelagic and benthic marine habitats, is scarcer. In summary, the increase in subsidized carrion has led to an increase in its representation in the diet of facultative species, and behavioral changes derived from the search for, exploitation of and competition for carrion.

Effects of Human-Subsidized Carrion on Scavenger Activity and Behavior

The availability of natural prey has varied in different regions of the world, and thus so has searching behavior, the size of the foraging areas and the likelihood of obtaining carrion. The abundance of prey generating greater carrion biomass, such as ungulates, has been altered by human activity. Thus, in vast areas of Africa, Asia and South America a sharp decline in populations of large herbivores has occurred as a result of direct human persecution, either by poaching for meat consumption or for illegal trafficking (Brashares et al. 2004; Ripple et al. 2015). On the contrary, the large herbivores in Europe and North America have experienced a recovery in recent decades, especially due to an interest in hunting and the application of conservation programs (Côté et al. 2004; Massei et al. 2015). As a result, the populations of vertebrate scavengers are encountering variations in the availability of carcasses of the large herbivores they exploit and, subsequently, are adapting their foraging patterns (Carrete and Donázar 2005). Nevertheless, the feeding behaviour of African vultures is not mediated by ungulate and antelope abundance (Kendall et al. 2014) so the impact of the alteration of their availability is not well known (Ogada et al. 2016b).

Humans competed with wild scavengers for resources such as carrion prior to hunting and livestock husbandry (Domínguez-Rodrigo and Pickering 2003; Moleón et al. 2014). This has changed and humans have shifted from consumers to suppliers of carrion, and thus competition, and even predation, by humans has changed to a relationship of mutualism and commensalism with wild scavengers (Gangoso et al. 2013). Some animals are increasingly frequent visitors to urban areas and other human infrastructures, searching for food, and resulting in a greater tolerance for these species. This is a relatively new phenomenon in regions such as Europe, while in Africa and Asia this behavior has been common at least since the early twentieth century due to the presence of carrion in towns and cities (Thiollay 2006; Gangoso et al. 2013). In the same way, animals tolerate shorter distances to people when carrion is provided. For example, animals will approach humans around supplementary feeding points, during hunting activities or in the surroundings of fishing boats (Skov and Durinck 2001; Mateo-Tomás and Olea 2010; Zuberogoitia et al. 2010). This greater proximity to people can generate negative effects both for the conservation of the species and for human safety. Thus, wild scavengers attracted to supplementary feeding points can be poached or hunted more easily, as in the case of the grey wolf *Canis lupus* (Mateo-Tomás et al. 2012). This could also trigger the implementation of deterrents towards large carnivores, such as the brown bear *Ursus arctos* due to their habituation to villages with food resources (Bateman and Fleming 2012).

As human-subsidized carrion is used by a great number of scavengers, dependence on this resource has emerged in different animals. For certain populations, food acquisition is linked to measures directly implemented by humans, such as supplementary feeding programs, hunting events, inputs to landfills or the disposal of fishing remains (Oro et al. 2013; Plaza and Lambertucci 2017). This dependence has increased in recent decades as the abundance of carrion has grown, allowing animals to adapt to a new exploitable resource (López-López et al. 2013). However, this dependence on anthropogenic carrion makes scavengers more sensitive to sudden changes in availability. For example, in Europe in the early 2000s regulations led to the control and elimination of livestock by-products due to the mad-cow disease crisis, which resulted in the closure of many traditional sites where carrion was provisioned and the mandatory removal of carcasses of all livestock species (Donázar et al. 2009; Margalida et al. 2010). Another example of this sudden shift occurred in several Spanish regions, which legislated for the removal from the field of all remains of ungulate shot during hunting activities due to alleged health risks (Margalida and Moleón 2016). Also in Europe, human-mediated changes in the availability of carrion include the closure of rubbish dumps or their transformation into organic matter processing plants using incineration, in which carrion is not accessible to animals, or the prohibition of fishing discards by European Union fisheries policy, which aimed to eliminate this resource completely (Condie et al. 2014; Real et al. 2017). The implementation of different policies, be they sanitary, public health, economic or environmental, produce the elimination of predictable sources of food in a very short time and do not allow wildlife to adapt to new resources without some detrimental consequences. For instance, changes in breeding parameters have been demonstrated (Margalida et al. 2014a; Hernández et al. 2018), as well as effects on the physical condition and survival of scavenging birds (Margalida et al. 2014a), spatial reconfiguration of breeding and a subsequent concentration around spots with available carrion. The latter causes an increase in competition for the resource and density-dependent fecundity regulation phenomena (Carrete et al. 2006). Other types of prey are selected including live animals and consequently, social effects ensue due to increased predation and agonistic interactions with livestock or wildlife (Margalida et al. 2014c; Newsome et al. 2015). The provision of anthropogenic carrion has changed the movements performed by wild scavengers. The decrease or increase in the availability of carcasses as well as the occurrence of areas with an abundant presence of carrion, may change foraging patterns. On the one hand, in places where resources have been reduced, scavengers may search more extensive areas for a longer time, which could result in an alteration in their fitness (Stephens et al. 2007). However, prospecting vast territories and the collaboration between individuals in the detection of carrion are evolutionarily acquired behaviors within the vulture guild. These animals are thus better adapted to searching for and finding food in the case of diminishing carrion resources (Ruxton and Houston 2004; Jackson et al. 2008). On the other hand, the appearance of carrion in larger quantities and in predictable places is more common, and animals know a priori where they can find food more easily. This has also led to changes in foraging patterns, reducing both the distance covered and time dedicated to searching for food (López-López et al. 2013). In this sense, routine and prefixed movements between breeding, roosting and feeding places have increased, as a consequence of learning the times and locations of carrion occurrence (Oro et al. 2013; López-López et al. 2014; Cortés-Avizanda et al. 2016; Margalida et al. 2016; Plaza and Lambertucci 2017). This shortening of movements generates energy savings and a consequent improvement in survival and reproduction prospects, although there is a resulting concentration of birds, greater inter- and intra-specific interactions and more competition (Oro et al. 2013; Newsome et al. 2015).

Soil characteristics are also altered by the continued contribution of nutrients from carrion provided by humans (Barton et al. 2013a, b). The composition of soil in a given territory is a consequence of its own lithological, meteorological and geomorphological traits, as well as the cycles of organic inputs (Ganz et al. 2012). However, a continuous supply of nutrients, especially nitrogen and carbon, modifies the acidity of the soil, alters the productivity patterns and changes the plant species (Strickland and Wickings 2016). In addition, it can also alter the exchange cycles of elements with the atmosphere or groundwater (Barton et al. 2013a, b).

Areas with continuous supplies of carrion attract a large number of both obligate and facultative scavengers. Among the latter, there are species that behave as predators, such as raptors (e.g. eagles, harriers or kites) and carnivorous mammals. As a consequence of this attraction, the abundance of predators increases as does predation pressure on prey species (Yarnell et al. 2015). This occurs in the surroundings of supplementary feeding points where carrion is continuously provided, and has demographic impacts, especially in ground-nesting birds and lagomorphs (Cortés-Avizanda et al. 2009).

Population Prospects

The food resources provided by humans have a dual effect on the demographic dynamics. First, a greater availability of carrion has led to an increase in numbers of wild scavengers, both obligate and facultative (Oro et al. 2013). Thus, in regions where subsidies increased throughout the twentieth century and where no severe threats occurred, such as massive poisonings, contamination with veterinary drugs or the disappearance of natural food sources (Prakash et al. 2003; Margalida et al. 2014a; Ogada et al. 2016a, b), populations of scavengers grew or remained stable (Donázar et al. 2009; Deinet et al. 2013). The increase in carcasses from cattle and game animals in temperate zones has favored the conservation of species such as the griffon and cinereous vultures in Europe and the black *Coragyps atratus* and turkey vultures *Cathartes aura* in North America. This is also true of medium-sized raptors and corvids, and canids like the red fox *Vulpes vulpes*, grey wolf *Canis lupus*, jackals *Canis aureus* and coyotes *Canis latrans* (Prugh et al. 2009; Ogada et al. 2012; Chapron et al. 2014). In the case of arthropods, positive effects have also been documented in some scarce necrophagous species, with supplementary feeding points serving as essential localities for their conservation (Martín-Vega and Baz 2011). Storks, gulls and even raptors, such as vultures and kites, also take advantage of the carrion deposited in garbage dumps, which has led to increases in their populations (Pons 1992; Tortosa et al. 2002). In the marine environment, the growing amounts of fishing discards are directly related to the recovery of threatened species, such as the Audouin's gull and the Balearic shearwater *Puffinus mauretanicus* (Oro et al. 1996; Bicknell et al. 2013).

Conversely, the exploitation of anthropogenic carrion can constitute an ecological trap with adverse demographic effects. Changes in the availability of subsidies, may revert population parameters. The interruption of the food supply can lead to a decrease in breeding success, fecundity or survival (Bicknell et al. 2013; Margalida et al. 2010, 2014b; Plaza and Lambertucci 2017). Thus, the almost exclusive dependence on anthropic resources by scavengers can lead to unexpected demographic changes. The quality of the carrion provided is also key to the survival of scavengers. Illegal or irresponsible actions have resulted in poisoning events impacting scavenger populations. The quasi-extinction of vultures in India due to the use of the anti-inflammatory veterinary drug Diclofenac is an example (Oaks et al. 2004; Green et al. 2004), as is the poaching of vultures in eastern and southern Africa (Ogada et al. 2016a, b) or lead poisoning in vultures due to the consumption of meat contaminated with lead ammunition from hunting activities (Fisher et al. 2006; Gangoso et al. 2009; Hernández and Margalida 2009a). Social conflicts due to complaints related to the predation on living livestock by vultures and other scavengers have emerged in countries with important numbers of necrophagous birds, such as Spain and France (Margalida et al. 2011a, b, 2014b). This predation has been exacerbated by food shortages from sanitary management policies in the European Union (Donázar et al. 2009; Margalida et al. 2010; Lagos and Barcenas 2015). Demands for compensation and a growing animosity toward scavengers have been triggered by changes in their feeding patterns. They may approach livestock for the purpose of consuming placentas or predating on dying, diseased or physically impeded animals (Margalida et al. 2011a, b, 2014b). An increase in these behaviors in vultures can then change the perception of these species as public health allies in the natural environment (Margalida et al. 2014c; Donázar et al. 2016). As a result, increased persecution toward these animals increase their risk of extinction, as occurs in Africa and Asia (Ogada et al. 2012, 2016a, b; Buechley and Şekercioğlu 2016).

Key Changes in Ecological Processes at Different Scales

The human-subsidized carrion changes the functioning of food webs, consequently triggering effects on ecological processes at the individual, population and ecosystem levels (Table 1; Oro et al. 2013). Given that food is a limiting factor in the wild that largely determines the fitness of individuals, the transition of this resource from being an object of competition to being available *ad libitum* impacts the life traits of scavenger species. Not all effects apply to all systems where carrion appears, nor to all populations or individuals. Nonetheless, these effects have been demonstrated and are therefore shown below.

Table 1 Effects of human-subsidized carrion on ecological processes at individual, population and ecosystem and community levels

Type of alteration	Effect on ecological process and/or display
Individual level	
Increase in body mass	Scavengers accessing more abundant human-mediated resources can display higher body sizes than those for whom food is a limiting factor (Yom-Tov 2003; Oro et al. 2013). Supplementary feeding programs for raptors provisioned with carrion reduces sibling aggression allowing a faster growth of the chicks in the nest due to an optimized gain of body mass (González et al. 2006). Moreover, the physical condition of adults also improves if food is abundant (Votier et al. 2010)
Increase in fecundity	The territories where carrion occurs in greater amounts have a higher quality and favor a better physical condition for successful reproduction. This has been shown in seabird populations that have been exposed to non-limiting carrion subsidies that later suddenly disappeared (Pons and Migot 1995)
Survival	Predictable carrion allows food to be obtained without competing with conspecifics and heterospecifics. It also mitigates the need for searching over wide areas. In this way, large carnivores or gregarious scavengers have increased their life expectancy (Oro et al. 2008)
Physiological changes	The less effort needed in the discovery and competition for food, the greater the energy savings. Predictable carrion contributes to the investment of energy in vital activities other than searching and fighting for food. In individuals benefitting from plentiful carrion, breeding success is increased and the display of sex traits is enhanced (Oro et al. 2013)
Dispersal alteration	The presence of predictable food sources reduces the size of dispersive movements in pre-adult ages (Margalida et al. 2013). The visitation of locations with carrion diminishes the risks of scanning unknown territories, which can be potentially hazardous due to poisoning, poaching, etc. This can contribute to reducing mortality rates (Oro et al. 2008). The concentration of immature individuals around predictable sources of carrion is higher than that of adults; this affects the selection of breeding areas, which also tend to be closer to these sources of food (Margalida et al. 2011a, b)
Changes in space use	Foraging areas in habitats that are little transformed by humans are being replaced by agriculture or semi-urban landscapes where finding carrion is more likely. The selection of feeding habitats has changed towards the inclusion of anthropic elements, both at a landscape level and related to human activity (Margalida et al. 2016, 2017b)

Table 1 (continued)

Type of alteration	Effect on ecological process and/or display
Population level	
Population size	Scavenger populations have increased as a result of the greater availability of food resources (Mundy et al. 1992; Donázar et al. 2009), especially in areas where subsidies are generated abundantly. The anthropogenic carrion favors a better physical condition of the individuals prior to reproduction, increasing fecundity, allows better development of the offspring and increases survival (Oro et al. 2008; Margalida et al. 2014a)
Population fluctuations and resilience	Species with short-life cycles (R-strategists) whose populations fluctuate according to changes in the availability of resources, in the prevalence of epizootics or in meteorology, have seen their population dynamics altered due to the absence of food shortage periods. Thus, exploitation of carrion by rodents, ungulates, passerines or fishes reduces periodic population fluctuations allowing greater resilience to stochastic environmental scenarios (Oro et al. 2013)
Demographic structure	The supply of subsidized carrion modifies the age ratios of avian scavenger populations. The population pyramid is modified, increasing the representation of pre-adult age classes (Oro et al. 2008; Margalida et al. 2011a, b)
Population density and concentration of breeding nuclei	In areas where carrion is abundant the density of scavengers increases (Parra and Tellería 2004). There are changes in emigration-immigration rates and an attraction to these areas from other patches that are not subsidized (Martínez-Abraín et al. 2012; Monsarrat et al. 2013). This provokes the packing of territories, a greater frequency of inter- and intraspecific competition and density-dependent phenomena (Carrete et al. 2006; Margalida et al. 2009)
Foraging behavior	This change is especially relevant in vultures, who are evolutionarily adapted to large prospective movements and visual communication over long distances (Jackson et al. 2008; Cortés-Avizanda et al. 2014). The foraging movements have decreased in length and the routes have been fixed in advance towards places where anthropogenic carrion appears
Migratory behavior	As one of the main drivers of migration is the limitation of trophic resources, the abundance of food throughout the annual cycle has reduced the size of the migratory movements and has even caused populations to stop migrating and remain close to breeding areas (Pulido 2007). In other cases, migrants relocate their wintering destinations to specific areas with subsidized carrion
Diet	The contribution of anthropogenic carrion to the environment has modulated the diet of different necrophagous species, making this biomass more represented (Donázar et al. 2010)
Ecosystem and community levels	
Competition	A higher availability of carrion relaxes the competition during its exploitation since a greater number of individuals can take advantage of it, especially pre-adult individuals (Margalida et al. 2011a, b, 2013; Moreno-Opo et al. 2015a, 2016). However, the hierarchies in its exploitation continue operating and the most dominant individuals obtain the food (Cortés-Avizanda et al. 2010; Moreno-Opo et al. 2016). The subsidized carrion can increase competition in other fields: aggressive behaviors for breeding territories close to predictable food and a subsequent decrease in the attention investment towards other vital activities and the reduction of habitat quality (Carrete et al. 2006; Robb et al. 2008)

(continued)

Table 1 (continued)

Type of alteration	Effect on ecological process and/or display
Predator-prey relationships	When scavengers have predatory ability, an increase in their population could generate hyperpredation on prey species in the surroundings of landfills or vulture restaurants (Cooper and Ginnett 2000; Cortés-Avizanda et al. 2009). The population increase of facultative necrophages also has consequences on other species that become prey when carrion begins to dwindle: some gulls that take advantage of fishing discards later prey on pulli of other seabirds in mixed colonies (Oro et al. 2013) or the increase of opportunistic mesocarnivores due to the high availability of carrion could exacerbate predation rates in threatened prey species (Moreno-Opo et al. 2015a, b)
Relationships within food-webs	Omnivorous or facultative scavenger species would be favored with respect to specialized species and those with less ecological plasticity as well as with regards to primary consumers, increasing their abundance and influence on the relationships at different levels of food webs (Margalef 1997)
Species composition	The homogenization of the characteristics of carrion occurring in the wild favors the most dominant and generalist scavenger species, outcompeting other more specialist species. This decreases species diversity at a local scale in carrion exploitation events (Cortés-Avizanda et al. 2012). Carrion may serve as an important resource for some alien species increasing their chances of establishment and their conversion into invasives (Benbow et al. 2016)
Habitat characteristics	The existence of an energetic element as powerful as carrion diminishes the diversity of behaviors and relationships between living beings, skewing towards the superabundance of certain animals with respect to others. This has long-term effects on the population dynamics, due to the variation in intensity and frequency in the individual animal-plant and animal–animal interactions (Bascompte et al. 2006). The simplification in habitats where subsidized carrion appears is produced either by reinforcement or relaxation of these interactions depending on their affinity for carrion or specialization with respect to other life strategies (Benbow et al. 2016)
Nutrient transfer	When carrion is abundant, alterations in soil conditions occur at a local scale due to the increase of different elements (nitrogen, potassium, sodium and sulfur) from both the carrion and the animals that are concentrated for its exploitation (Barton et al. 2013a, b). In addition to pH modification, bacterial and fungal activity increases due to an increase in nutrient availability (Leff et al. 2015)
Human-wildlife conflicts	Anthropogenic carrion can trigger an increase in the interactions of wildlife with livestock, agriculture or even human security. These interactions derive mainly from the concentration of animals around specific points (dumps and/or vulture restaurants) that could interact with human uses or enhance the perception of increased predations on livestock (Margalida et al. 2011a, b, 2014a; Donázar et al. 2016; Morales-Reyes et al. 2018)

Conclusions and Future Perspectives

Patterns of carrion occurrence are being modified at a global scale and in a short-period of time, shifting from a pulsed random resource to a more predictable food source depending on human activities. As a result, wild scavengers are adapting their behavioural and feeding ecology to the different types of human-mediated carrion subsidies: livestock carcasses, hunting remains, fishing discards and rubbish dumps are the main sources of non-natural carrion exploited both, by obligate and facultative scavenger species, especially in the most industrialized countries. Otherwise, the decline of wildlife populations as well as the implementation of some sanitary regulations are reducing the availability of natural and human-mediated carrion resources, respectively, worldwide. All these alterations are provoking different effects on scavengers at different levels.

From an individual standpoint, the biological fitness of scavengers is affected by the increase or decrease of human-mediated carrion, including changes in parameters such as fertility, survival, and space use. To these individual effects, population implications have been observed, including demographic (changes in population size and density-dependence phenomena), behavioural (intra- and interspecific competition and migration), and distribution (modulation of distribution range) modifications. At the community level, anthropogenic carrion subsidies have also produced alterations, especially in food web performance, in predator-prey relationships, and other related issues like species richness and even in characteristics of the landscape.

The influence of human activities on carrion-related ecological processes requires a further greater research effort. It is also important to deepen this analysis from a conservation biology perspective. In this sense, analyzing the ecological requirements of the different scavengers and their communities should be prioritized. Once this knowledge is thoroughly adquired, it is necessary to avoid the negative impacts from the anthropogenization on the occurrence of an essential resource as carrion for the most specialized and sensitive species. Accordingly, management measures determining carrion availability should be adapted to natural scenarios in which the carrying capacity of the environment is mediated by ecological pressures, being influenced as little as possible by human activity. This should be a conservation target for policy-makers and managers with the purpose of ensuring a key ecosystem service provided by scavengers and, in parallel, the monitoring and conservation needs of threatened populations.

Acknowledgements We are indebted to several colleagues involved in the projects performed to assess the role of avian scavengers and carrion consumption: L. M. González, R. Casanovas, D. García, J. Caldera, A. Aranda, A. Trujillano, J. J. García, A. Arredondo, R. Higuero. We also thank the editors for the opportunity in taking part of this book and the reviews of drafts of our manuscript. This research was supported by project CGL2015-66966-C2-2-R2.

References

Allen B (2016) Animals in religion: devotion, symbol and ritual. Reaktion Books, London

Barton PS, Cunningham SA, Lindenmayer DB, Manning AD (2013a) The role of carrion in maintaining biodiversity and ecological processes in terrestrial ecosystems. Oecologia 171(4):761–772

Barton PS, Cunningham SA, Macdonald BC et al (2013b) Species traits predict assemblage dynamics at ephemeral resource patches created by carrion. PLoS One 8(1):e53961

Bascompte J, Jordano P, Olesen JM (2006) Asymmetric coevolutionary networks facilitate biodiversity maintenance. Science 312:431–433

Bateman PW, Fleming PA (2012) Big city life: carnivores in urban environments. J Zool 287(1):1–23

Bellan SE, Turnbull PCB, Beyer W, Getz WM (2013) Effects of experimental exclusion of scavengers from carcasses of anthrax-infected herbivores on *Bacillus anthracis* sporulation, survival, and distribution. Appl Environ Microbiol 79:3756–3761

Benbow ME, Tomberlin JK, Tarone AM (2016) Carrion ecology, evolution and their applications. CRC Press, Taylor and Francis Group, Boca Raton

Benskin CM, Wilson K, Jones K, Hartley IR (2009) Bacterial pathogens in wild birds: a review of the frequency and effects of infection. Biol Rev 84(3):349–373

Berny P, Vilagines L, Cugnasse JM et al (2015) Vigilance Poison: illegal poisoning and lead intoxication are the main factors affecting avian scavenger survival in the Pyrenees (France). Ecotoxicol Environ Saf 118:71–82

Bicknell AW, Oro D, Camphuysen KC, Votier SC (2013) Potential consequences of discard reform for seabird communities. J Appl Ecol 50(3):649–658

Bignal EM, McCracken DI (2000) The nature conservation value of European traditional farming systems. Environ Rev 8:149–171

Blázquez M, Sánchez-Zapata JA (2009) The role of wild ungulates as a resource for the community of vertebrate scavengers. In: Donázar JA, Margalida A, Campión D (eds) Vultures, feeding stations and sanitary legislation: a conflict and its consequences from the perspective of conservation biology, Munibe 29 (Suppl). Sociedad de Ciencias Aranzadi San Sebastián, Spain, pp 308–327

Blázquez M, Sánchez-Zapata JA, Botella F, Carrete M, Eguía S (2009) Spatio-temporal segregation of facultative avian scavengers at ungulate carcasses. Acta Oecol 35:645–650

Blench R (2000) Extensive pastoral livestock systems: issues and options for the future. FAO-Japan Cooperative Project "Collection of Information on Animal Production and Health". FAO, Rome

Bodey TW, Jessopp MJ, Votier SC et al (2014) Seabird movement reveals the ecological footprint of fishing vessels. Curr Biol 24:R514–R515

Bozzano A, Sarda F (2002) Fishery discard consumption rate and scavenging activity in the northwestern Mediterranean Sea. ICES J Mar Sci 59(1):15–28

Brashares JS, Arcese P, Sam MK et al (2004) Bushmeat hunting, wildlife declines, and fish supply in West Africa. Science 306(5699):1180–1183

Breck SW, Kluever BM, Panasci M et al (2011) Domestic calf mortality and producer detection rates in the Mexican wolf recovery area: implications for livestock management and carnivore compensation schemes. Biol Conserv 144(2):930–936

Brown CJ, Plug I (1990) Food choice and diet of the bearded vulture *Gypaetus barbatus* in southern Africa. S Afr J Zool 25(3):169–177

Buechley ER, Şekercioğlu ÇH (2016) The avian scavenger crisis: looming extinctions, trophic cascades, and loss of critical ecosystem functions. Biol Conserv 198:220–228

Carrasco-García R, Barroso P, Montoro V et al (2018) Consumption of big game remains by scavengers: a potential risk as regards disease transmission in Central Spain. Front Vet Sci 5:4

Carrete M, Donázar JA (2005) Application of central-place foraging theory shows the importance of Mediterranean dehesas for the conservation of the cinereous vulture, *Aegypius monachus*. Biol Conserv 126(4):582–590

Carrete M, Donázar JA, Margalida A (2006) Density-dependent productivity depression in pyrenean bearded vultures: implications for conservation. Ecol Appl 16(5):1674–1682

Casas-Díaz E, Cristòfol C, Cuenca R et al (2016) Determination of fluoroquinolone antibiotic residues in the plasma of Eurasian griffon vultures (*Gyps fulvus*) in Spain. Sci Total Environ 557:620–626

Catchpole TL, Elliott S, Peach D, Mangi SC, Gray TS (2017) How to deal with the EU landing obligation: lessons from an English discard ban sea trial. ICES J Mar Sci 75(1):270–278

Chamberlain CP, Waldbauer JR, Fox-Dobbs K et al (2005) Pleistocene to recent dietary shifts in California condors. Proc Natl Acad Sci U S A 102:16707–16711

Chapron G, Kaczensky P, Linnell J et al (2014) Recovery of large carnivores in Europe's modern human-dominated landscapes. Science 346:1517–1519

Coleman FC, Williams SL (2002) Overexploiting marine ecosystem engineers: potential consequences for biodiversity. Trends Ecol Evol 17(1):40–44

Condie HM, Grant A, Catchpole TL (2014) Incentivising selective fishing under a policy to ban discards; lessons from European and global fisheries. Mar Policy 45:287–292

Cooper SM, Ginnett TF (2000) Potential effects of supplemental feeding of deer on nest predation. Wildl Soc Bull 28:660–666

Cortés-Avizanda A, Carrete M, Serrano D, Donázar JA (2009) Carcasses increase the probability of predation of ground-nesting birds: a caveat regarding the conservation value of vulture restaurants. Anim Conserv 12(1):85–88

Cortés-Avizanda A, Carrete M, Donázar JA (2010) Managing supplementary feeding for avian scavengers: guidelines for optimal design using ecological criteria. Biol Conserv 143(7):1707–1715

Cortés-Avizanda A, Jovani R, Carrete M, Donázar JA (2012) Resource unpredictability promotes species diversity and coexistence in an avian scavenger guild: a field experiment. Ecology 93(12):2570–2579

Cortés-Avizanda A, Jovani R, Donázar JA, Grimm V (2014) Bird sky networks: how do avian scavengers use social information to find carrion. Ecology 95:1799–1808

Cortés-Avizanda A, Blanco G, DeVault TL et al (2016) Supplementary feeding and endangered avian scavengers: benefits, caveats, and controversies. Front Ecol Environ 14(4):191–199

Costillo E, Corbacho C, Morán R, Villegas A (2007a) Diet plasticity of Cinereous Vulture *Aegypius monachus* in different colonies in the Extremadura (SW Spain). Ardea 95(2):201–211

Costillo E, Corbacho C, Morán R, Villegas A (2007b) The diet of the black vulture Aegypius monachus in response to environmental changes in Extremadura (1970–2000). Ardeola 54:197–204

Côté SD, Rooney TP, Tremblay JP et al (2004) Ecological impacts of deer overabundance. Annu Rev Ecol Evol Syst 35:113–147

Darimont CT, Fox CH, Bryan HM, Reimchen TE (2015) The unique ecology of human predators. Science 349(6250):858–860

Darimont CT, Codding BF, Hawkes K (2017) Why men trophy hunt. Biol Lett 13:20160909

Davies RWD, Cripps SJ, Nickson A, Porter G (2009) Defining and estimating global marine fisheries bycatch. Mar Policy 33(4):661–672

Davis SE, Nager RG, Furness RW (2005) Food availability affects adult survival as well as breeding success of parasitic jaegers. Ecology 86(4):1047–1056

Deinet S, Ieronymidou C, McRae L et al (2013) Wildlife comeback in Europe. The recovery of selected mammal and bird species. ZSL, BirdLife International and the European Bird Census Council, London

Delibes-Mateos M, Farfán MA, Olivero J et al (2009) Long-term changes in game species over a long period of transformation in the Iberian Mediterranean landscape. Environ Manag 43:1256–1268

DeVault TL, Rhodes OE Jr, Shivik JA (2003) Scavenging by vertebrates: behavioral, ecological, and evolutionary perspectives on an important energy transfer pathway in terrestrial ecosystems. Oikos 102(2):225–234

Deygout C, Gault A, Duriez O et al (2010) Impact of food predictability on social facilitation by foraging scavengers. Behav Ecol 21(6):1131–1139

Di Minin E, Leader-Williams N, Bradshaw CJ (2016) Banning trophy hunting will exacerbate biodiversity loss. Trends Ecol Evol 31(2):99–102

Domínguez-Rodrigo M (2001) A study of carnivore competition in riparian and open habitats of modern savannas and its implications for hominid behavioral modelling. J Hum Evol 40:77–98

Domínguez-Rodrigo M, Pickering TR (2003) Early hominid hunting and scavenging: a zooarcheological review. Evol Anthropol 12(6):275–282

Donázar JA (1992) Muladares y basureros en la biología y conservación de las aves en España. Ardeola 39(2):29–40

Donázar JA (1993) Los buitres ibéricos: biología y conservación. JM Reyero, Madrid

Donázar JA, Fernández C (1990) Population trends of Griffon Vultures (*Gyps fulvus*) in northern Spain between 1969 and 1989 in relation to conservation measures. Biol Conserv 53:83–91

Donázar JA, Naveso MA, Tella JL, Campión D (1997) Extensive grazing and raptors in Spain. In: Pain D, Pienkowski M (eds) Farming and birds in Europe: the common agricultural policy and it's implications for bird conservation. Academic, London, pp 117–149

Donázar JA, Margalida A, Campión D (2009) Vultures, feedings stations and sanitary legislation: a conflict and its consequences from the perspective of conservation biology, Munibe 29 (Suppl). Sociedad de Ciencias Aranzadi San Sebastián, Spain

Donázar JA, Cortés-Avizanda A, Carrete M (2010) Dietary shifts in two vultures after the demise of supplementary feeding stations: consequences of the EU sanitary legislation. Eur J Wildl Res 56(4):613–621

Donázar JA, Cortés-Avizanda A, Fargallo JA et al (2016) Roles of raptors in a changing world: from flagships to providers of key ecosystem services. Ardeola 63(1):181–234

Dubois S, Fraser D (2013) A framework to evaluate wildlife feeding in research, wildlife management, tourism and recreation. Animals 3:978–994

FAO (2016) The State of World Fisheries and Aquaculture 2016. Contributing to food security and nutrition for all. Food and Agriculture Organization-United Nations, Rome

Fernández-Giménez ME, Fillat F (2012) Pyrenean pastoralists' ecological knowledge: documentation and application to natural resource management and adaptation. Hum Ecol 40(2):287–300

Fielding D, Newey S, van der Wal R, Irvine RJ (2014) Carcass provisioning to support scavengers: evaluating a controversial nature conservation practice. Ambio 43(6):810–819

Fisher IJ, Pain DJ, Thomas VG (2006) A review of lead poisoning from ammunition sources in terrestrial birds. Biol Conserv 131(3):421–432

Flanders F, Gillespie JR (2015) Modern livestock and poultry production. Cengage Learning, Boston

Fryxell JM (1991) Forage quality and aggregation by large herbivores. Am Nat 138(2):478–498

Gangoso L, Álvarez-Lloret P, Rodríguez-Navarro AA et al (2009) Long-term effects of lead poisoning on bone mineralization in vultures exposed to ammunition sources. Environ Pollut 157(2):569–574

Gangoso L, Agudo R, Anadón JD et al (2013) Reinventing mutualism between humans and wild fauna: insights from vultures as ecosystem services providers. Conserv Lett 6(3):172–179

Ganz HH, Karaoz U, Getz WM et al (2012) Diversity and structure of soil bacterial communities associated with vultures in an African savanna. Ecosphere 3(6):47

García-Heras MS, Cortés-Avizanda A, Donázar JA (2013) Who are we feeding? Asymmetric individual use of surplus food resources in an insular population of the endangered Egyptian vulture *Neophron percnopterus*. PLoS One 8(11):e80523

García-Tarrasón M, Bécares J, Bateman S et al (2015) Sex-specific foraging behavior in response to fishing activities in a threatened seabird. Ecol Evol 5:2348–2358

Garthe S, Camphuysen K, Furness RW (1996) Amounts of discards by commercial fisheries and their significance as food for seabirds in the North Sea. Mar Ecol Progr Ser 136:1–11

Gavashelishvili A, McGrady MJ (2006) Breeding site selection by bearded vulture (*Gypaetus barbatus*) and Eurasian griffon (*Gyps fulvus*) in the Caucasus. Anim Conserv 9:159–170

Gompper ME, Vanak AT (2008) Subsidized predators, landscapes of fear and disarticulated carnivore communities. Anim Conserv 11(1):13–14

González LM, Margalida A, Sánchez R, Oria J (2006) Supplementary feeding as an effective tool for improving breeding success in the Spanish imperial eagle (Aquila adalberti). Biol Conserv 129(4):477–486

Gortazar C, Diez-Delgado I, Barasona JA et al (2015) The wild side of disease control at the wildlife-livestock-human interface: a review. Front Vet Sci 1:27

Green RE, Newton IAN, Shultz S et al (2004) Diclofenac poisoning as a cause of vulture population declines across the Indian subcontinent. J Appl Ecol 41:793–800

Guil F, Fernández-Olalla M, Moreno-Opo R et al (2011) Minimising mortality in endangered raptors due to power lines: the importance of spatial aggregation to optimize the application of mitigation measures. PLoS One 6(11):e28212

Gunson KE, Mountrakis G, Quackenbush LJ (2011) Spatial wildlife-vehicle collision models: a review of current work and its application to transportation mitigation projects. J Environ Manag 92(4):1074–1082
Hahn GL (1981) Housing and management to reduce climatic impacts on livestock. J Anim Sci 52:175–186
Heath MR, Cook RM, Cameron AI et al (2014) Cascading ecological effects of eliminating fishery discards. Nature 5:3893
Hernández M, Margalida A (2008) Pesticide abuse in Europe: effects on the Cinereous vulture (*Aegypius monachus*) population in Spain. Ecotoxicology 17(4):264–272
Hernández M, Margalida A (2009a) Assessing the risk of lead exposure for the conservation of the endangered Pyrenean bearded vulture (*Gypaetus barbatus*) population. Environ Res 109(7):837–842
Hernández M, Margalida A (2009b) Poison-related mortality effects in the endangered Egyptian vulture (*Neophron percnopterus*) population in Spain. Eur J Wildl Res 55(4):415–423
Hernández M, Colomer MA, Pizarro M, Margalida A (2018) Changes in eggshell thickness and ultrastructure in the Bearded Vulture (*Gypaetus barbatus*): a long-term analysis. Sci Total Environ 624:713–721
Herrero M, Havlík P, Valin H et al (2013) Biomass use, production, feed efficiencies, and greenhouse gas emissions from global livestock systems. Proc Natl Acad Sci U S A 110(52): 20888–20893
Hertel F (1994) Diversity in body size and feeding morphology within past and present vulture assemblages. Ecology 75(4):1074–1084
Hobbs RJ, Higgs E, Harris JA (2009) Novel ecosystems: implications for conservation and restoration. Trends Ecol Evol 24(11):599–605
Houston DC, Cooper JE (1975) The digestive tract of the white-back griffon vulture and its role in disease transmission among wild ungulates. J Wildl Dis 11:306–313
Hulme-Beaman A, Dobney K, Cucchi T, Searle JB (2016) An ecological and evolutionary framework for commensalism in anthropogenic environments. Trends Ecol Evol 31(8):633–645
Jackson AL, Ruxton GD, Houston DC (2008) The effect of social facilitation on foraging success in vultures: a modelling study. Biol Lett 4(3):311–313
Jayathilakan K, Sultana K, Radhakrishna K, Bawa AS (2012) Utilization of byproducts and waste materials from meat, poultry and fish processing industries: a review. J Food Sci Technol 49:278–293
Kane A, Healy K, Guillerme T et al (2017) A recipe for scavenging in vertebrates–the natural history of a behaviour. Ecography 40(2):324–334
Kelleher K (2005) Discards in the world's marine fisheries. An update. FAO Fisheries Technical Paper. No. 470. FAO, Rome
Kelly TR, Grantham J, George D et al (2014) Spatiotemporal patterns and risk factors for lead exposure in endangered California condors during 15 years of reintroduction. Conserv Biol 28:1721–1730
Kendall CJ (2013) Alternative strategies in avian scavengers: how subordinate species foil the despotic distribution. Behav Ecol Sociobiol 67(3):383–393
Kendall CJ, Virani MZ, Hopcraft JGC et al (2014) African vultures don't follow migratory herds: scavenger habitat use is not mediated by prey abundance. PLoS One 9(1):e83470
Lagos L, Bárcena F (2015) EU sanitary regulation on livestock disposal: implications for the diet of wolves. Environ Manag 56(4):890–902
Lambertucci SA, Speziale KL, Rogers TE, Morales JM (2009a) How do roads affect the habitat use of an assemblage of scavenging raptors? Biodivers Conserv 18(8):2063–2074
Lambertucci SA, Trejo A, Di Martino S et al (2009b) Spatial and temporal patterns in the diet of the Andean condor: ecological replacement of native fauna by exotic species. Anim Conserv 12(4):338–345
Leff JW, Jones SE, Prober SM et al (2015) Consistent responses of soil microbial communities to elevated nutrient inputs in grasslands across the globe. Proc Natl Acad Sci U S A 112:10967–10972

Lindsey P, Roulet P, Romanach S (2007) Economic and conservation significance of the trophy hunting industry in sub-Saharan Africa. Biol Conserv 134:455–469

López-López P, Benavent-Coral J, García-Ripollés C, Urios V (2013) Scavengers on the move: behavioural changes in foraging search patterns during the annual cycle. PLoS One 8(1):e54352

López-López P, García-Ripollés C, Urios V (2014) Food predictability determines space use of endangered vultures: implications for management of supplementary feeding. Ecol Appl 24(5):938–949

Loss SR, Will T, Loss SS, Marra PP (2014a) Bird–building collisions in the United States: estimates of annual mortality and species vulnerability. Condor 116(1):8–23

Loss SR, Will T, Marra PP (2014b) Refining estimates of bird collision and electrocution mortality at power lines in the United States. PLoS One 9(7):e101565

Lucas RE (2003) The industrial revolution. Federal Reserve Bank of Minneapolis

Margalef R (1997) Our biosphere. Ecology Institute, Oldendorf/Luhe

Margalida A (2008) Bearded vultures (*Gypaetus barbatus*) prefer fatty bones. Behav Ecol Sociobiol 63(2):187–193

Margalida A (2010) Supplementary feeding during the chick-rearing period is ineffective in increasing the breeding success in the bearded vulture (*Gypaetus barbatus*). Eur J Wildl Res 56(4):673–678

Margalida A (2012) Baits, budget cuts: a deadly mix. Science 338(6104):192–192

Margalida A, Colomer MA (2012) Modelling the effects of sanitary policies on European vulture conservation. Sci Rep 2:753

Margalida A, Moleón M (2016) Toward carrion-free ecosystems? Front Ecol Environ 14(4):183–184

Margalida A, Mañosa S, Bertran J, Garcia D (2007) Biases in studying the diet of the Bearded Vulture. J Wildl Manag 71(5):1621–1625

Margalida A, Bertran J, Heredia R (2009) Diet and food preferences of the endangered Bearded Vulture *Gypaetus barbatus*: a basis for their conservation. Ibis 151(2):235–243

Margalida A, Donázar JA, Carrete M, Sánchez-Zapata JA (2010) Sanitary versus environmental policies: fitting together two pieces of the puzzle of European vulture conservation. J Appl Ecol 47(4):931–935

Margalida A, Campión D, Donázar JA (2011a) European vultures' altered behaviour. Nature 480:457

Margalida A, Colomer MA, Sanuy D (2011b) Can wild ungulate carcasses provide enough biomass to maintain avian scavenger populations? An empirical assessment using a bio-inspired computational model. PLoS One 6(5):e20248

Margalida A, Carrete M, Hegglin D et al (2013) Uneven large-scale movement patterns in wild and reintroduced pre-adult bearded vultures: conservation implications. PLoS One 8(6):e65857

Margalida A, Bogliani G, Bowden CG et al (2014a) One Health approach to use of veterinary pharmaceuticals. Science 346(6215):1296–1298

Margalida A, Colomer MA, Oro D (2014b) Man-induced activities modify demographic parameters in a long-lived species: effects of poisoning and health policies. Ecol Appl 24(3):436–444

Margalida A, Campión D, Donázar JA (2014c) Vultures *vs.* livestock: conservation relationships in an emerging conflict between humans and wildlife. Oryx 48(02):172–176

Margalida A, Pérez-García JM, Afonso I, Moreno-Opo R (2016) Spatial and temporal movements in Pyrenean bearded vultures (*Gypaetus barbatus*): integrating movement ecology into conservation practice. Sci Rep 6:35746

Margalida A, Martínez JM, Gómez de Segura A et al (2017a) Supplementary feeding and young extraction from the wild are not a sensible alternative to captive breeding for reintroducing bearded vultures *Gypaetus barbatus*. J Appl Ecol 54:334–340

Margalida A, Pérez-García JM, Moreno-Opo R (2017b) European policies on livestock carcasses management did not modify the foraging behavior of a threatened vulture. Ecol Indic 80:66–83

Margalida A, Oliva-Vidal P, Llamas A, Colomer, MA (2018) Bioinspired models for assessing the importance of transhumance and transboundary management in the conservation of European avian scavengers. Biol Conserv 228:321–330

Martínez-Abraín A, Tavecchia G, Regan HM et al (2012) Effects of wind farms and food scarcity on a large scavenging bird species following an epidemic of bovine spongiform encephalopathy. J Appl Ecol 49:109–117
Martín-Vega D, Baz A (2011) Could the 'vulture restaurants' be a lifeboat for the recently rediscovered bone-skippers (*Diptera*: *Piophilidae*)? J Insect Conserv 15(5):747
Massé S, Dussault C, Dussault C, Ibarzabal J (2014) How artificial feeding for tourism-watching modifies black bear space use and habitat selection. J Wildl Manag 78:1228–1238
Massei G, Kindberg J, Licoppe A et al (2015) Wild boar populations up, numbers of hunters down? A review of trends and implications for Europe. Pest Manag Sci 71(4):492–500
Mateo-Tomás P, Olea PP (2010) When hunting benefits raptors: a case study of game species and vultures. Eur J Wildl Res 56:519–528
Mateo-Tomás P, Olea PP, Sánchez-Barbudo IS, Mateo R (2012) Alleviating human–wildlife conflicts: identifying the causes and mapping the risk of illegal poisoning of wild fauna. J Appl Ecol 49:376–385
Mateo-Tomás P, Olea PP, Moleón M et al (2015) From regional to global patterns in vertebrate scavenger communities subsidized by big game hunting. Divers Distrib 21:913–924
Mateo-Tomás P, Olea PP, Jiménez-Moreno M et al (2016) Mapping the spatio-temporal risk of lead exposure in apex species for more effective mitigation. Proc R Soc Lond B Biol Sci 283:20160662
Mithen S (1999) The hunter-gatherer prehistory of human-animal interactions. Anthrozoös 12:195–204
Moleón M, Sánchez-Zapata JA, Margalida A et al (2014) Humans and scavengers: the evolution of interactions and ecosystem services. Bioscience 64:394–403
Monsarrat S, Benhamou S, Sarrazin F et al (2013) How predictability of feeding patches affects home range and foraging habitat selection in avian social scavengers? PLoS One 8:e53077
Morales-Reyes Z, Martín-López B, Moleón M et al (2018) Farmer perceptions of the ecosystem services provided by scavengers: what, who, and to whom. Conserv Lett 11(2):e12392
Morelle K, Lehaire F, Lejeune P (2013) Spatio-temporal patterns of wildlife-vehicle collisions in a region with a high-density road network. Nat Conserv 5:53–73
Moreno-Opo R, Margalida A (2013) Carcasses provide resources not exclusively to scavengers: patterns of carrion exploitation by passerine birds. Ecosphere 4(8):1–15
Moreno-Opo R, Margalida A, Garcia F et al (2012) Linking sanitary and ecological requirements in the management of avian scavengers: effectiveness of fencing against mammals in supplementary feeding sites. Biodivers Conserv 21:1673–1685
Moreno-Opo R, Trujillano A, Arredondo A et al (2015a) Manipulating size, amount and appearance of food inputs to optimize supplementary feeding programs for European vultures. Biol Conserv 181:27–35
Moreno-Opo R, Trujillano A, Margalida A (2015b) Optimization of supplementary feeding programs for European vultures depends on environmental and management factors. Ecosphere 6(7):1–15
Moreno-Opo R, Trujillano A, Margalida A (2016) Behavioral coexistence and feeding efficiency drive niche partitioning in European avian scavengers. Behav Ecol 27:1041–1052
Mundy P, Butchart D, Ledger J, Piper S (1992) The vultures of Africa. Academic, London
Nardone A, Ronchi B, Lacetera N et al (2010) Effects of climate changes on animal production and sustainability of livestock systems. Livest Sci 130:57–69
Neumann W, Ericsson G, Dettki H et al (2012) Difference in spatiotemporal patterns of wildlife road-crossings and wildlife-vehicle collisions. Biol Conserv 145(1):70–78
Newsome TM, Dellinger JA, Pavey CR et al (2015) The ecological effects of providing resource subsidies to predators. Glob Ecol Biogeogr 24(1):1–11
Oaks JL, Gilbert M, Virani MZ et al (2004) Diclofenac residues as the cause of vulture population decline in Pakistan. Nature 427(6975):630–633
Ogada DL, Keesing F, Virani MZ (2012) Dropping dead: causes and consequences of vulture population declines worldwide. Ann N Y Acad Sci 1249(1):57–71

Ogada D, Botha A, Shaw P (2016a) Ivory poachers and poison: drivers of Africa's declining vulture populations. Oryx 50(4):593–596
Ogada D, Shaw P, Beyers RL et al (2016b) Another continental vulture crisis: Africa's vultures collapsing toward extinction. Conserv Lett 9(2):89–97
Olea PP, Mateo-Tomás P (2009) The role of traditional farming practices in ecosystem conservation: the case of transhumance and vultures. Biol Conserv 142(8):1844–1853
Oro D, Genovart X, Ruiz X et al (1996) Differences in diet, population size and reproductive performance between two colonies of Audouin's Gull *Larus audouinii* affected by a trawling moratorium. J Avian Biol 27(3):245–251
Oro D, Margalida A, Carrete M et al (2008) Testing the goodness of supplementary feeding to enhance population viability in an endangered vulture. PLoS One 3(12):e4084
Oro D, Genovart M, Tavecchia G et al (2013) Ecological and evolutionary implications of food subsidies from humans. Ecol Lett 16(12):1501–1514
Parra J, Tellería JL (2004) The increase in the Spanish population of Griffon Vulture *Gyps fulvus* during 1989–1999: effects of food and nest site availability. Bird Conserv Int 14(1):33–41
Peris J (2003) Feeding in urban refuse dumps: ingestion of plastic objects by the White Stork (*Ciconia ciconia*). Ardeola 50:81–84
Plaza PI, Lambertucci SA (2017) How are garbage dumps impacting vertebrate demography, health, and conservation? Glob Ecol Conserv 12:9–20
Ponce C, Alonso JC, Argandoña G et al (2010) Carcass removal by scavengers and search accuracy affect bird mortality estimates at power lines. Anim Conserv 13(6):603–612
Pons JM (1992) Effects of changes in the availability of human refuse on breeding parameters in a herring gull. Ardea 80:143–150
Pons JM, Migot P (1995) Life-history strategy of the herring gull: changes in survival and fecundity in a population subjected to various feeding conditions. J Anim Ecol 64:592–599
Prakash V, Pain DJ, Cunningham AA et al (2003) Catastrophic collapse of Indian white-backed *Gyps bengalensis* and long-billed *Gyps indicus* vulture populations. Biol Conserv 109(3):381–390
Prugh LR, Stoner CJ, Epps CW et al (2009) The rise of the mesopredator. Bioscience 59(9):779–791
Pulido F (2007) The genetics and evolution of avian migration. Bioscience 57:165–174
Real E, Oro D, Martínez-Abraín A et al (2017) Predictable anthropogenic food subsidies, density-dependence and socio-economic factors influence breeding investment in a generalist seabird. J Avian Biol 48(11):1462–1470
Rioux S, Savard JPL, Gerick AA (2013) Avian mortalities due to transmission line collisions: a review of current estimates and field methods with an emphasis on applications to the Canadian electric network. Avian Conserv Ecol 8(2):7
Ripple WJ, Newsome TM, Wolf C et al (2015) Collapse of the world's largest herbivores. Sci Adv 1:e140010
Robb GN, McDonald RA, Chamberlain DE, Bearhop S (2008) Food for thought: supplementary feeding as a driver of ecological change in avian populations. Front Ecol Environ 6:476–484
Ruiz M, Ruiz JP (1986) Ecological history of transhumance in Spain. Biol Conserv 37:73–86
Ruxton GD, Houston DC (2004) Obligate vertebrate scavengers must be large soaring fliers. J Theor Biol 228(3):431–436
Rytwinski T, Soanes K, Jaeger JA et al (2016) How effective is road mitigation at reducing road-kill? A meta-analysis. PLoS One 11(11):e0166941
Sardà F, Coll M, Heymans JJ, Stergiou KI (2015) Overlooked impacts and challenges of the new European discard ban. Fish Fish 16:175–180
Sarrazin F, Lecomte J (2016) Evolution in the Anthropocene. Science 351(6276):922–923
Sarrazin F, Bagnolin C, Pinna JL et al (1994) High survival estimates of griffon vultures (*Gyps fulvus fulvus*) in a reintroduced population. Auk 111(4):853–862
Schutgens M, Shaw JM, Ryan PG (2014) Estimating scavenger and search bias for collision fatality surveys of large birds on power lines in the Karoo, South Africa. Ostrich 85(1):39–45
Selva N, Fortuna MA (2007) The nested structure of a scavenger community. Proc R Soc Lond B Biol Sci 274(1613):1101–1108

Sesé J A, Antor RJ, Alcántara M, Ascaso JC, Gil JA (2005) La alimentación suplementaria en el quebrantahuesos: estudio de un comedero del Pirineo occidental aragonés. Biología de la conservación del Quebrantahuesos *Gypaetus barbatus* en España. Organismo Autónomo Parques Nacionales, Madrid, pp 279–304

Shester GG, Micheli F (2011) Conservation challenges for small-scale fisheries: bycatch and habitat impacts of traps and gillnets. Biol Conserv 144(5):1673–1681

Sigurðardóttir S, Stefánsdóttir EK, Condie H et al (2015) How can discards in European fisheries be mitigated? Strengths, weaknesses, opportunities and threats of potential mitigation methods. Mar Policy 51:366–374

Skov H, Durinck J (2001) Seabird attraction to fishing vessels is a local process. Mar Ecol Progr Ser 214:289–298

Stara K, Sidiropoulos L, Tsiakiris R (2016) Bound eagles, evil vultures and cuckoo horses. Preserving the bio-cultural diversity of carrion eating birds. Hum Ecol 44(6):751–764

Stephens DW, Brown JS, Ydenberg RC (eds) (2007) Foraging: behavior and ecology. University of Chicago Press, Chicago

Strickland MS, Wickings K (2016) Carrion effects on belowgrounds communities and consequences for soil processes. In: Benbow ME, Tomberlin JK, Tarone AM (eds) Carrion ecology, evolution and their applications. CRC Press, Boca Ratón

Tauler-Ametller H, Hernández-Matías A, Pretus JL, Real J (2017) Landfills determine the distribution of an expanding breeding population of the endangered Egyptian Vulture *Neophron percnopterus*. Ibis 159:757–768

Teixeira FZ, Coelho AVP, Esperandio IB, Kindel A (2013) Vertebrate road mortality estimates: effects of sampling methods and carcass removal. Biol Conserv 157:317–323

Thiollay J (2006) The decline of raptors in West Africa: long-term assessment and the role of protected areas. Ibis 148(2):240–254

Thoma VG (2013) Lead-free hunting rifle ammunition: product availability, price, effectiveness, and role in global wildlife conservation. Ambio 42(6):737–745

Tortosa FS, Caballero JM, Reyes-López J (2002) Effect of rubbish dumps on breeding success in the white stork in southern Spain. Waterbirds 25(1):39–43

Vicente J, Carrasco R, Acevedo P, Montoro V, Gortazar C (2011) Big game waste production: sanitary and ecological implications. In: Integrated waste management—volume II. IntechOpen, Rijeka

Virani MZ, Kendall C, Njoroge P, Thomsett S (2011) Major declines in the abundance of vultures and other scavenging raptors in and around the Masai Mara ecosystem, Kenya. Biol Conserv 144:746–752

Votier SC, Furness RW, Bearhop S et al (2004) Changes in fisheries discard rates and seabird communities. Nature 427(6976):727–730

Votier SC, Bearhop S, Witt MJ et al (2010) Individual responses of seabirds to commercial fisheries revealed using GPS tracking, stable isotopes and vessel monitoring systems. J Appl Ecol 47:487–497

Wilmers CC, Stahler DR, Crabtree RL, Smith DW, Getz WM (2003) Resource dispersion and consumer dominance: scavenging at wolf- and hunter-killed carcasses in Greater Yellowstone, USA. Ecol Lett 11:993–1006

Wilson EE, Wolkovich EM (2011) Scavenging: how carnivores and carrion structure communities. Trends Ecol Evol 26(3):129–135

Yarnell RW, Phipps WL, Dell S et al (2015) Evidence that vulture restaurants increase the local abundance of mammalian carnivores in South Africa. Afr J Ecol 53:287–294

Yom-Tov Y (2003) Body sizes of carnivores commensal with humans have increased over the past 50 years. Funct Ecol 17:323–327

Zorrilla I, Martinez R, Taggart MA, Richards N (2015) Suspected flunixin poisoning of a wild Eurasian Griffon Vulture from Spain. Conserv Biol 29(2):587–592

Zuberogoitia I, Martínez JE, Margalida A et al (2010) Reduced food availability induces behavioural changes in Griffon Vulture *Gyps fulvus*. Ornis Fennica 87(2):52

What Makes Carrion Unsafe for Scavengers? Considerations for Appropriate Regulatory Policies and Sound Management Practices

Darcy Ogada, Ngaio Richards, and Shannon Behmke

Contents

D. Ogada (✉)
The Peregrine Fund, Boise, ID, USA

National Museums of Kenya, Nairobi, Kenya
e-mail: ogada.darcy@peregrinefund.org

N. Richards
Working Dogs for Conservation (WD4C), Bozeman, MT, USA
e-mail: ngaio@wd4c.org

S. Behmke
Division of Forestry and Natural Resources, West Virginia University, Morgantown, WV, USA
e-mail: sbehmke@mix.wvu.edu

P. P. Olea et al. (eds.), *Carrion Ecology and Management*, Wildlife Research Monographs 2, https://doi.org/10.1007/978-3-030-16501-7_9

Introduction

At present, carrion is managed almost exclusively for the benefit of people, i.e., to minimize potential repercussions to human health, with little thought afforded to the important role this resource plays in supporting an array of the earth's biodiversity (or, for that matter, its contribution to ecosystem nutrient cycling) (chapter "Ecological Functions of Vertebrate Scavenging"). As mentioned in previous chapters, carrion is a critically important food resource for vultures, and scavenging mammals such as foxes, jackals, and hyenas (see chapter "Vertebrate Scavenging Communities"). It also serves as the breeding and foraging grounds for a host of invertebrate species (chapter "Invertebrate Scavenging Communities") whose micro-scavenging ecosystem services are still not commonly recognized and, as such, remain largely undervalued. Due to its unsavory nature, carrion is above all regarded as something that must be disposed of quickly. And while vultures are nature's most highly evolved scavengers, with an ability to rapidly strip a carcass bare—often in a matter of minutes (Ruxton and Houston 2004), the increasing interference in the scavenger food web through a host of anthropogenic activities has correspondingly severely comprised the availability and quality of carrion. In stark contrast to the pre-Anthropocene era, it is now rare that carrion from wild animals can be found on the landscape anywhere in the world, except for that killed along roads or shot by hunters (chapter "Human-Mediated Carrion: Effects on Ecological Processes"). As populations of scavengers likely collapsed in response to this decrease in wild carrion, some species simultaneously evolved to feed on the deceased domestic animals that have followed in man's footsteps across the planet (Oro et al. 2013).

It is from this co-evolution of scavengers and domestic animals that the decreasing quality of carrion comes into focus as its management moves away from traditional systems and shifts increasingly, and sometimes controversially, to the center of policy debates regarding its disposal (Margalida et al. 2010).

Although dwindling in scale, the practice of transhumance (i.e., the seasonal movement of people and their livestock) is a traditional form of livestock management that in turn provides carrion for an array of scavenging species (Olea and Mateo-Tomás 2009). In some areas of the world, notably India, religion exerts a powerful influence on the management of, and even arguably perceptions towards, carrion. The sacred cows of Hindu culture have ensured an unending supply of carrion, especially for vultures. Here, the resultant carcass dumps provided an abundant, steady and geographically static food source that artificially inflated vulture populations (Taggart et al. 2015) (see chapter "Vertebrate Scavenging Communities", Box 1). Similarly, in southern Europe, feeding stations synonymous with vulture restaurants, have for decades operated so that farmers could cheaply and efficiently dispose of their domestic carrion (Dupont et al. 2012). Across the world, there has been a long-standing symbiosis whereby scavengers consume the gut piles and other remains left behind by hunters (Mateo-Tomás et al. 2015; chapter "Human-Mediated Carrion: Effects on Ecological Processes").

Traditional management of carrion has relied on the ecosystem services provided by scavengers, especially vultures. However, alongside this, the development of pharmaceutical drugs and pesticides, with the concurrent rise and evolution in their usage, and the long-standing presence and availability of lead in the environment, have all introduced a wide range of toxic, although largely invisible, threats that have devastated many populations of carrion feeders (particularly vultures) throughout the world. An increased human demand for meat and meat-based products has led to a corresponding rise in livestock rearing (Margalida et al. 2014a), which should theoretically benefit scavengers. Instead, the increased use of pharmaceutical drugs to treat livestock has made scavengers more vulnerable than ever (Taggart et al. 2015).

In this chapter we will review historic practices related to carrion management and how these have impacted more recent management practices and perceptions, as well as scavengers themselves. Drawing from our own perspectives and experiences in field research, wildlife biology and ecotoxicology, we will outline ways in which current aspects of carrion management and the carrion now offered to scavengers, especially vultures, is either inherently unsafe or carries a high potential risk, largely due to widespread assumptions made about its safety and insufficient safety verification practices. We will start by discussing how the widespread use of non-steroidal anti-inflammatory drugs (NSAIDs) in livestock has altered the traditional management of carrion in South East Asia. We will evaluate how human-predator conflict in Africa and Europe shapes the mismanagement of carrion, leading to large casualties due to non-targeted poisoning. Similarly, we will discuss the myriad cases of secondary poisoning related to agricultural use of pesticides and anticoagulant rodenticides. The impacts that arise from lead poisoning in hunter-killed carcasses, mainly for vultures, and the implications of lead contamination for ecosystems, will also be discussed.

More recently, a series of disease outbreaks in both domestic and wild animal populations have thrust the issue of carrion management into sharper focus (see chapter "Human-Mediated Carrion: Effects on Ecological Processes"), often to the detriment of scavengers. Outbreaks of bovine spongiform encephalopathy (BSE) or 'mad cow disease' and chronic wasting disease have raised concerns over the possibility of further transmission from infected carcasses and, ultimately, as we underlined at the beginning of this section, the potential repercussions to human health (see chapter "The Role of Scavenging in Disease Dynamics"). In the case of BSE these concerns have resulted in a series of regulatory actions for the disposal of carrion in the European Union (EU). In light of this, sanitary policies that restrict carrion management and limit its provision to scavengers will also be considered. Throughout this chapter, we reflect not only upon the role but on the effectiveness of legislation in the ongoing management of carrion as well. We also discuss the pros and cons of provisioning food at feeding stations for improving the survival of threatened scavengers. A number of innovative approaches being taken to reduce the contamination of carcasses and carrion, thereby ensuring their safety for scavengers and the surrounding ecosystems, are also presented, in tandem with a general 'forensic approach' with emphasis on feeding habits and dietary predilections. Wherever possible, areas where we believe more vigilance and attention is required for safe, effective carrion management are highlighted.

Widespread Contamination of Carrion with NSAIDs (and other Veterinary Agents) and Ecosystem-Level Effects in Asia: with Perspectives from Africa, North America and Europe

The threat posed by the NSAID diclofenac to Old World vultures on the Indian subcontinent (a.k.a. the 'Asian Vulture Crisis') has been thoroughly established (see, for example Taggart et al. 2015; Richards et al. 2018; Zorrilla et al. 2015). There, three species of *Gyps* vultures, namely the: Oriental White-backed Vulture *Gyps bengalensis*, Long-billed Vulture *G. indicus*, and Slender-billed Vulture *G. tenuirostris*, once among the most populous on earth, were nearly brought to extinction following exposure to residues of the drug in dead livestock left at carcass dumps expressly for vultures to dispose of (Prakash 1999; Gilbert et al. 2002; Oaks et al. 2004). The plummet in the number of vultures and the unprecedented speed at which it came about—in less than a decade—precipitated an influx of vectors such as feral dogs, which, to compound matters, cannot rival the proficiency and speed of vultures in disposing of a carcass (Pain et al. 2003; Ogada et al. 2012). Many of these animals are also aggressive towards people, with an increase in bite incidents and subsequent rabies transmission rates being of concern (Markandya et al. 2008). Tanners and bone collectors—who benefitted from the thoroughness with which vultures clean a carcass and its hide—saw their livelihoods adversely affected too. True, the vulture population on the Indian subcontinent was artificially inflated due to the presence of the carcass dumps (Taggart et al. 2015). But the existence of the dumps, the rapid cleanup provided by the vultures (and their deterrence of feral dogs) as well as the facilitation of the tanning and bone collection livelihood constitutes an interesting (and, unfortunately all too rare) example of symbiosis, where humans and wildlife co-exist closely in a largely urbanized environment without being in conflict.

Further investigation since then has revealed that additional NSAIDs (e.g., ketoprofen, ibuprofen, flunixin) are also likely harmful—if not outright toxic—to Old World vultures and other avian scavengers (Cuthbert et al. 2007, 2015). The necessity of determining presence, identifying exposure sources and considering potential repercussions of other veterinary agents like antimicrobials/antibiotics and euthanasia drugs also continues to be highlighted (e.g., by Taggart et al. 2015; Wells et al. 2019). In 2016, a movement began to ban the NSAID aceclofenac, which metabolizes in vivo to diclofenac in cattle, and as such is considered to pose an equal danger (Galligan et al. 2016), the toxicity and threat of aceclofenac having first been raised in 2012 (by Sharma 2012). Several studies (including that of Casas-Díaz et al. 2016) have demonstrated that (Eurasian Griffon) Vultures are indeed accessing antibiotic residues in livestock carcasses provided at feeding stations, and that such exposure—at concentrations deemed too low to be toxic, but that could lead to antibiotic resistance given frequent exposure—is widespread.

Nonetheless, there remains a troubling discord with actual implementation of tangible steps to safeguard beneficial scavengers from contaminated carrion or in some cases, even with an acknowledgement that such threats do indeed exist. In

2014, diclofenac was registered for veterinary use in Spain, despite that country hosting over 95% of Europe's vulture population, an important number of Red Kites *Milvus milvus*, and the entire population of the threatened Spanish Imperial Eagle *Aquila adalberti* (Margalida et al. 2014b). The registration was achieved under the premise that the conditions, which prevailed on the Indian subcontinent (i.e., intensive use of the drug in livestock and an extensive network of carcass dumps) are not applicable in a European context.

In western Europe, for example in Spain, veterinary drug withdrawal periods prior to slaughter are not imposed for carcasses provided to scavengers, as is required with those slated for human consumption (Casas-Díaz et al. 2016). Indeed, the presence/absence of veterinary residues in carcasses offered to vultures at feeding stations and in animals that are extensively reared (i.e., outside) still remains largely unclear and the same holds true for carcass meat provided at captive breeding facilities (Richards et al. 2014; Zorrilla et al. 2018). As yet, no government body can provide reliable statistics on the actual incidences of animal exposures (Ruiz-Suárez et al. 2015).

The availability of residues of other harmful agents like the euthanasia drug sodium pentobarbital in livestock and domestic animal carcasses, which have already claimed the lives of many carrion feeders, must also be thoroughly investigated. Numerous studies—largely based in North America (e.g. Hayes 1988; Krueger and Krueger 2002; Langelier 1993; O'Rourke 2002) have shown both that sodium pentobarbital residues are highly persistent, and they have reflected that the management of euthanized animal carcasses is neither universally regulated nor transparent at the moment—an issue highlighted in Taggart et al. (2015). Payne et al. (2015) examined decomposition of euthanized horses composted in wood chips and detected residues of sodium pentobarbital over a year (367 days) later. A review conducted by Wells et al. (2019) uncovered more than 400 related animal poisonings—including of Griffon vulture and eagles—and of which 75% of cases were not part of the published literature.

Another increasingly well studied area that must now receive broader consideration within management and regulatory circles is the potential repercussion of such residues to micro-scavengers, in terms of disrupting the various life cycle stages, further halting the decomposition of carcasses and encouraging the propagation of disease. As such, these findings must also be assimilated and embraced by the macro-scavenger research community for their complementarity. Indeed, human forensic research has shown that residues of recreational drugs (e.g., heroin and methamphetamine) in a cadaver can slow larval development, which in turn affects the rate of decomposition (not to mention the determination of the time of death; Verma and Paul 2013). An ample breadth of research has examined the potential for veterinary agent residues (e.g., endectocides, antimicrobials) in the dung of livestock to be harmful to coprophagous beetles (e.g., Lumaret and Errouissi 2002; Floate 2006; Tixier et al. 2015; Verdú et al. 2015). A comprehensive risk assessment to evaluate the toxicity of residues in cattle dung was recently developed (Floate et al. 2015). We note that such a risk assessment would also be relevant for coprophagous (and opportunistic) scavengers like the Egyptian Vulture *Neophron percnop-*

terus. Even if coprophagy is not their primary dietary predilection, and though it is not strictly tied to their consumption of carrion, their population status is such that any threat must be considered and explored (Richards et al. 2014), especially in light of how it could undermine those regulations and management practices enacted to protect the species. To our knowledge, no studies have yet examined which veterinary agents are available to scavenging insects when they feed upon the carcasses of previously medicated livestock, and whether such exposure impacts the (currently undervalued) nutrient cycling ecosystem service they provide.

In summary, researchers and managers should consider the likelihood that, broadly speaking:

a. Residues of NSAIDs and other veterinary agents are more prevalent in the carcasses of livestock animals (see, for example, Casas-Díaz et al. 2016; Wells et al. 2019) than is current widely accepted;
b. Vultures therefore face ongoing exposure to sub-lethal (if not lethal—see Zorrilla et al. 2015) levels of veterinary agents.

We believe that the practice of deliberately lacing livestock carcasses with acutely toxic agents—often organophosphorus or carbamate pesticides—to kill predators, is a contributing factor to the current dearth in information regarding presence/absence of veterinary agent residues in carcasses, and to a lack of routine monitoring for evidence of chronic exposure in vultures (see Richards and Ogada 2014). Following an incident of vulture/scavenger mortality in which a poisoned carcass has been implicated, and where residues have been detected in deceased wildlife, poisoning is determined to have caused death, and no additional inquiry is carried out. However, it would also be appropriate to opportunistically carry the investigation one step further, to verify the presence of relevant veterinary agents both in the implicated livestock carcass(es) and in the deceased animals that fed upon them. This is a critical piece of information that is not currently being recorded or factored in to the bigger carcass management and safety assessment picture.

The Role of Human-Predator Conflict in the Illicit Use of Carrion-Based Poison Baits in Africa and Europe

Whether scavengers are attracted by smell or strictly by sight (e.g., Old World vultures), carrion is an irresistible lure for both wild and domestic carnivores (e.g., lions, hyenas, domestic dogs and cats). It is impossible to overstate the gravity of the use of carrion-based poisoned baits to eliminate predators on a global scale. While there are a number of motives for using carrion to attract or to bait carnivores, the overwhelming one is malicious (Hernández and Margalida 2008; Giorgi and Mengozzi 2011; Richards 2012). However, baits are also used (mostly legally) by national wildlife authorities to control wildlife populations (e.g., Lewis 1968; Jarvis and La Grange 1982; Linhart et al. 1993; Whitfield et al. 2003; Olea et al. 2009).

Indeed, carrion baits have been used to administer contraceptives or vaccinations, and as part of sanctioned predator elimination measures (Linhart et al. 1993). While many countries worldwide formerly instituted poisoning campaigns to eliminate predators (Leopold et al. 1964; Martínez-Haro et al. 2008; Ogada 2014), the current legal usages of poison baits against predators are largely confined to island nations seeking to control or eradicate those that have been introduced (e.g., Alterio 2000; Glen et al. 2007). However, the bulk of legal, sanctioned use of poison baits does not fail to prevent secondary poisoning of predators, most notably mustelids and raptors (Kaukeinen 1982; Berny et al. 1997; Gillies and Pierce 1999; Eason et al. 2002; Brakes and Smith 2005; Olea et al. 2009). Household pets are also at risk (Meenken and Booth 1997). Increasingly, however, the use of specially trained conservation detection dogs to pinpoint the whereabouts of the predator species of interest may considerably reduce secondary and unintentional poisonings, by helping managers confine eradication efforts only to those specific areas (see for e.g. http://www.doc.govt.nz/our-work/conservation-dog-programme/).

The predominant use of poisoned baits stems from conflicts between humans and predators, and this widescale and illegal use occurs in every conceivable part of the world. The use of poison baits represents the principal non-natural factor in the extinction of several European vertebrate megafauna over the previous two centuries, and it can be considered the most significant cause of non-natural mortality of wildlife in Spain, one of Europe's main vulture strongholds (Margalida 2012; Margalida et al. 2014c; Mateo-Tomás et al. 2012). Poisoned baits targeting predators also affect many other species, most notably obligate scavengers. In the following sections, we discuss the implications of the illegal use of carrion-based poison baits on populations of predators and scavengers in Africa and Europe and the related implications for carrion management.

Africa

As in many other parts of the world, the use of poisons for lethal predator control to eliminate mammals such as lions, hyenas, wild dogs, and jackals has also been prevalent in a large number of African countries (Ogada 2014). Most of these campaigns, instigated by national wildlife authorities and often lasting for decades, involved baiting carrion with strychnine (Ogada 2014). Consequently, populations of predators and scavengers, particularly long-lived and slowly reproducing species, such as vultures, have been devastated. Jackson (1938) wrote of the demise of Rüppell's Vultures *G. rueppellii* in the Kenyan highlands due to strychnine poisoning during the early 1900s. While most government-led poisoning campaigns have since ended, using poisons to control predators and other animals perceived to be 'damage-causing' has become an almost reflexive practice entrenched in many African countries such that today wildlife poisoning remains one of the most serious threats to all predators and scavengers, in addition to a host of other 'problem' animals e.g. monkeys, baboons, seed-eating birds (Fig. 1; Ogada 2014).

Fig. 1 Intentionally poisoned spotted hyena *Crocuta crocuta* whose carcass subsequently poisoned a critically endangered White-headed Vulture *Trigonoceps occipitalis* and a Bateleur *Terathopius ecaudatus* in the Masai Mara, Kenya. Photo by H. Courant

The use of baited carrion to poison lions, hyenas and other large mammalian predators has had substantial impacts on their populations throughout Africa. Lions are particularly vulnerable to baited carrion because they readily scavenge, and whole prides can be wiped out at a single baited carcass (Frank 2011). Already lions are extinct in North Africa, and only a few relic populations currently exist in West and Central Africa, while in East Africa lion populations have been devastated by widespread poisoning. Three species of African hyena (spotted; *Crocuta crocuta*, brown; *Hyaena brunnea*, and striped; *H. hyaena*) are the biggest target of poisoning campaigns (Ogada 2014). Their opportunism and, hence, their indiscriminate food choices, make them particularly vulnerable and their extinction across large parts of their former African range has doubtless arisen due to poisoning (Fig. 1; Mills and Hofer 1998; St John et al. 2011). Wild dogs *Lycaon pictus*, leopards *Panthera pardus*, jackals (*Canis* sp.) and caracals *Caracal caracal* also join the list of unfortunate mammalian predators that have been widely persecuted in Africa through the use of baited carcasses (Ogada 2014).

Populations of African raptors, and particularly those that scavenge as a major portion of their diet, have been hard hit by the indiscriminate use of baited carrion, most notably Tawny Eagles *Aquila rapax*, Bateleurs *Terathopius ecaudatus*, and vultures. There can be no denying the massive impact of poisoned carcasses on populations of African vultures. The vast majority of vulture poisonings have resulted from livestock farmers baiting carrion or meat pieces in order to eliminate mammalian predators (Ogada et al. 2016a). However, some vultures, particularly Lappet-faced Vultures *Torgos tracheliotos* in southern Africa, have been intentionally poisoned by livestock farmers who allege that the birds kill newborn lambs

(Mundy et al. 1992; Bridgeford 2001). More recently, poachers have taken to lacing elephant carcasses with pesticides to deliberately kill vultures whose overhead circling gives away the location of their illicit activities (Ogada et al. 2016b).

Beginning as early as the 1900s, entire populations of some species of vulture were notably declining due to the indiscriminate use of poisons to kill mammalian predators (Jackson 1938; Boshoff and Vernon 1980). By mid-century, lethal predator control campaigns were well underway across the continent (Thévenot et al. 1985, 2003; Toure 1988; Cherkaoui 2005; Chardonnet et al. 2005) with devastating impacts for vultures. During a survey of wildlife in Zaire (Democratic Republic of Congo) lions were poisoned at baited carcasses, which also caused significant declines of vultures in several parts of the country (Verschuren 1975). A recent review of African vulture declines indicated that 61% of mortalities during 1961–2014 were caused by poisoning and this was the single biggest factor behind plummeting populations, estimated to have declined by 62% on average for eight species of vulture over the past 30 years (Ogada et al. 2016a).

Europe

Across Europe, the use of poisoned baits to control predators, such as wolves and jackals, was widespread and legal until the mid-to-late twentieth century, e.g., in Spain it was legal until 1983 (Tavares 2013). Perhaps due to this long legacy of legal use, deliberate poisoning of predators to protect livestock and wild game continues illegally in many countries, particularly in Spain, Scotland, France, Greece and Italy (Guitart et al. 2010; Tingay et al. 2012), although illegal poisoning probably occurs in every country in Europe (see, for e.g., Richards 2012). Here, as in Africa, illegal poisoning is most commonly carried out by lacing baits such as carcasses or meat with toxic pesticides (Tavares 2013).

Southern European vulture populations have been devastated by the illicit use of carrion-based poisoned baits. For example, the use of poison is thought to be the main cause of the local extinction of several species (e.g., *G. fulvus*) in Romania that arose in the 1970s (Mateo-Tomás et al. 2012). Population declines of Bearded, Egyptian and Cinereous *Aegypius monachus* vultures in Spain have all been linked to illegal poisoning, particularly of adult birds, which could have major implications on the population dynamics (and general recovery) of these species (Hernández and Margalida 2008; 2009; Margalida et al. 2008, 2014a). In the French Pyrenees, 25% of Bearded Vulture mortalities were caused by poisoning (Berny et al. 2015). The use of poisoned baits targeting predators, mainly Red Fox *Vulpes vulpes,* is also an important factor in the decline of Spanish Imperial Eagle populations (González et al. 2007).

In the UK, poisoning is the most prevalent form of illicit control of birds of prey, particularly in Scotland (Whitfield et al. 2003; Tingay et al. 2012). Carcasses, or other bait laced with poison, are used to kill predators, typically foxes and birds of prey, often in areas with a high prevalence of game hunting. The persecution of Golden Eagles *A. chrysaetos*, mainly on grouse moors, probably has national-level impacts and is estimated to cause 3–5% annual adult mortality, a level which predisposes their Scottish populations to decline (Whitfield et al. 2004).

Legal Use of Insecticides and Rodenticides Contaminates Carrion and Secondarily Poisons Scavengers

During the legal application of pesticides, particularly to control agricultural pests, carrion often becomes a toxic by-product, or 'casualty'. Pest species such as rodents and Quelea birds are often targeted for removal using anticoagulant rodenticides and organophosphate pesticides (Bruggers et al. 1989; Thomas et al. 2011). Invariably, the contaminated carcasses not found during removal efforts are readily consumed by a wide variety of predators and scavengers with likely enormous—but as yet unquantified—consequences for the populations of those species, which exist in naturally low numbers and occupy a position at the top of the food chain.

More than 30 pesticides registered in North America and Europe have been known to cause the death of wild terrestrial vertebrates, *even when used according to the relatively stringent regulations in force in those countries* (Mineau 2013). In 2016, a newly released draft evaluation from the US Environmental Protection Agency maintained that three organophosphate pesticides: malathion, chlorpyrifos and diazinon pose an especially significant threat to numerous endangered species in North America (U.S. Environmental Protection Agency (USEPA) 2016a, b, c). And, where toxic carrion does not kill a scavenger outright, a high proportion of currently registered pesticides have the potential to disrupt reproduction by reducing egg production, hatching, or fledging success (Mineau 2013).

Over the years, many secondary poisoning incidents have involved raptors ingesting carrion contaminated by organophosphate and carbamate pesticides (Elliott et al. 1996; Mineau et al. 1999; Elliott et al. 1997, 2008). Particularly worrying have been single incidents where alarmingly high levels of mortality have been recorded. For example, an estimated 400 diurnal raptors and owls perished in Israel after monocrotophos, an organophosphorous pesticide, was applied for the purpose of killing voles (Mendelssohn and Paz 1977). Although the difficulties in quantifying the effects of pesticide-related mortality on raptor populations has been noted by Henny et al. (1999), attempts have nonetheless been made to enumerate the impacts of secondary poisoning due to the consumption of contaminated small mammals and birds. Balcomb (1983) noted that the then current 3,000,000 ha of corn treated annually with granular carbofuran in the United States had the capacity of killing or incapacitating several thousand raptors via secondary poisoning. In Kenya, after two nesting colonies of Red-billed Quelea *Quelea quelea* were sprayed (50 ha), 70% of the raptor population was found to be exposed to the avicide and two individual raptors found dead, indicating both lethal and sub-lethal effects to predatory birds (Bruggers et al. 1989). A study of Bald Eagles *Haliaeetus leuco-*

cephalus along the James River, Virginia, USA (USEPA 1989) found that secondary poisoning by carbofuran may have accounted for 6.5–8.3% of eagle mortality during 1985 and concluded that this level of pesticide-related mortality would likely result in a noticeable decrease in the population of breeding eagles.

Anticoagulants are toxic to all vertebrates. Second generation anticoagulant rodenticides (e.g., difenacoum, bromadiolone, brodifacoum) are more acutely toxic at lower doses than their first generation predecessors (e.g., warfarin, pindone) and are more persistent in vertebrate livers (Parmar et al. 1987; Stone et al. 1999). This extreme toxicity and longer persistence increases their potency as an agent of secondary poisoning (Thomas et al. 2011). An array of literature that spans the globe recounts the hazards of secondary poisoning due to second generation anticoagulant rodenticides, and identifies birds of prey, particularly owls, as the species most affected (Eason et al. 1999; Newton et al. 1999; Stone et al. 1999, 2003; Berny and Gaillet 2008).

Forensic analysis determined that a majority of Great Horned Owls *Bubo virginianus* and about 50% of Red-tailed Hawks *Buteo jamaicensis* in New York State were exposed to anticoagulants (Stone et al. 2003). In Spain, similar forensic data and analysis showed that 61% of owls, 37% of carnivores and 27% of diurnal raptors had been exposed to anticoagulants (Sánchez-Barbudo et al. 2012). In Britain, 20% of Tawny Owls *Strix aluco* had been exposed (Walker et al. 2008). Despite a known high prevalence of exposure to anticoagulants, the effects to the overall fitness of wild populations are difficult to measure (Rattner et al. 2014). Although Smith (1999) found no evidence that rodenticide use causes large-scale population declines of predatory and scavenging birds, a recent study by Thomas et al. (2011) estimated that a minimum of 11% of Canada's Great Horned Owl population is at fatal risk of poisoning due to anticoagulants, a figure which does not take into account sub-lethal exposure that could indirectly contribute to mortality.

Contamination from Lead Ammunition in Hunter-Killed Carrion and its Effects on Scavengers and Ecosystems

Naturally found in the Earth's crust, lead is released into both terrestrial and aquatic environments through various anthropogenic sources. These sources include coal-fired power plants, mining activities and smelting (among other forms of industry), as well as leaded gasoline, lead-based paints and ammunition. Hunters using lead ammunition to harvest game and nuisance animals generate lead-fragment-contaminated carrion. Lead fragments in the carcass pose a risk to the people or wildlife consuming it (Hunt et al. 2006; Knott et al. 2010; Haig et al. 2014), whether the carcasses are collected for human consumption or left or lost in the natural environment.

Lead enters the bodies of living organisms as a nonessential and toxic element through inhalation of lead dust or ingestion of paint chips or lead fragments. Lead poisoning has been documented in numerous obligate and opportunistic scavengers—

many already identified in previous sections of this chapter as being susceptible to other toxicants in carrion. Several scavengers evaluated for lead exposure include the Egyptian Vulture (*Neophron percnopterus*; Gangoso et al. 2009), Griffon Vulture (*Gyps fulvus*; Espín et al. 2014), Common Raven (*Corvus corax*; Craighead and Bedrosian 2008), Bald Eagle (*Haliaeetus leucocephalus;* Scheuhammer and Templeton 1998), White-Tailed Sea Eagle (*Haliaeetus albicilla;* Helander et al. 2009), White-Backed Vultures (*Gyps africanus*; Kenny et al. 2015) Black Vulture (*Coragyps atratus*; Behmke et al. 2015), Turkey Vulture (*Cathartes aura*; Kelly and Johnson 2011), and California Condor (*Gymnogyps californianus;* Finkelstein et al. 2012).

The consequences of lead range in avian scavengers from noticeable symptoms (e.g., lack of coordination, "droop-winged"; Sileo and Fefer 1987; Carpenter et al. 2003) to systemic effects (e.g., decreased haemoglobin synthesis, decreased bone mineralization; Locke and Thomas 1996; Gangoso et al. 2009). Lead poisoning even affects wildlife at the population level by impeding recovery of critically endangered species such as the California Condor (Church et al. 2008; Finkelstein et al. 2012). The consequences of lead exposure, however, vary at both the individual and species levels (Pokras and Kneeland 2009; Haig et al. 2014). While wildlife such as California Condors, Golden Eagles and Bald Eagles often exhibit symptoms of lead poisoning that result in morbidity or mortality (e.g., crop stasis resulting in starvation; Aguilar et al. 2012), other scavenging species seldom are found dead or dying from this heavy metal. For the avian scavengers not commonly debilitated by lead, such as the Turkey Vulture, experimental lead dosing studies have determined the lead threshold for the species (Carpenter et al. 2003). Despite the many studies evaluating the point at which lead poisoning affects various species, surprisingly little is known about the effects of sub-lethal lead levels in wildlife populations. To assess the potential effects of sub-lethal lead levels, such as possible behavioural or demographic changes, it may be valuable to study wildlife in a controlled laboratory environment. Studies in controlled environments have been used to evaluate the effects of sub-lethal mercury exposure in wildlife (Lewis et al. 2013; Varian-Ramos et al. 2014), and may prove useful in examining the effects of sub-lethal lead levels as well.

While it is challenging to assess the importance and effects of sub-lethal lead exposure in wildlife, there is an extensive list of vertebrate taxa affected by lead poisoning (Pokras and Kneeland 2009). Additionally, 63 avian species are documented as having ingested or been exposed to lead from ammunition sources (Pain et al. 2009). In some cases, such as with Griffon Vultures, birds change their movements and roosts to be closer to hunters (Mateo-Tomás and Olea 2010)—a behavioural observation that allowed researchers to estimate that nearly 2000 vultures over a 5- to 6-month period could be fed with the remains of hunted Red Deer (*Cervus elaphus*) and Wild Boar (*Sus scrofa*) carcasses (Mateo-Tomás and Olea 2010). Consequently, lead ammunition represents a significant carrion management issue.

Research and management often considers carrion in at least two parts—the eviscerated carcass and the viscera, the latter commonly discarded and left for scavengers. When an animal is shot with lead ammunition, lead fragments from the bullet and disperses along wound channels through the meat (Hunt et al. 2006). In a study that evaluated the presence and dispersal of lead fragments in carrion, the average number of fragments in the eviscerated carcass was over 350 (1.20 g), while the viscera contained approximately 180 (1.48 g, Knott et al. 2010). These findings closely matched that of Hunt et al. (2006), which estimated 213 fragments per carcass and 160 in the viscera. The dispersal of lead fragments within an eviscerated carcass were on average 24 cm apart, with the most widely separated lead fragments 43 cm apart (Knott et al. 2010).

Managing carrion for possible lead contamination is essential to reducing the impact that it has on scavenging populations and, ultimately, ecosystem function. In contrast to other toxic threats discussed in this chapter, visual inspections of carcasses can aide in identifying the presence of lead. However, the presence of an entry wound does not conclusively indicate an animal was shot with lead ammunition. This is because several forms of non-lead ammunition, such as copper bullets, now exist for hunters to use. Even with a portable x-ray machine, it is still possible to miss lead fragments embedded in a carcass (Knott et al. 2010). While Nadjafzadeh et al. (2015) found that avian scavengers, such as White-tailed Eagles, can deliberately avoid fragments as small as 9 mm in size, Hunt et al. (2006) and Knott et al. (2010) demonstrated the likelihood that many smaller and, hence, less easily avoidable fragments are present in a carcass and likely to be consumed. Here again there exists a dilemma of leaving out carcasses containing potentially harmful traces of lead, or removing a very important food source.

The scientific community has studied extensively the negative consequences of lead in the bodies of wildlife and humans (Cook and Trainer 1966; Pattee et al. 1981; Batuman et al. 1983; Locke and Thomas 1996; Dietrich et al. 2001; Canfield et al. 2003; Carpenter et al. 2003; Needleman 2004; Nevin 2007; Gangoso et al. 2009; Mielke and Zahran 2012; Aguilar et al. 2012). Despite the understanding that lead is hazardous when ingested or inhaled, there seems to be a disconnect regarding lead's toxicity and using lead ammunition. To better understand this situation in the hunting community, Sieg et al. (2009) published the results of the Arizona Game and Fish Department's efforts to create an effective communication strategy to promote voluntary lead reduction within the range of the critically endangered California Condor. These efforts employed the use of phone surveys and focus groups to design and modify their campaign. With 205 hunters surveyed, a surprisingly low 23% were "aware that lead poisoning is a problem faced by [California] condors," while an even lower percentage of hunters (9%) were "aware of any educational efforts to try and raise awareness of this issue" (Sieg et al. 2009). Despite the low awareness rate among hunters that lead poisoning is harmful to scavengers, the majority of hunters surveyed (between 83% and 97%) said they were willing to take action to assist in the recovery of the California Condor. This research presents gaps between the scientific community and the target audience, but details strategies to improve education and outreach programs to promote voluntary lead reduction efforts. In effectively using marketing and social psychology principles, the Arizona Game and Fish Department's voluntary lead reduction program resulted in a greater than 80% hunter compliance rate, no lead poisoning fatalities during that hunting season and a declining lead exposure rate in California condors (Sieg et al. 2009).

Strengths and Limitations of Regulatory Policies and Voluntary Programs Enacted to Improve Carrion Quality and Facilitate the Implementation of Sound Management Practices

As a result of the population crashes of Old World vultures and California Condors, regulatory policies and voluntary programs have increasingly been enacted to safeguard carrion quality.

In 2006, the governments of India, Pakistan and Nepal withdrew the manufacturing licenses for veterinary diclofenac. In India, this directive was further strengthened in 2008, when it was made an imprisonable offence to manufacture, retail or use diclofenac for veterinary purposes. These actions have proven effective in gradually reducing the prevalence and concentration of diclofenac in carcasses (Cuthbert et al. 2011), while in Pakistan this legislation has been attributed to the (albeit slow) reversal of declines of Long-billed Vultures (Chaudhry et al. 2012). The NSAID meloxicam has been confirmed to be a vulture-safe alternative to diclofenac (Swarup et al. 2007). In a further step to combat ongoing diclofenac-related mortality in Indian vultures, the government of India also banned the use of multi dose vials of diclofenac intended for human use, because they were being illicitly used for veterinary purposes, i.e., to treat livestock (Cuthbert et al. 2015). A paper reported on experimental trials that confirmed the NSAID aceclofenac metabolized to diclofenac in cattle and called for a ban on the drug (Galligan et al. 2016). This concern had initially been raised several years before (Sharma 2012), and may now in hindsight explain some of the observed slow recovery of vulture populations (Cuthbert et al. 2015). Other NSAIDs have been identified as being harmful to *Gyps* vultures, however, due to their infrequent use or short half-life in cattle, there have been relatively few regulatory policies enacted to limit their use in order to improve carrion quality (Cuthbert et al. 2014). Still, "Vulture Safe Zones" have been created in some places wherein ketoprofen usage is banned (e.g., in Bangladesh and Tamil Nadu, since 2017 and 2015 respectively).

Similarly, several successful steps have been taken to address and reduce lead originating from ammunition in the environment (Pokras and Kneeland 2009). For example, the United States Army has moved from lead to "green bullets" with a copper core, and many national parks worldwide have banned the use of lead shot in waterfowl hunting. Yet few countries have banned lead ammunition for all forms of hunting (Haig et al. 2014). For example, in Spain, lead ammunition is banned in wetlands, to hunt waterfowl, but is still used to hunt big and small game (Espín et al. 2014). Similarly, only lead shot for waterfowl hunting has been banned in the United States and Canada since the 1990s (Haig et al. 2014). Clearly, hunters are key to the survival of susceptible scavengers. However, merely eliminating lead ammunition in big game hunters does not represent a workable solution (Haig et al. 2014). Regulatory policies and voluntary programs to ban or reduce the prevalence of the use of lead ammunition (at least within the US) have proven somewhat difficult to implement. Getting hunters to switch to non-lead ammunition has yet to fully take into consideration the following, (1) finding effective alternatives, such as copper and copper alloy bullets, (2) economic costs (lead shots currently remains one of the least expensive options, but see Thomas 2013), and (3) whether bans would be enforceable or effective (Epps 2014). Although a ban on lead ammunition was enacted in California in 2013, the benefits to California Condors remain to be seen (Finkelstein et al. 2012). However, voluntary programs such as those already implemented by the State of Arizona and the Yurok Tribe in northern California have shown high levels of compliance (Sieg et al. 2009; Haig et al. 2014). Facilitating constructive relationships between hunters that use lead ammunition and those that have switched to non-toxic shot may be one way to promote voluntary compliance. Practical measures within the ballistics realm, such as the development (and use of) bullets that only fragment into particles that are large enough for avian scavengers to avoid (Nadjafzadeh et al. 2015) may also provide a temporary solution.

Attempts to regulate access to highly toxic pesticides has proven more difficult, especially in Africa where government cooperation has been minimal. In 2009 following numerous high profile lion poisonings, the manufacturers of carbofuran, FMC Corporation, visited Kenya and immediately began a buy-back program to remove all remaining stocks from the shelves of agro-vet shops (Kahumbu 2012).

Finally, the ubiquitous but invisible presence of rodenticides within habitats, and residues in deceased prey, remains not only a significant threat to a number of scavenging species but also to the related conservation efforts underway. However, of the threats we have discussed, rodenticides seem to have received the least attention. Somewhat ironically, assessing the extent and repercussions of sublethal exposure, in *living* animals, will considerably inform managers regarding the degree of pervasiveness of rodenticides within a given habitat, as well as appropriate mitigation and remedial measures to protect scavengers. The proactive, non-invasive collection of faecal matter from a representative selection of susceptible species for rodenticide analysis may be of particular value (e.g., following analytical methods developed by Sage et al. 2010).

Regulatory Policy Affects the Quantity of Food Available for Scavengers

Scavengers, particularly obligate scavengers, currently face three major carrion-related threats: (1) the consumption of an abundant food source that may be contaminated by multiple possible toxins, (2) an often abrupt withdrawal of food sources, or (3) changes to the temporal and spatial availability of food sources (Morales-Reyes et al. 2015). Broadly speaking, the modernization of livestock management, improved sanitation at slaughterhouse and legislation on the destruction of organic waste have all reduced the quantity of carrion available for scavengers.

In Europe, an outbreak of bovine spongiform encephalopathy (BSE) between 1996 and 2000 caused the EU to regulate the disposal of livestock carcasses through mandatory incineration at designated plants (Tella 2001; Margalida et al. 2010). These actions were taken despite the likelihood of their exacting serious consequences/repercussions on endangered vultures, including a halt in population growth, a decrease in breeding success and an apparent increase in the mortality of young age-classes (Donázar et al. 2009; Margalida et al. 2010). Morales-Reyes et al. (2015) have pointed out that these measures in fact contravened the obligation that EU member states had pledged to safeguard scavenging species—and furthermore had the unintended consequence of increasing greenhouse gas emissions. Concurrently, the number of farmer complaints about vultures (mostly Griffon Vultures) attacking and killing cattle increased exponentially (Margalida et al. 2014a). Fortunately for vultures, new EU regulations were adopted in 2011 and the Spanish government subsequently has allowed farmers to abandon downed livestock carcasses in the wild or to dispose of them at feeding stations, subject to *some* sanitary and administrative restrictions (Margalida et al. 2012).

The disposal of carcasses through removal and incineration is a costly endeavour. Morales-Reyes et al. (2015) estimated annual costs in the range of 50 million USD in payouts to insurance companies, with over 77,000 metric tonnes of CO_2 emitted into the atmosphere per year, and this attached *solely* to the transport of carcasses, even prior to their actual disposal and incineration. In southern France, Dupont et al. (2012) recently developed a model to assess the environmental and economic consequences of two methods of carrion removal, by private companies or by vultures. The model explored the risks and benefits associated with each method, namely reducing monetary costs and carbon emissions associated with carcass disposal by vultures, but also recognizing that there remains a sanitary risk if vultures fail to remove carrion efficiently (Dupont et al. 2012). The study found that the best results were obtained by a compromise solution whereby *both* private companies *and* vultures were used to dispose of carcasses. This strategy took into consideration both the high economic costs of using private companies, and also the temporal mismatch between livestock mortality and food requirements of vultures (Dupont et al. 2012). The results of this study stress the need to involve multiple stakeholders (e.g., farmers, wildlife managers, sanitary authorities, industrial companies and governmental institutions) in decisions about carrion management to maximize environmental and economic benefits (Dupont et al. 2012).

Improving Carrion Quantity and Quality Through Feeding Stations

Vultures remain the most viable means of actively disposing of carrion, and it is due to concerns over their often drastic population declines that feeding stations (e.g., vulture restaurants) continue to be the main conservation management strategy. As a management tool, feeding stations are intended to supplement scarce food resources (e.g., Wilbur et al. 1974; Piper et al. 1999; García-Ripollés et al. 2004) and they are increasingly used under the auspices of providing an alternative safe food supply that is free from contaminants (e.g., Gilbert et al. 2007). Feeding stations have also been used to increase the success of vulture reintroduction programs (e.g., Sarrazin et al.

1994) and to facilitate the recolonization of abandoned breeding sites (Mundy et al. 1992). Whether using vultures to remove carrion or using carrion to augment the food supply of vultures, managers must ensure that provision of this resource coincides both with the geographic range and foraging predilections of the targeted scavengers.

Feeding stations have been in long-term use primarily in southern Africa, Spain, France, Israel and the western US. The pros and cons of their use has been widely debated (Fielding et al. 2014). Feeding stations provide threatened vultures with food, facilitate the monitoring of tagged birds, provide eco-tourism opportunities, and increase public awareness—among other benefits (Piper 2005; Anderson and Anthony 2005). Despite this, feeding stations can also have a number of detrimental and unintentional consequences (Piper 2004, 2005; Anderson and Anthony 2005). Foremost among these are the attraction of mammalian scavengers to feeding stations that may contribute to the spread of disease (Moreno-Opo et al. 2012; Yarnell et al. 2015), the possibility of increasing vulture mortality if carrion placed at feeding stations is not free of contaminants such as barbiturates, lead or antiparasitics used to treat livestock (Piper 2004; Mateo et al. 2015), and the role of provisioned feeding on decreased vulture foraging ability and efficiency of ecosystem service provision (Deygout et al. 2009). A recent review of the pros and cons of feeding stations gives this topic even greater consideration (Cortés-Avizanda et al. 2016).

Ideally, feeding stations have a net positive impact on threatened vulture populations, particularly if those populations are under severe and imminent threat. Four studies implicitly examined the effect of supplementary feeding on vulture survival (Piper et al. 1999; Gilbert et al. 2007; Oro et al. 2008; Margalida 2010). In three of these studies, results showed either increased survival in sub-adult birds (Piper et al. 1999), or a delay in the extinction of populations under imminent threat due to poisoning (Gilbert et al. 2007; Oro et al. 2008). The only exception was in the preliminary results of Margalida (2010), which suggested that the breeding success of an obligate brood reducer (the Bearded Vulture) were not improved through supplemental feeding.

The primary and unanimous factor in provisioning carrion at a feeding station is careful, vigilant operation. Most of the detrimental consequences of feeding stations can be eliminated or reduced through proper and continuous management (Moreno-Opo et al. 2012, 2015). A well-managed feeding station can also enhance foraging of threatened species by reducing competition from more abundant vulture species. For example, Moreno-Opo et al. (2015) found that Bearded, Cinereous and Egyptian vultures preferred when small parts of ungulates (e.g., sheep and goats) were provided in abundance and scattered, rather than having to tackle a carcass in its entirety.

Conclusions and Future Perspectives

The examples in this chapter clearly illustrate there is an urgent need to start managing carrion for the benefit of biodiversity and in recognition of the ecosystem services provided by vultures. Globally, carrion has become a reservoir, even a conduit, for residues of pharmaceutical drugs, pesticides and other poisons, and heavy metals such as lead. The consumption of carrion by predators and scavengers has given these toxins a pathway to contaminate the larger food web. There are few populations of predators and scavengers worldwide that have not been affected by at least one or more of these toxins emanating from carrion. Population-level effects as a result of consuming contaminated carrion have been documented for many species of vultures, eagles, and large mammalian predators. Most predators and scavengers occur at the top of the food chain and are long-lived, which increases their vulnerability to bioaccumulation and its sub-lethal effects. The abundance of these species is naturally limited and any loss may constitute a significant reduction in local populations and impair the functioning of ecosystems. For example, avian scavengers likely limit the transmission of diseases at carcasses and recycle nutrients back into soils. Predators and scavengers are also important cultural icons and vultures, for example, are crucial recyclers of the dead within the Zoroastrian-practicing Parsi community of India.

Our goal has been to provide comprehensive information for managers and regulators/policy-makers regarding the array of harmful compounds that may be present in carrion, and that may compromise its quality and safety as a food resource for scavengers. From a conservation perspective, it is catastrophically counter-productive to draw in vulnerable scavenging species, especially vultures, only to provide them with a tainted food source. Margalida et al. (2014b) have discussed some of the inherent flaws in the current risk assessment process regarding the evaluation of potential toxicants, including loopholes that would enable known toxic substances to be registered or to remain below the radar.

Given the need to ensure the overall safety of carcasses and be able to trace carrion/carcasses back to their source, it is concerning that very little of such data exists (Mateo et al. 2015; Zorilla et al. 2018). Monitoring must be routinely conducted, on a region by region basis. We also believe that too many assumptions of compliance are made, and that it must be regularly investigated and ascertained. For example, when assessing potential sources of lead exposure and resultant stress markers in Griffon Vultures, Espín et al. (2014) did not analyze the metal content of provisioned pig carcasses, stating that "…the lead levels are expected to be low in the carcass in accordance with legislation…". Our concern about compliance, or lack thereof, stems from studies where, for example, nearly 30% of the injection sites sampled from animals rendered at a slaughterhouse in Belgium tested positive for a veterinary product in excess of the established maximum residue limits safety parameter, with roughly 7% containing residues of the NSAIDs flunixin, meloxicam and tolfenamic acid (Van Hoof et al. 2004).

To ensure a scenario where scavengers no longer come into contact with contaminated carcasses, additional research (along the lines of Mateo et al. 2015; Casas-Díaz et al. 2016) and educational outreach (e.g., SASOL feeding station guidelines; Fig. 2) will be required. We encourage managers to seek out the existing literature, with emphasis on studies such as Hunt et al. (2006), which attempt to quantify a known threat (in this case the number of lead shot fragments in

Fig. 2 The SASOL guide to the establishment and operation of supplementary feeding sites for vultures, developed by the Endangered Wildlife Trust (EWT 2011) and other collaborators (https://www.ewt.org.za/eBooks/booklets/Vulture%20Restuarant%20booklet.pdf)

different edible parts of carcass) and in so doing illustrate the fundamental ingredients required to assess risk. Some studies have already established the presence of residues in carcass components known to be favoured by scavenging species. For example, since lamb's feet represent one of the main food components provided at supplemental feeding stations for the imperilled Bearded Vulture, Mateo et al. (2015) recovered a selection of these items from slaughter houses, feeding stations and carcasses that would have been consumed in the wild, and found that the topical antiparasitic agents (some having organophosphorus insecticides as their active ingredient) used widely on sheep were present therein. Studies like that conducted by Casas-Díaz et al. (2016) are invaluable in very clearly reporting on presence of residues in provisioned carcasses and evidence of scavenger exposure, contrasted against the actual implications of such exposure (e.g., too low to be toxic but sufficiently chronic to result in antibiotic resistance). The authors also highlighted a simple but fundamental principle, that "…wildlife should not have access to food contaminated with pharmaceuticals" (Casas-Díaz et al. 2016).

For species whose populations are already fragile, we recommend scrutinizing and following up on studies that have identified dietary predilections, to elucidate potential sources of exposure at an even finer scale. Houston and Copsey's (1994) examination of bone digestion and intestinal morphology of the Bearded Vulture provides a case in point. The authors found that, given the choice, individuals would prefer bones over meat, and favour old, dried bones rather than fresh, presenting a range of different exposure scenarios for exploration. Along similar lines, Bearded Vultures might be facing incidental exposure to residues of veterinary drugs present on the surface of sheep's wool used to line nests (e.g., Richards et al. 2011a), and coprophagues like the Egyptian Vulture may be encountering such residues in livestock dung. Additional possible routes for testing veterinary agent residues and heavy metals include: retrieval of wool, hair, regurgitated pellets and bones from vulture nests, and selective collection of livestock feces to either identify or rule out the many potential sources of exposure in the wild (e.g., Richards et al. 2011a, b, 2014). Considered separately, incidental sources of exposure may seem of negligible concern. These should therefore be weighed together for each species to obtain an overall comprehensive sense of risk—or lack thereof.

In terms of pesticides and other poisons in use, regulations regarding their accessibility must be tightened, particularly in developing countries. Pesticides that have been banned for use in the majority of developed countries (where, reprehensively, their manufacture sometimes still remains legal) should also be banned in developing countries. Law enforcement and judicial proceedings regarding the illegal use of poisons needs to be strengthened, particularly in Africa (Ogada 2014). Conversely, harmful residues need not have been introduced via illicit means, or with deliberate intent to poison scavengers (e.g., poachers poisoning an elephant carcass to kill vultures). As we have discussed, some deliberate poisoning is entirely legal and government sanctioned (e.g., quelea culls). Residues of different veterinary agents (or, as evidence continues to show, their equally toxic metabolites) can be present following various therapeutic treatments, and not necessarily administered close to the time of death. Likewise, lead ammunition remains a legal means for hunters to capture their quarry.

It is imperative that managers and regulators think "outside the box" whenever possible and reach out to other concerned stakeholders for additional information and insight. Conferring more closely with veterinarians, eco-toxicologists, conservationists, wildlife agents and hunters (often one and the same) and policy-makers would help bring to light outstanding factors and exposure sources within the immediate environment that could be undermining the precautions being implemented to ensure the safety of the carrion provided. In parallel, the tremendous value of the complimentary carcass clean up and nutrient cycling performed by micro-scavengers must be fully recognized and the potential repercussions of toxic compound residues in carcasses to these organisms has to be acknowledged and incorporated into mitigative measures. As we continue to dispense toxins into the environment, just like scavengers, we will eventually pay the ultimate price. The onus is on each and every one of us to ensure the safest possible conditions under which nature can, very simply, be allowed to do its job.

Acknowledgements SB thanks Dr. Todd Katzner for his helpful review. We appreciate all those working to protect our vultures and scavengers, whether through research or careful policy-making.

References

Aguilar RF, Yoshicedo JN, Parish CN (2012) Ingluviotomy tube placement for lead-induced crop stasis in the California condor (*Gymnogyps californianus*). J Avian Med Surg 26:176–181

Alterio N (2000) Controlling small mammal predators using sodium monofluoroacetate (1080) in bait stations along forestry roads in a New Zealand beech forest. N Z J Ecol 24(1):3–9

Anderson MD, Anthony A (2005) The advantages and disadvantages of vulture restaurants versus simply leaving livestock (and game) carcasses in the veldt. Vult News 53(1):42–45

Balcomb R (1983) Secondary poisoning of red-shouldered hawks with carbofuran. J Wildl Manag 47(4):1129–1132

Batuman V, Landy E, Maesaka J, Wedeen R (1983) Contribution of lead to hypertension with renal impairment. N Engl J Med 309:17–21

Behmke S, Fallon J, Duerr AE et al (2015) Chronic lead exposure is epidemic in obligate scavenger populations in eastern North America. Environ Int 79:51–55

Berny P, Gaillet JR (2008) Acute poisoning of Red Kites (Milvus milvus) in France: data from the SAGIR network. J Wildl Dis 44(2):417–426

Berny PJ, Buronfosse T, Buronfosse F et al (1997) Field evidence of secondary poisoning of foxes (*Vulpes vulpes*) and buzzards (*Buteo buteo*) by bromadiolone, a 4-year survey. Chemosphere 35(8):1817–1829

Berny P, Vilagines L, Cugnasse JM et al (2015) VIGILANCE POISON: illegal poisoning and lead intoxication are the main factors affecting avian scavenger survival in the Pyrenees (France). Ecotoxicol Environ Saf 118:71–82

Boshoff AF, Vernon CJ (1980) The past and present distribution and status of the Cape Vulture in the Cape Province. Ostrich 51(4):230–250

Brakes CR, Smith RH (2005) Exposure of non-target small mammals to rodenticides: short-term effects, recovery and implications for secondary poisoning. J Appl Ecol 42(1):118–128

Bridgeford P (2001) More vulture deaths in Namibia. Vult News 44:22–26

Bruggers RL, Jaeger MM, Keith JO et al (1989) Impact of fenthion on nontarget birds during quelea control in Kenya. Wildl Soc Bull 17(2):149–160

Canfield RL, Henderson CR, Cory-Sletchta DA et al (2003) Intellectual impariment in children with blood lead concentrations below 10 μg per deciliter. N Engl J Med 348:1517–1526
Carpenter JW, Pattee OH, Fritts SH et al (2003) Experimental lead poisoning in turkey vultures (Cathartes aura). J Wildl Dis 39:96–104
Casas-Díaz E, Cristòfol C, Cuenca R et al (2016) Determination of fluoroquinolone antibiotic residues in the plasma of Eurasian griffon vultures (Gyps fulvus) in Spain. Sci Total Environ 557–558:620–626
Chardonnet P, Belemsobgo U, Crosmary W et al (2005) Influences directes et indirectes sur la conservation du lion en Afrique de l'Ouest et en Afrique Centrale. Atelier sur la Conservation du Lion d'Afrique de l'Ouest et d'Afrique Centrale, Douala, Cameroon, 5–7 October
Chaudhry MJI, Ogada DL, Malik RN et al (2012) First evidence that populations of the critically endangered Long-billed Vulture Gyps indicus in Pakistan have increased following the ban on diclofenac in South Asia. Bird Conserv Int 22(4):389–397
Cherkaoui I (2005) The Bearded Vulture Gypaetus barbatus in Morocco. Vult News 52:37
Church ME, Gwiazda R, Risebrough R et al (2008) Ammunition is the principal source of lead accumulated by California condors re-introduced to the wild. Environ Sci Technol 42:1809–1811
Cook RC, Trainer DO (1966) Experimental lead poisoning of Canada geese. J Wildl Manag 30:1–8
Cortés-Avizanda A, Blanco G, DeVault TL et al (2016) Supplementary feeding and endangered avian scavengers: benefits, caveats, and controversies. Front Ecol Environ 14(4):191–199
Craighead D, Bedrosian B (2008) Blood lead levels of common ravens with access to big-game offal. J Wildl Manag 72(1):240–245
Cuthbert R, Parry-Jones J, Green RE, Pain DJ (2007) NSAIDs and scavenging birds: potential impacts beyond Asia's critically endangered vultures. Biol Lett 3(1):90–93
Cuthbert R, Taggart MA, Prakash V et al (2011) Effectiveness of action in India to reduce exposure of Gyps Vultures to the toxic veterinary drug diclofenac. PLoS One 6:e19069
Cuthbert RJ, Taggart MA, Prakash V et al (2014) Avian scavengers and the threat from veterinary pharmaceuticals. Philos Trans R Soc Lond Ser B Biol Sci 369(1656). pii: 20130574
Cuthbert RJ, Taggart MA, Saini M et al (2015) Continuing mortality of vultures in India associated with illegal veterinary use of diclofenac and a potential threat from nimesulide. Oryx 50:104–112
Deygout C, Gault A, Sarrazin F, Bessa-Gomes C (2009) Modeling the impact of feeding stations on vulture scavenging service efficiency. Ecol Model 220(15):1826–1835
Dietrich K, Ris M, Succop P et al (2001) Early exposure to lead and juvenile delinquency. Neurotoxicol Teratol 23:511–518
Donázar JA, Margalida A, Campión D (2009) Vultures, feeding stations and sanitary legislation: a conflict and its consequences from the perspective of conservation biology. Munibe 29:1–551
Dupont H, Mihoub JB, Bobbe S, Sarrazin F (2012) Modelling carcass disposal practices: implications for the management of an ecological service provided by vultures. J Appl Ecol 49(2):404–411
Eason CT, Milne L, Potts M et al (1999) Secondary and tertiary poisoning risks associated with brodifacoum. N Z J Ecol 23(2):219–224
Eason CT, Murphy EC, Wright GR, Spurr EB (2002) Assessment of risks of brodifacoum to non-target birds and mammals in New Zealand. Ecotoxicology 11(1):35–48
Elliott JE, Langelier KM, Mineau P, Wilson LK (1996) Poisoning of bald eagles and red-tailed hawks by carbofuran and fensulfothion in the Fraser Delta of British Columbia, Canada. J Wildl Dis 32(3):486–491
Elliott JE, Wilson LK, Langelier KM et al (1997) Secondary poisoning of birds of prey by the organophosphorus insecticide, phorate. Ecotoxicology 6(4):219–231
Elliott JE, Birmingham AL, Wilson LK et al (2008) Fonofos poisons raptors and waterfowl several months after granular application. Environ Toxicol Chem 27(2):452–460
Endangered Wildlife Trust (2011) The Sasol guide to the establishment and operation of supplementary feeding sites for vultures. http://www.ewt.org.za/eBooks/booklets/Vulture%20Restuarant%20booklet.pdf. Accessed 10 Apr 2016

Epps CW (2014) Considering the switch: challenges of transitioning to non-lead hunting ammunition. Condor 116(3):429–434
Espín S, Martínez-López E, Jiménez P et al (2014) Effects of heavy metals on biomarkers for oxidative stress in Griffon vulture (Gyps fulvus). Environ Res 129:59–68
Fielding D, Newey S, van der Wal R, Irvine RJ (2014) Carcass provisioning to support scavengers: evaluating a controversial nature conservation practice. Ambio 43(6):810–819
Finkelstein ME, Doak DF, George D et al (2012) Lead poisoning and the deceptive recovery of the critically endangered California condor. Proc Natl Acad Sci U S A 109(28):11449–11454
Floate K (2006) Endectocide use in cattle and fecal residues: environmental effects in Canada. Can J Vet Res 70(1):1–10
Floate KD, Düring RA, Hanafi J et al (2015) Validation of a standard field test method in four countries to assess the toxicity of residues in dung of cattle treated with veterinary medical products. Environ Toxicol Chem 35:1934–1946
Frank L (2011) Living with lions: lessons from Laikipia. In: Georgiadis NJ (ed) Conserving wildlife in African landscapes: Kenya's Ewaso ecosystem. Smithsonian Contributions to Zoology No. 632, pp 73–84
Galligan TH, Taggart MA, Cuthbert RJ et al (2016) Metabolism of aceclofenac in cattle to vulture-killing diclofenac. Conserv Biol 30(5):1122–1127
Gangoso L, Álvarez-Lloret P, Rodríguez-Navarro A et al (2009) Long-term effects of lead poisoning on bone mineralization in vultures exposed to ammunition sources. Environ Pollut 157(2):569–574
García-Ripollés C, López-López P, García-López F (2004) Management and monitoring of a vulture restaurant in Castellón Province, Spain. Vult News 50:5–14
Gilbert M, Watson RT, Ahmed S et al (2002) Breeding and mortality of Oriental White-backed Vulture Gyps bengalensis in Punjab Province, Pakistan. Bird Conserv Int 12(4):311–326
Gilbert M, Watson RT, Ahmed S et al (2007) Vulture restaurants and their role in reducing diclofenac exposure in Asian vultures. Bird Conserv Int. 17(01):63–77
Gillies CA, Pierce RJ (1999) Secondary poisoning of mammalian predators during possum and rodent control operations at Trounson Kauri Park, Northland, New Zealand. N Z J Ecol 23(2):183–192
Giorgi M, Mengozzi G (2011) Malicious animal intoxications: poisoned baits. Vet Med 56(4):173–179
Glen AS, Gentle MN, Dickman CR (2007) Non-target impacts of poison baiting for predator control in Australia. Mammal Rev 37(3):191–205
González LM, Margalida A, Mañosa S et al (2007) Causes and spatio-temporal variations of non-natural mortality in the vulnerable Spanish imperial eagle *Aquila adalberti* during a recovery period. Oryx 41(04):495–502
Guitart R, Sachana M, Caloni F et al (2010) Animal poisoning in Europe. Part 3: wildlife. Vet J 183(3):260–265
Haig SM et al (2014) The persistent problem of lead poisoning in birds from ammunition and fishing tackle. Condor 116(3):408–428
Hayes B (1988) British Columbia: deaths caused by barbiturate poisoning in bald eagles and other wildlife. Can Vet J 29(2):173–174
Helander B, Axelsson J, Borg H et al (2009) Ingestion of lead from ammunition and lead concentrations in white-tailed sea eagles (Haliaeetus albicilla) in Sweden. Sci Total Environ 407(21):5555–5563
Henny CJ, Mineau P, Elliott JE, Woodbridge B (1999) Raptor poisonings and current insecticide use: what do isolated kill reports mean to populations. In: Proceedings of the 22nd international ornithological congress, University of Natal, Durban
Hernández M, Margalida A (2008) Pesticide abuse in Europe: effects on the Cinereous vulture (Aegypius monachus) population in Spain. Ecotoxicology 17(4):264–272
Hernández M, Margalida A (2009) Poison-related mortality effects in the endangered Egyptian vulture (Neophron percnopterus) population in Spain. Eur J Wildl Res 55(4):415–423

Houston DC, Copsey JA (1994) Bone digestion and intestinal morphology of the Bearded Vulture. J Raptor Res 28(2):73–78

Hunt WG, Burnham W, Parish CN et al (2006) Bullet fragments in deer remains: implications for lead exposure in avian scavengers. Wildl Soc Bull 34(1):167–170

Jackson FJ (1938) The birds of Kenya colony and the Uganda protectorate. Gurney and Jackson, London

Jarvis MJ, La Grange M (1982) Problem vertebrate management in Zimbabwe. In: Proceedings of the tenth vertebrate pest conference, paper 25. http://digitalcommons.unl.edu/vpc10/25. Downloaded 18 Dec 2013

Kahumbu P (2012) Banned in America killing in Kenya: the story of a poison. Swara 3:30

Kaukeinen D (1982) A review of the secondary poisoning hazard potential to wildlife from the use of anticoagulant rodenticides. In: Proceedings of the tenth vertebrate pest conference, p 27. http://digitalcommons.unl.edu/vpc10/25. Accessed 10 Apr 2016

Kelly TR, Johnson CK (2011) Lead exposure in free-flying turkey vultures is associated with big game hunting in California. PLoS One 6:e15350

Kenny D, Reading R, Maude G et al (2015) Blood lead levels in White-Backed Vultures (Gyps africanus) from Botswana, Africa. Vult News 68:25–31

Knott J, Gilbert J, Hoccom DG, Green RE (2010) Implications for wildlife and humans of dietary exposure to lead from fragments of lead rifle bullets in deer shot in the UK. Sci Total Environ 409(1):95–99

Krueger B, Krueger KA (2002) U.S. fish and wildlife service fact sheet: secondary pentobarbital poisoning of wildlife. Habitat and Animal Health Concern, Inc., Stafford, VA. http://www.fws.gov/southeast/news/2002/12-03SecPoisoningFactSheet.pdf Accessed 6 Mar 2016

Langelier KM (1993) Barbiturate poisoning in twenty-nine bald eagles. In: Redig PT, Hunter B (eds) Raptor biomedicine. University of Minnesota Press, Minneapolis, MN, pp 231–232

Leopold AS, Cain SA, Cottam CM et al (1964) Predator and rodent control in the United States. US Fish and Wildlife Publications. Paper 25. http://digitalcommons.unl.edu/usfwspubs/254. Downloaded 15 Oct 2015

Lewis JC (1968) Use of poison bait to control rabies in Tennessee wildlife. Public Health Rep 83(1):69–74

Lewis CA, Cristol DA, Swaddle JP, Varian-Ramos CW, Zwollo P (2013) Decreased immune response in zebra finches exposed to sublethal doses of mercury. Arch Environ Contam Toxicol 64:327–336

Linhart SB, Kappeler A, Windberg LA (1993) A review of baits and bait delivery systems for free-ranging carnivores and ungulates. Contraception in wildlife: Animal and Plant Health Inspection Service. United States Department of Agriculture, Oakland, CA, pp 69–132

Locke LN, Thomas NJ (1996) Lead poisoning of waterfowl and raptors. In: Fairbrother A, Locke LN, Hoff GL (eds) Noninfectious diseases of wildlife. Manson, London, pp 108–117

Lumaret JP, Errouissi F (2002) Use of anthelmintics in herbivores and evaluation of risks for the nontarget fauna of pastures. Vet Res 33(5):547–562

Margalida A (2010) Supplementary feeding during the chick-rearing period is ineffective in increasing the breeding success in the bearded vulture (Gypaetus barbatus). Eur J Wild Res 56(4):673–678

Margalida A (2012) Baits, budget cuts: a deadly mix. Science 338(6104):192–192

Margalida A, Heredia R, Razin M, Hernández M (2008) Sources of variation in mortality of the Bearded vulture Gypaetus barbatus in Europe. Bird Conserv Int 18(01):1–10

Margalida A, Donázar JA, Carrete M, Sánchez-Zapata JA (2010) Sanitary versus environmental policies: fitting together two pieces of the puzzle of European vulture conservation. J Appl Ecol 47(4):931–935

Margalida A, Carrete M, Sánchez-Zapata JA, Donázar JA (2012) Good news for European vultures. Science 335(6066):284–284

Margalida A, Campión D, Donázar JA (2014a) Vultures *vs.* livestock: conservation relationships in an emerging conflict between humans and wildlife. Oryx 48(02):172–176

Margalida A, Sánchez-Zapata JA, Blanco G et al (2014b) Diclofenac approval as a threat to Spanish vultures. Conserv Biol 28(3):631–632
Margalida A, Colomer MA, Oro D (2014c) Man-induced activities modify demographic parameters in a long-lived species: effects of poisoning and health policies. Ecol Appl 24(3):436–444
Markandya A, Taylor T, Longo A et al (2008) Counting the cost of vulture decline—an appraisal of the human health and other benefits of vultures in India. Ecol Econ 67(2):194–204
Martínez-Haro M, Mateo R, Guitart R et al (2008) Relationship of the toxicity of pesticide formulations and their commercial restrictions with the frequency of animal poisonings. Ecotoxicol Environ Saf 69(3):396–402
Mateo R, Sánchez-Barbudo IS, Camarero PR, Martínez JM (2015) Risk assessment of bearded vulture (Gypaetus barbatus) exposure to topical antiparasitics used in livestock within an ecotoxicovigilance framework. Sci Total Environ 536:704–712
Mateo-Tomás P, Olea PP (2010) When hunting benefits raptors: a case study of game species and vultures. Eur J Wildl Res 56:519–528
Mateo-Tomás P, Olea PP, Sánchez-Barbudo IS, Mateo R (2012) Alleviating human-wildlife conflicts: identifying the causes and mapping the risk of illegal poisoning of wild fauna: mapping illegal poisoning of wildlife. J Appl Ecol 49(2):376–385
Mateo-Tomás P, Olea PP, Moleón M et al (2015) From regional to global patterns in vertebrate scavenger communities subsidized by big game hunting. Divers Distrib 21(8): 913–924
Meenken D, Booth LH (1997) The risk to dogs of poisoning from sodium monofluoroacetate (1080) residues in possum (Trichosurus vulpecula). N Z J Agric Res 40(4): 573–576
Mendelssohn H, Paz U (1977) Mass mortality of birds of prey caused by Azodrin, an organophosphorus insecticide. Biol Conserv 11(3):163–170
Mielke HW, Zahran S (2012) The urban rise and fall of air lead (Pb) and the latent surge and retreat of societal violence. Environ Int 43:48–55
Mills MGL, Hofer H (1998) Hyaenas: status survey and conservation action plan. IUCN/SSC Hyaena Specialist Group. IUCN, Gland, Switzerland and Cambridge
Mineau P (2013) Bird: pesticide use impacts. Encycl Environ Manag. https://doi.org/10.1081/E-EEM-120046191
Mineau P, Fletcher MR, Glaser LC et al (1999) Poisoning of raptors with organophosphorus and carbamate pesticides with emphasis on Canada, US and UK. J Raptor Res 33(1):1–37
Morales-Reyes Z, Pérez-García JM, Moleón M et al (2015) Supplanting ecosystem services provided by scavengers raises greenhouse gas emissions. Sci Rep 5:7811
Moreno-Opo R, Margalida A, García F et al (2012) Linking sanitary and ecological requirements in the management of avian scavengers: effectiveness of fencing against mammals in supplementary feeding sites. Biodivers Conserv 21(7):1673–1685
Moreno-Opo R, Trujillano A, Arredondo A et al (2015) Manipulating size, amount and appearance of food inputs to optimize supplementary feeding programs for European vultures. Biol Conserv 181:27–35
Mundy P, Butchart D, Ledger J, Piper S (1992) The vultures of Africa. Academic, London
Nadjafzadeh M, Hofer H, Krone O (2015) Lead exposure and food processing in white-tailed eagles and other scavengers: an experimental approach to simulate lead uptake at shot mammalian carcasses. Eur J Wildl Res 61(5):763–774
Needleman H (2004) Lead poisoning. Ann Rev Med 55:209–222
Nevin R (2007) Understanding international crime trends: the legacy of preschool lead exposure. Environ Res 104:315–336
Newton I, Shore RF, Wyllie I et al (1999) Empirical evidence of side-effects of rodenticides on some predatory birds and mammals. In: Cowan DP, Feare CJ (eds) Advances in vertebrate pest management. Filander, Fürth, pp 347–367
O'Rourke K (2002) Euthanized animals can poison wildlife: veterinarians receive fines. J Am Vet Med Assoc 220:146–147

Oaks JL, Gilbert M, Virani MZ et al (2004) Diclofenac residues as the cause of vulture population decline in Pakistan. Nature 427:630–633

Ogada DL (2014) The power of poison: pesticide poisoning of Africa's wildlife. Ann N Y Acad Sci 1322:1–20

Ogada DL, Torchin ME, Kinnaird MF, Ezenwa VO (2012) Effects of vulture declines on facultative scavengers and potential implications for mammalian disease transmission. Conserv Biol 26(3):453–460

Ogada D, Botha A, Shaw P (2016b) Ivory poachers and poisons; drivers of Africa's declining vulture populations. Oryx 50:593–596

Ogada D, Shaw P, Beyers RL et al (2016a) Another continental vulture crisis: Africa's vultures collapsing toward extinction. Conserv Lett 9:89–97

Olea PP, Mateo-Tomás P (2009) The role of traditional farming practices in ecosystem conservation: the case of transhumance and vultures. Biol Conserv 142(8):1844–1853

Olea PP, Sánchez-Barbudo IS, Viñuela J et al (2009) Lack of scientific evidence and precautionary principle in massive release of rodenticides threatens biodiversity: old lessons need new reflections. Environ Conserv 36(01):1–4

Oro D, Margalida A, Carrete M et al (2008) Testing the goodness of supplementary feeding to enhance population viability in an endangered vulture. PLoS One 3(12):e408

Oro D, Genovart M, Tavecchia G et al (2013) Ecological and evolutionary implications of food subsidies from humans. Ecol Lett 16:1501–1514

Pain DJ, Cunningham AA, Donald PF et al (2003) Causes and effects of temporospatial declines of *Gyps* Vultures in Asia. Conserv Biol 17(3):661–671

Pain DJ, Fisher IJ, Thomas VG (2009) A global update of lead poisoning in terrestrial birds from ammunition sources. In: Richard T, Watson RT, Pokras M, Hunt G (eds) Ingestion of lead from spent ammunition: implications for wildlife and humans, May 2008. The Peregrine Fund, Boise, ID, pp 99–118

Parmar G, Bratt H, Moore R, Batten PL (1987) Evidence for a common binding- site in vivo for the retention of anticoagulants in rat liver. Hum Toxicol 6:431–432

Pattee OH, Wiemeyer SN, Mulhern BM et al (1981) Experimental lead-shot poisoning in bald eagles. J Wildl Manag 45:806–810

Payne J, Farris R, Parker G et al (2015) Quantification of sodium pentobarbital residues from equine mortality compost piles. J Anim Sci 93(4):1824–1829

Piper SE (2004) Vulture restaurants-conflict in the midst of plenty. In: Chancellor RD, Meyburg B-U (eds) Raptors worldwide. WWGBP/MME, Budapest, pp 341–349

Piper SE (2005) Supplementary feeding programs: how necessary are they for the maintenance of numerous and healthy vultures populations? In: Houston DC, Piper SE (eds) Proceedings of the international conference on conservation and management of vulture populations. Natural History Museum of Crete & WWF Greece, Thessaloniki, pp 41–50

Piper SE, Boshoff AF, Scott HA (1999) Modelling survival rates in the Cape Griffon Gyps coprotheres, with emphasis on the effects of supplementary feeding. Bird Study 46(S1): S230–S238

Pokras MA, Kneeland MK (2009) Understanding lead uptake and effects across species lines: a conservation medicine approach. In: Watson RT, Fuller M, Pokras M, Hunt WG (eds) Ingestion of lead from spent ammunition: implications for wildlife and humans. The Peregrine Fund, Boise, ID. https://doi.org/10.4080/ilsa.2009.0101

Prakash V (1999) Status of vultures in Keoladeo National Park, Bharatpur, Rajasthan, with special reference to population crash in Gyps species. J Bombay Nat Hist Soc 96:365–378

Rattner BA, Lazarus RS, Elliott JE et al (2014) Adverse outcome pathway and risks of anticoagulant rodenticides to predatory wildlife. Environ Sci Technol 48(15):8433–8445

Richards N (ed) (2012) Carbofuran and wildlife poisoning: global perspectives and forensic approaches. Wiley, Chichester

Richards NL, Ogada D (2014) The relative risks of veterinary agents and toxic compounds to avian scavengers and other wildlife within the European Union and Africa. Royal Society of Chemistry Bulletin September Issue

Richards N, Hall S, Scott K, Harrison N (2011a) First detection of an NSAID, flunixin, in sheep's wool using GC/MS. Environ Pollut 159(5):1446–1450
Richards NL, Cook G, Simpson V et al (2011b) Qualitative detection of the NSAIDs diclofenac and ibuprofen in the hair of Eurasian otters (Lutra lutra) occupying UK waterways with GC-MS. Eur J Wildl Res 57(5):1107–1114
Richards NL, Hall SW, Harrison NM et al (2014) Merging wildlife and environmental monitoring approaches with forensic principles: application of unconventional and non-invasive sampling in eco-pharmacovigilance. J. Forensic Res 5:228
Richards NL, Gilbert M, Taggart M, Naidoo V (2018) A cautionary tale: diclofenac and its profound impact on vultures. In: Dellasala DA, Goldstein MI (eds) Encyclopedia of the Anthropocene. Elsevier, Oxford, pp. 247–255
Ruiz-Suárez N, Boada LD, Henríquez-Hernández LA et al (2015) Continued implication of the banned pesticides carbofuran and aldicarb in the poisoning of domestic and wild animals of the Canary Islands (Spain). Sci Total Environ 505:1093–1099
Ruxton GD, Houston DC (2004) Obligate vertebrate scavengers must be large soaring fliers. J Theor Biol 228:431–436
Sage M, Fourel I, Coeurdassier M et al (2010) Determination of bromadiolone residues in fox faeces by LC/ESI-MS in relationship with toxicological data and clinical signs after repeated exposure. Environ Res 110(7):664–674
Sánchez-Barbudo IS, Camarero PR, Mateo R (2012) Primary and secondary poisoning by anticoagulant rodenticides of non-target animals in Spain. Sci Total Environ 420:280–288
Sarrazin F, Bagnolini C, Pinna JL et al (1994) High survival estimates of griffon vultures (Gyps fulvus fulvus) in a reintroduced population. Auk 111(4):853–862
Scheuhammer AM, Templeton DM (1998) The use of stable isotope ratios to distinguish sources of lead exposure in wild birds. Ecotoxicology 7:37–42
Sharma P (2012) Aceclofenac as a potential threat to critically endangered vultures in India: a review. J Raptor Res 46(3):314–318
Sieg R, Sullivan KA, Parish CN (2009) Voluntary lead reduction efforts within the northern Arizona range of the California Condor. In: Watson RT, Fuller M, Pokras M et al (eds) Ingestion of lead from spent ammunition: implications for wildlife and humans. The Peregrine Fund, Boise
Sileo L, Fefer S (1987) Paint chip poisoning of Laysan albatross at Midway Atoll. J Wildl Dis 23:432–437
Smith RH (1999) Population biology and non-target effects of rodenticides: trying to put the eco into ecotoxicology. In: Cowan DP, Feare CJ (eds) Advances in vertebrate pest management. Filander, Furth, pp 331–346
St John FAV, Keane AM, Edwards-Jones G et al (2011) Identifying indicators of illegal behaviour: carnivore killing in human-managed landscapes. Proc R Soc Lond B Biol Sci 279:804–812
Stone WB, Okoniewski JC, Stedelin JR (1999) Poisoning of wildlife with anticoagulant rodenticides in New York. J Wildl Dis 35(2):187–193
Stone WB, Okoniewski JC, Stedelin JR (2003) Anticoagulant rodenticides and raptors: recent findings from New York, 1998-2001. Bull Environ Contam Toxicol 70:34–40
Swarup DPRC, Patra RC, Prakash V et al (2007) Safety of meloxicam to critically endangered Gyps vultures and other scavenging birds in India. Anim Conserv 10(2):192–198
Taggart MA, Richards NL, Kinney CA (2015) Impacts of pharmaceuticals on terrestrial wildlife. In: Hester RE, Harrison RM (eds) Pharmaceuticals in the environment. Royal Society of Chemistry, pp 216–254
Tavares J (2013) Overview of the main threats to vultures in Europe today. In: Papazoglou C, Charalambous C (eds) Proceedings of the Griffon Vulture conference. BirdLife Cyrpus, Nicosia, pp 77–88
Tella JL (2001) Action is needed now, or BSE crisis could wipe out endangered birds of prey. Nature 410(6827):408–408
Thévenot M, Bergier P, Beaubrun P (1985) Present distribution and status of raptors in Morocco. In: Newton I, Chancellor RD (eds) Conservation studies on raptors. International Council for Bird Preservation, Cambridge, pp 83–101

Thévenot M, Vernon R, Bergier P (2003) The birds of Morocco. British Ornithologist's Union. p 594
Thomas VG (2013) Lead-free hunting rifle ammunition: product availability, price, effectiveness, and role in global wildlife conservation. Ambio 42(6):737–745
Thomas PJ, Mineau P, Shore RF et al (2011) Second generation anticoagulant rodenticides in predatory birds: probabilistic characterisation of toxic liver concentrations and implications for predatory bird populations in Canada. Environ Int 37(5):914–920
Tingay RE, McAdam D, Taylor MJ (2012) Perspectives on wildlife poisoning by carbofuran in the United Kingdom and Republic of Ireland–With a particular focus on Scotland. In: Carbofuran and wildlife poisoning: global perspectives and forensic approaches, pp 171–188
Tixier T, Blanckenhorn WU, Lahr J et al (2015) A four-country ring test of non-target effects of ivermectin residues on the function of coprophilous communities of arthropods in breaking down livestock dung. Environ Toxicol Chem 35(8):1953–1958
Toure O (1988) The pastoral environment of northern Senegal. Rev Afr Polit Econ 15(42):32–39
U.S. Environmental Protection Agency (USEPA) (1989) Carbofuran—special review technical support document. Office of Pesticides and Toxic Substances, Washington, DC
U.S. Environmental Protection Agency (USEPA) (2016a) Biological evaluation chapters for Chlorpyrifos. Biological evaluation chapters for chlorpyrifos. www.epa.gov/endangered-species/biological-evaluation-chapters-chlorpyrifos. Accessed 9 Apr 2016
U.S. Environmental Protection Agency (USEPA) (2016b) Biological evaluation chapters for diazinon. www.epa.gov/endangered-species/biological-evaluation-chapters-diazinon. Accessed 9 Apr 2016
U.S. Environmental Protection Agency (USEPA) (2016c) Biological evaluation chapters for malathion. www.epa.gov/endangered-species/biological-evaluation-chapters-malathion. Accessed 9 Apr 2016
Van Hoof N, De Wasch K, Poelmans S et al (2004) Multi-residue liquid chromatography /tandem mass spectrometry method for the detection of non-steroidal anti-inflammatory drugs in bovine muscle: optimisation of ion trap parameters. Rapid Commun Mass Spectrom 18(23):2823–2829
Varian-Ramos CW, Swaddle JP, Cristol DA (2014) Mercury reduces avian reproductive success and imposes selection: an experimental study with adult- or lifetime-exposure in zebra finch. PLoS One 9(4):e95674
Verdú JR, Cortez V, Ortiz AJ et al (2015) Low doses of ivermectin cause sensory and locomotor disorders in dung beetles. Sci Rep 5:13912
Verma K, Paul R (2013) Assessment of post mortem interval, (PMI) from forensic entomotoxicological studies of larvae and flies. Entomol Ornithol Herpetol 2:104–108
Verschuren J (1975) Wildlife in zaire. Oryx 13:149–163
Walker LA, Turk A, Long SM et al (2008) Second generation anticoagulant rodenticides in tawny owls (Strix aluco) from Great Britain. Sci Total Environ 392(1):93–98
Wells K, Butterworth A, Richards NL (2019) A review of secondary pentobarbital poisoning in wild, domestic and captive carnivores. J Veterinary Forensic Sci (Inaugural Issue). In Press
Whitfield DP, McLeod DR, Watson J et al (2003) The association of grouse moor in Scotland with the illegal use of poisons to control predators. Biol Conserv 114(2):157–163
Whitfield DP, Fielding AH, McLeod DRA, Haworth PF (2004) Modelling the effects of persecution on the population dynamics of golden eagles in Scotland. Biol Conserv 119(3):319–333
Wilbur SR, Carrier WD, Borneman JC (1974) Supplemental feeding program for California condors. J Wildl Manag 38(2):343–346
Yarnell RW, Phipps WL, Dell S et al (2015) Evidence that vulture restaurants increase the local abundance of mammalian carnivores in South Africa. Afr J Ecol 53(3):287–294
Zorrilla I, Taggart M, Martinez R, Richards NL (2015) Suspected flunixin poisoning of a wild Eurasian Griffon Vulture from Spain. Conserv Biol 29(2):587–592
Zorrilla I, Richards NL, Benítez JR et al (2018) Case study: detection of two non-steroidal anti-inflammatory drugs (NSAIDs) in the eggs of captive-reared Bearded Vultures at a breeding centre in southern Spain. J Wildlife Rehab. 38(1):15–27.

Part III
Methodological Approaches: Description of Some Methods Used to Study Carrion Ecology

Methods for Monitoring Carrion Decomposition in Aquatic Environments

Gail S. Anderson and John R. Wallace

Contents

Introduction

In aquatic ecosystem biological monitoring (biomonitoring), living organisms, predominantly benthic invertebrates or those invertebrates living in the benthos of fresh and marine waters, have been used to determine the quality of the aquatic environment as well as the impact of disturbance or toxicants in such environments (Van Urk et al. 1993; Maret et al. 2003; Rosenberg et al. 2008). Because terrestrial invertebrates have evolved to feed on carrion as described in chapters "Invertebrate Scavenging Communities" and "Carrion Decomposition", there is a predictable assemblage of taxa, whose community is associated with carrion that is intimately tied to spatial and temporal constraints (Merritt and De Jong 2016). However, similar convergent evolutionarily patterns of aquatic organisms with carrion have not been well studied, and therefore, aquatic (specifically freshwater) organisms offer

G. S. Anderson (✉)
School of Criminology, Centre for Forensic Research, Simon Fraser University, Burnaby, BC, Canada
e-mail: ganderso@sfu.ca

J. R. Wallace
Department of Biology, Millersville University, Millersville, PA, USA
e-mail: John.Wallace@millersville.edu

P. P. Olea et al. (eds.), *Carrion Ecology and Management*, Wildlife Research Monographs 2, https://doi.org/10.1007/978-3-030-16501-7_10

up few indicators of comparable relationships as terrestrial carrion decomposers (Merritt and Wallace 2010). For this reason, monitoring carrion in aquatic environments is much more complex than in terrestrial habitats.

The difficulties in using benthic organisms in the biomonitoring of carrion vary from constraining factors such as accessibility to the carrion of interest and dealing with abiotic features in these aqueous environs to study design and analysis (Rosenberg and Resh 2001). For example, most carcasses are inaccessible except by divers or remotely operated vehicles such as those used to study whale falls in deep abyssal oceanic environments (Smith et al. 2002; Smith and Baco 2003). In deeper waters there are considerably more safety issues associated with aquatic environs which impact access; for example, depth and weather conditions impact diver safety and equipment suitability. If divers are used, more than one diver is required as a safety requirement and most access requires vessel time that greatly increases costs. The variability of abiotic factors in both freshwater and marine environments such as the dramatic changes in temperature, dissolved oxygen, pH, salinity gradients between freshwater, estuarine and marine ecosystems, current and tidal strength, pressure over just a few meters in deep water oceanic environs, play an important role in carrion decomposition and the rate of decomposition (see chapter "Invertebrate Scavenging Communities"), but also influence the dominant pathways of decomposition especially those mediated by biological mechanisms such as with microbial, invertebrate and vertebrate decomposers of carrion (Merritt and Wallace 2010; Wallace 2016). Therefore, in a traditional monitoring sense, physical/chemical monitoring of abiotic features such as temperature, current, pH and salinity offer up instantaneous snapshots of the conditions existing where the carrion are found (Rosenberg et al. 2008). Whereas, the use of those organisms feeding on carrion exposed to this abiotic variation provide temporal and spatial components associated with carrion decomposition (Rosenberg et al. 2008; Wallace 2016).

Regarding study design and analysis, because the abiotic features mentioned above can affect the distribution and abundance of microbial and invertebrate carrion decomposers, this interaction requires the need for ecological knowledge of all species involved and judiciously considered in the empirical design in carrion monitoring studies (Rosenberg and Resh 2001). Cost can be prohibitive with regards to quantitative aspects of sampling e.g., adequate replication and high sample sizes required for statistical precision especially with molecular analyses of microbe diversity on carrion and in the monitoring carrion in deep, abyssal environments in oceans. These studies can also be time consuming as well (Rosenberg and Resh 2001). Taxonomic analyses of organisms colonizing carrion in both freshwater and marine ecosystems can be difficult e.g., midges (Chironomidae) in freshwater, and some species of amphipods (Amphipoda) in marine settings and microbial diversity in both environments. Lastly, while there are many biotic and diversity indices available for biomonitoring the water quality, to our knowledge, there are few if any designed to evaluate carrion decomposition via organismal colonization. However, the use of microbial diversity in such studies may provide datasets for which these traditional metrics can be applied (Benbow et al. 2015).

A comprehensive review of the abiotic parameters within freshwater and marine ecosystems and their use in water quality evaluations in these systems is presented in Hauer and Lamberti (1996) and Twomey et al. (2009). While physical/chemical parameters and biological sampling are not mutually exclusive in the understanding of carrion decomposition in aqueous environments, the focus of this chapter will be on the methods used to sample or collect biological organisms associated with carrion decomposition in freshwater and marine ecosystems.

Freshwater

In manipulative studies where carrion or carcasses are placed in either lentic (ponds, lakes) or lotic (streams, rivers) systems, it may be desired to understand the complete picture of macroinvertebrate diversity in these systems for comparative purposes to those that might be found on the carcass itself. With that goal in mind, sampling freshwater organisms including plants, microbes and macroinvertebrates associated with carrion is generally based on the habitat and community sampled; however, the ultimate criterion for which sampling device is used depends on the objectives of the study (Merritt et al. 2008). A thorough understanding of the habitat to be sampled is necessary due to the non-random and clumped distribution patterns of these organisms associated with microhabitat preferences, food etc. in freshwater systems (Resh 1979). Whether the objectives of a particular carrion study are qualitative or quantitative plays a role in which sampling device is selected. A table listing the sampling devices and applicable habitats is described in depth in Merritt et al. (2008). A qualitative or semi quantitative approach (an approximate area that is delineated) to sample the macroinvertebrates present in a system where carrion remains might be studied in freshwater systems would include a kick net or seine (Merritt et al. 2008). A kick seine would provide relative abundances of organisms present in a specific area for comparative purposes to those found on carrion. If sampling directly from carrion is desired in a manipulative fashion, carcasses, such as stillborn pigs could be placed in cages and anchored to stream bottoms with rebar as shown in Fig. 1. When cages are removed from the streams for invertebrate removal, D-frame nets placed behind or on the downstream end of the cage can be used to collect any macroinvertebrates lost or released from the carrion upon removal (Fig. 1). There are numerous sampling devices for quantitative sampling of freshwater streams e.g., Surber, T or Hess samplers that are described in Merritt et al. (2008). Water depth and velocity will be determining factors on which device is most appropriate.

Other approaches to study macroinvertebrate diversity on submerged carcasses have involved the use of a specifically constructed sampling apparatus designed to lift the carcass out of a lentic system such as a pond and allow manual removal of invertebrates (Fig. 2).

Macroinvertebrate samples can be sorted using a number of procedures depending on the time and cost issues. There are a number of issues with these procedures that may influence the quality of data obtained, for further elaboration, consult Table 3G in Merritt et al. (2008). In terms of preserving freshwater macroinvertebrates, current recommendations would depend on the objectives for the study. For example, for simple macroinvertebrate taxonomic evaluation, 70–80% ethanol is adequate (Merritt et al. 2008).

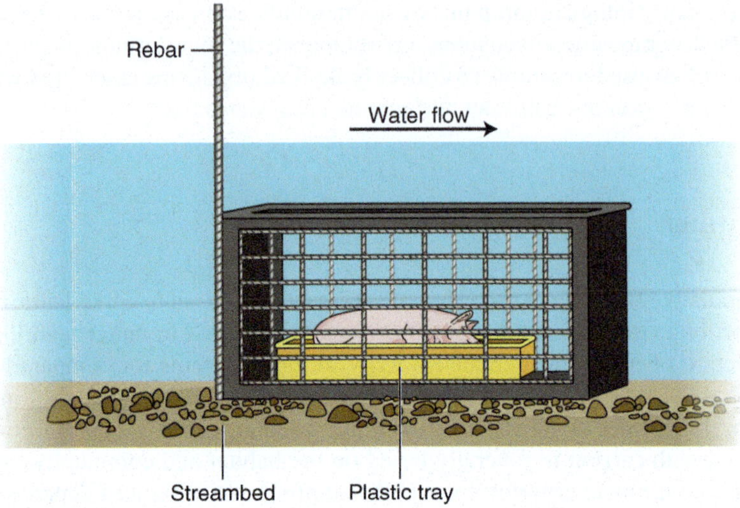

Fig. 1 Experimental cage design to study macroinvertebrate colonization on carrion and placement of D-frame nets to collect aquatic organisms. Credits: the chapter authors

Fig. 2 A sampling unit that holds the carcass in a basket and allows for it to be lifted from the water as described by Vance et al. (1995) (With Permission of the American Academy of Forensic Sciences as they hold the copyright)

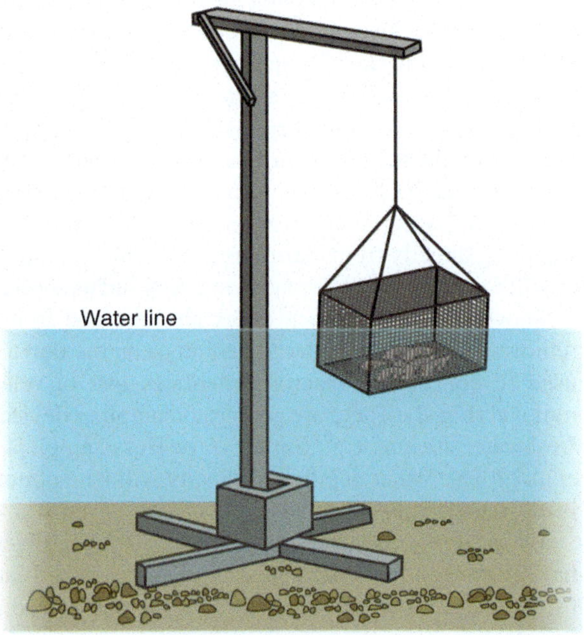

Although not generally considered, algae will colonize carrion as described in several studies focused on forensic applications (Haefner et al. 2004; Zimmerman and Wallace 2008). Because aquatic plants such as algae may utilize a carcass as nutrients for growth, their order of appearance or biomass may influence faunal colonization of carrion remains in these environments (Keiper and Casamatta 2001). Therefore, sampling techniques from carrion would be important in understanding the role of algae in carrion decomposition. Using a similar design to Fig. 1, algal samples can be collected for either qualitative (species composition) or quantitative (biomass) analyses using a modified 2—syringe periphyton sampler as described by Aloi (1990) (Fig. 3). Using a periphyton sampler, samples can be scraped from carrion as well as artificial tile substrates for comparison (Fig. 4). Chlorophyll a concentrations can be determined from scraped samples as described by Haefner et al. (2004) to quantify the amount of algae on a carcass. Similarly, scraped samples can be analyzed for algal diversity along a carrion decomposition gradient as described by Zimmerman and Wallace (2008).

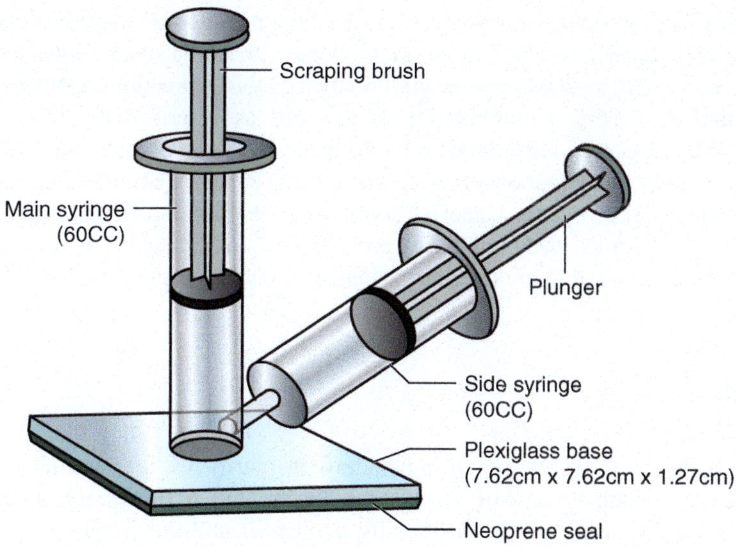

Fig. 3 Algal sampler as described by Aloi (1990) and illustrated by Haefner et al. (2004) (With Permission of the American Academy of Forensic Sciences as they hold the Copyright)

Fig. 4 Stillborn pig carcass with algal sample removed from the right front leg using a periphyton sampler (Photo Credit: J. Wallace)

Preservation techniques for algal samples vary according to study objectives. Previous studies suggest that samples used for chlorophyll analyses should be stored in a cool, dark environment post sampling in order to preserve chlorophyll for quantification (Haefner et al. 2004). If assessing algal diversity, samples should be stored in a glass vial and preserved with concentrated HCl and Lugol's solution and maintained in the dark to avoid degradation of the solution (Zimmerman and Wallace 2008).

The application of microbial communities to track temporal aspects of carrion decomposition has recently received increased attention (Benbow et al. 2015). The use of metagenomic tools to characterize the epinecrotic bacterial communities on vertebrate carrion in aquatic systems has allowed a more thorough treatise of the bacterial diversity present on carrion remains through decomposition (Benbow et al. 2015; Lang et al. 2016). These studies have employed similar carcass submersion methods as described above by using stillborn pigs in cages. However, sampling the epinecrotic habitats for microbes involves swabbing predetermined areas on pigs after designated time periods. The swabbings are done with sterile swabs. After swabbing, swabs are stored in a –20 °C freezer until ready for DNA extractions. DNA extraction is accomplished using a combination of methods as described by Benbow et al. (2015). The epinecrotic bacterial communities can either be characterized via pyrosequencing techniques (Benbow et al. 2015) or via ARISA techniques that generate a unique "fingerprint" of the bacterial community (Fisher and Triplett 1999). This method produces microbial community profiles that can be visualized with nonmetric multidimensional scaling multivariate statistical analyses (Lang et al. 2016).

Marine

Carrion studies in the marine environment have primarily revolved around extremely large carcasses such as those of whales which are natural to the marine environment (either naturally discovered or artificially deployed) and much smaller carcasses, deployed as human proxies, which are not natural to the environment. As the size differences are so great between the two extremes, monitoring is quite different, with larger carcasses taking decades to completely break down (Smith and Baco 2003) and human proxies taking only days or weeks to skeletonization (Anderson and Bell 2014, 2016). Much of the data collected in the ocean are photographic, with sometimes limited physical sampling. Most studies involve either observations made from a submersible vehicle or from a baited camera.

Early whale falls were serendipitous discoveries of natural deaths in various stages of decomposition, with the exact time of death unknown. In the last few years, naturally deceased whale carcasses have been deliberately deployed and observed over time. In such studies, the usual method for monitoring is to use manned or remotely operated underwater vehicles (ROV) (Smith et al. 2002) or towed cameras (Smith et al. 2014) to view, record and sample the carcass and water conditions on infrequent occasions. Most data are recorded photographically, but, depending on the submersible used, bones and fauna can also be sampled using the submersibles manipulators and sweep nets (Bennett et al. 1994). Due to the prohibitive costs, most whale falls are sampled only once (Bennett et al. 1994) or over a lengthy time series (Smith and Baco 2003) with a great deal of data being collected over very short periods in single dives over several years.

Baited cameras and traps are frequently used to record the presence and changes in opportunistic scavengers over a range of habitats and depths. Many such studies use a small fish or meat bait placed in front of a time-lapse submersible camera situated on an autonomous lander frame, and are usually monitored over a short time frame of one to two days, as the bait is rapidly consumed (Jones et al. 2003). Such experiments can be conducted over a large range of depths and provide data on changes in the necrophagous fauna that can be expected to scavenge fresh bait in various habitats and depths. Most fauna are identified visually but baited funnel traps can also be used (Janssen et al. 2000).

Baited cameras have also been used to study larger baits over longer time frames. Dolphin and porpoise carcasses have been monitored under baited cameras supplemented with baited funnel traps over a period of up to two weeks followed by carcass recovery using a box trap that closed over the carcass to bring the entire carcass to the surface for further study (Jones et al. 1998). In most circumstances, abiotic data can only be recorded at certain points or has not been recorded at all.

Ideally, monitoring of carcasses should include regular and frequent observations together with frequent recording of all abiotic parameters. The Victoria Experimental Network under Sea (VENUS) observatory (Ocean Networks Canada) is an elaborate cabled underwater laboratory in the Salish Sea, on the west coast of Canada and the US (http://dmas.uvic.ca/). High-speed, real-time data are provided from more than 50 oceanographic instruments gathering acoustic, physical, chemical and photographic data simultaneously. The instruments are connected via fibre optic cables to a science instrument interface module and then to the Node, a power and communication hub, which connects to a station on shore, which links to the University of Victoria. At the University, DMAS, a data managing and archiving system, receives and processes the data, allowing researchers to access their data via the internet (http://dmas.uvic.ca/). VENUS provides researchers with an ideal environment in which to study the ocean, including carrion, and a number of pig carcasses have been deployed in several habitats, seasons and depths since 2006 (Anderson and Bell 2014, 2016). Such deployments attract not only invertebrates but also large vertebrates such as sharks (Fallows et al. 2013), which can destroy the experiment. When invertebrates are the target, it is necessary to use a cage system to protect the carcasses from shark removal but allow full access for invertebrates (Fig. 5).

In these studies a camera platform was deployed by an ROV, under which pig carcasses were placed. The camera, originally an Olympus C8080, 8 mp, later replaced by a digital webcam, recorded a scan of the carcasses every 15 min, for 2 min, during which time a range of lights were turned on, then off to reduce light pollution. All carrion fauna and decompositional changes were recorded visually, as once placed, there was no further direct contact until recovery. Instruments recording abiotic factors were placed, ideally, on the camera platform, and also on an instrument panel ~100 m from the carcasses. Instruments recorded dissolved oxygen,

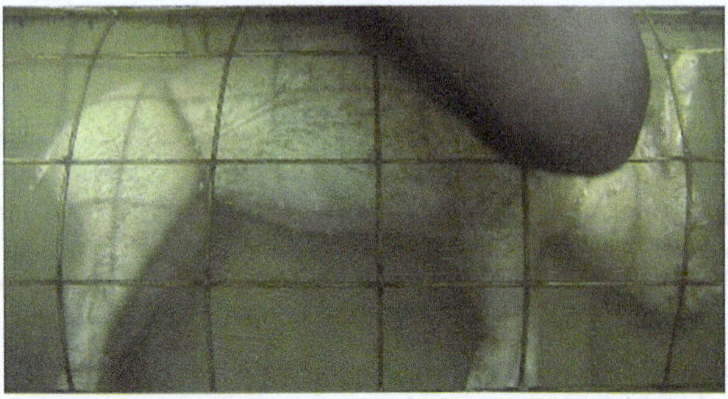

Fig. 5 Twenty-four kg pig carcass at a depth of 300 m in the Strait of Georgia, hours after placement, caged to prevent shark access. Sixgill shark (*Hexanchus griseus* (Bonneterre)) swimming over the carcass (Photo credits: Ocean Networks Canada VENUS Observatory)

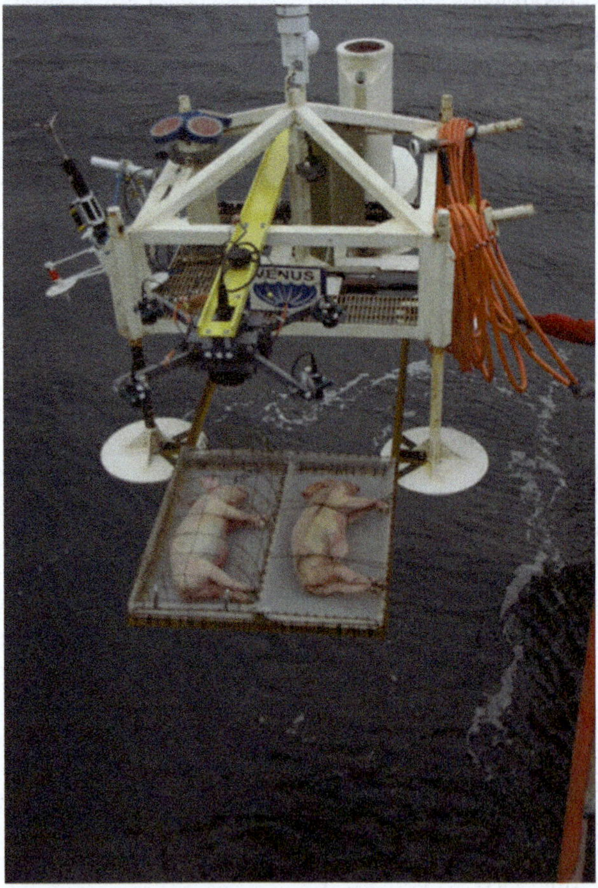

Fig. 6 Two twenty-four kg pig carcasses placed under a digital webcam on a platform and being lowered into the ocean from a research vessel. Instruments measuring abiotic parameters are mounted on the camera platform. Photo by G. S. Anderson. (Equipment Ocean Networks Canada, VENUS Observatory)

temperature, salinity, density, turbidity, conductivity and pressure at per second intervals (Fig. 6). This allowed abiotic parameters to be linked to changes in faunal scavenging. This, however, is an ideal arrangement and is rarely available. The only disadvantage is the lack of ability to sample fauna except when the carcasses are recovered, six months after deployment (http://dmas.uvic.ca/).

In shallower waters, it is possible to deploy divers to observe and sample carrion directly or bring a carcass to the surface for brief examination (Anderson 2010) but this limits the depth and habitats that can be studied.

Although most carrion research in the ocean relates to whale falls or human proxies, recently other carrion types have been explored including natural jellyfish die-offs. Most falls have so far been monitored by cameras towed behind a vessel with some core sampling (Billett et al. 2006) or baited cameras (Sweetman et al. 2014). In a review of jelly-fall observations, Lebrato et al. (2012) recommend several future sampling and monitoring techniques. Regional (large scale studies) methods recommended include Deep ocean observatory networks and collaborations with industry, such as off-shore oil and gas, to locate and monitor jelly-falls whereas, ROVs, towed cameras, time lapse cameras, fishery trawlers and tagging individuals acoustically were recommended for medium scale studies with sediment traps, water column profilers and sediment sampling for genetic profiling recommended for small scale studies (Lebrato et al. 2012).

Conclusions and Future Perspectives

Overall, there are many techniques available for studying aquatic carrion ecology, although the techniques employed vary depending on the research question and desired data. Studying carrion in aquatic environs is much more demanding than on land and involves unique challenges. There are many limitations and deficits in our empirical knowledge of carrion decomposition in freshwater systems. For example, carcasses are difficult to access and observe without anthropogenic disturbance due to proximity of human recreational activity. In some cases, permits may be required thereby slowing or prohibiting natural experiments. Manipulative empirical studies are lacking in deep freshwater lakes and estuarine systems and need to be addressed to understand carrion ecology in one of the most productive ecosystems on the planet. The advent of unique underwater laboratories such as Ocean Networks Canada's VENUS Observatory used in oceanic systems may be of some use in deep freshwater lakes.

In marine systems, physical aspects of oceanic systems may deter or make difficult the studying of carrion decomposition. For example, greater depths, tides, currents and distance from shore create increased safety concerns for researchers, and greatly increase costs. The cabled underwater laboratory, Ocean Networks Canada's VENUS Observatory, offers tremendous opportunities for researchers and, as almost all work, except deployment and recovery, are remote, removes the normal safety concerns. As well, it allows the possibility of continuously observable experiments at great depths and challenging environments not previously studied. Unfortunately, this facility has not yet been duplicated elsewhere in the world. Underwater carrion ecology is an area little studied and much more research needs to be carried out. Underwater cabled laboratories such as VENUS should be established and supported in other areas of the globe.

References

Aloi JE (1990) A critical review of recent freshwater periphyton field methods. Can J Fish Aquat Sci 47:656–670

Anderson GS (2010) Decomposition and invertebrate colonization of cadavers in coastal marine environments. In: Amendt J, Campobasso CP, Grassberger M, Goff ML (eds) Current concepts in forensic entomology. Springer, Berlin

Anderson GS, Bell LS (2014) Deep coastal marine taphonomy: investigation into carcass decomposition in the Saanich inlet, British Columbia using a baited camera. PLoS One 9(10):e110710. https://doi.org/10.1371/journal.pone.0110710

Anderson GS, Bell LS (2016) Impact of marine submergence and season on faunal colonization and decomposition of pig carcasses in the Salish Sea. PLoS One 11(3):e0149107. https://doi.org/10.1371/journal.pone.0149107

Benbow ME, Pechal JL, Lang JM, Erb R, Wallace JR (2015) The potential of high-throughput metagenomic sequencing of aquatic bacterial communities to estimate the postmortem submersion interval. J Forensic Sci 60(6):1500–1510. https://doi.org/10.1111/1556-4029.12859

Bennett BA, Smith CR, Glaser B, Maybaum HL (1994) Faunal community structure of a chemoautotrophic assemblage on whale bones in the deep Northeast Pacific Ocean. Mar Ecol Prog Ser 108:205–223

Billett DSM, Bett BJ, Jacons CL, Rouse IP, Wigham BD (2006) Mass deposition of jellyfish in the deep Arabian Sea. Limnol Oceanogr 51(5):2077–2083

Haefner JN, Wallace JR, Merritt RW (2004) Pig decomposition in lotic aquatic systems: the potential use of algal growth in establishing a postmortem submersion interval (PMSI). J Forensic Sci 49(2):330–336

Hauer FR, Lamberti GA (1996) Methods in stream ecology. Academic, New York, p 674

Janssen F, Treude T, Witte U (2000) Scavenger assemblages under differing trophic conditions: a case study in the deep Arabian Sea. Deep Sea Res II 47:2999–3026

Jones EG, Collins MA, Bagley PM, Addison S, Priede IG (1998) The fate of cetacean carcasses in the deep sea: observations on consumption rates and succession of scavenging species in the abyssal north-East Atlantic Ocean. Proc R Soc Lond B 265(1401):1119–1127

Jones EG, Tselepides A, Bagley PM, Collins MA, Priede IG (2003) Bathymetric distribution of some benthic and benthopelagic species attracted to baited cameras and traps in the deep eastern Mediterranean. Mar Ecol Prog Ser 251:75–86

Lang JM, Erg R, Pechal JL, Wallace JR, McEwan RW, Benbow ME (2016) Microbial biofilm community variation in flowing habitats: potential utility as bioindicators of postmortem submersion intervals. Microorganisms 4(1):1. https://doi.org/10.3390/microorganisms4010001

Lebrato M, Pitt KA, Sweetman AK, Jones DOB, Cartes JE, Oschlies A, Condon RH, Molinero JC, Adler L, Gaillard C, Lloris D, Billett DSM (2012) Jelly-falls historic and recent observations: a review to drive future research directions. Hydrobiologia 690(1):227–245. https://doi.org/10.1007/s10750-012-1046-8

Maret TR, Cain DJ, MacCoy DE, Short TM (2003) Response of benthic invertebrate assemblages to metal exposure and bioaccumulation associated with hard-rockmining in northwestern streams, USA. J N Am Benthol Soc 22:598–620

Merritt RW, Wallace JR (2010) The role of aquatic insects in forensic investigations. In: Byrd JH, Castner JL (eds) Forensic entomology: the utility of arthropods in legal investigations. CRC Press, Boca Raton, pp 271–319

Merritt RW, Cummins KW, Resh VH, Batzer DP (2008) Sampling aquatic insects: collection devices, statistical considerations and rearing procedures. In: Merritt RW, Cummins KW, Berg MB (eds) An introduction to the aquatic insects of North America. Kendall Hunt, Dubuque, pp 15–37

Resh VH (1979) Sampling variability and life history features: basic considerations in the Design of Aquatic Insect Studies. J Fish Res Bd Can 36:290–311

Rosenberg DM, Resh VH (2001) Introduction to freshwater biomonitoring and benthic macroinvertebrates. In: Rosenberg DM, Resh VH (eds) Freshwater biomonitoring and benthic macroinvertebrates. Kluwer Academic Publishers, Boston, p 5

Rosenberg DM, Resh VH, King RS (2008) Use of aquatic insects in biomonitoring. In: Merritt RW, Cummins KW, Berg MB (eds) An introduction to the aquatic insects of North America. Kendall Hunt, Dubuque, pp 123–137

Smith CR, Baco AR (2003) Ecology of whale falls at the deep-sea floor. Oceanogr Mar Biol Ann Rev 41:311–354

Smith CR, Baco AR, Glover AG (2002) Faunal succession on replicate deep-sea whale falls: time scales and vent-seep affinities. Cah Biol Mar 43:293–297

Smith KE, Thatje S, Singh H, Amsler MO, Vos SC, McClintock JB, Brothers CJ, Brown A, Ellis D, Anderson JS, Aronson RB (2014) Discovery of a recent, natural whale fall on the continental slope off Anvers Island, western Antarctic peninsula. Deep Sea Res 90:76–80. https://doi.org/10.1016/j.dsr.2014.04.013

Sweetman AK, Smith CR, Dale T, Jones DO (2014) Rapid scavenging of jellyfish carcasses reveals the importance of gelatinous material to deep-sea food webs. Proc Biol Sci 281(1796):20142210. https://doi.org/10.1098/rspb.2014.2210

Twomey LJ, Piehler MF, Paerl HW (2009) Priority parameters for monitoring of freshwater and marine systems and their measurement. In: Inyang II, Daniels JL (eds) Environmental monitoring. EOLSS, Kingston, pp 318–338

Van Urk GF, Kerkum F, Leeuwen CJV (1993) Insects and insecticides in the lower Rhine. Wat Res 27:205–213

Vance GM, VanDyk JK, Rowley WA (1995) A device for sampling aquatic insects associated with carrion in water. J Forensic Sci 40(3):479–482

Wallace JR (2016) Aquatic vertebrate carrion decomposition. In: Benbow ME, Tomberlin JK, Tarone AM (eds) Carrion ecology, evolution and their applications. CRC Press, Boca Raton, pp 247–271

Zimmerman K, Wallace JR (2008) Estimating a postmortem submersion interval using algal diversity on mammalian carcasses in brackish marshes. J Forensic Sci 53(4):935–941

Studying Movement of Avian Scavengers to Understand Carrion Ecology

Olivier Duriez, Roi Harel, and Ohad Hatzofe

Contents

Introduction

Studying carrion ecology and their importance in the ecosystem is difficult in the field, because carrions presence is usually unpredictable in time and space. Moreover, carrion is a pulsed resource that can be quite ephemeral in the ecosystems as they are usually rapidly eliminated by scavengers. Therefore, studying scavengers' movement ecology can help to a better understanding of the scavenging-related patterns and processes. The role of vultures in the ecosystems are of prime importance because they are the most specialized terrestrial vertebrate scavengers (Ruxton and Houston 2004) and they are highly skilled movers able to transport scavenging services over large areas (DeVault et al. 2016). In this chapter we will focus on the methods to study vulture movement ecology, and review recent results about their

O. Duriez (✉)
CEFE UMR 5175, CNRS, Université de Montpellier, Université Paul-Valéry Montpellier, EPHE, Montpellier, France
e-mail: olivier.duriez@cefe.cnrs.fr

R. Harel
Movement Ecology Lab, Department of Ecology, Evolution and Behavior, Alexander Silberman Institute of Life Sciences, The Hebrew University of Jerusalem, Jerusalem, Israel
e-mail: roi.harel@mail.huji.ac.il

O. Hatzofe
Science Division, Israel Nature and Parks Authority, Jerusalem, Israel
e-mail: ohad@npa.org.il

P. P. Olea et al. (eds.), *Carrion Ecology and Management*, Wildlife Research Monographs 2, https://doi.org/10.1007/978-3-030-16501-7_11

foraging ecology and movements. Tracking vultures can help identifying mortality of them as well as of both wild and domesticated ungulates. Thus, tracking scavengers can be highly important in the context of global environmental change for detecting alterations in the ecosystems derived from changes in the carrion availability.

Trapping Vultures

Trapping vultures is generally not an easy task and must be performed professionally in order to minimize the possibility of injury or stress to birds. When planning trapping efforts, the time of day, season (breeding status of the birds), weather, number of birds to be captured, staff required, and the possibility of predation should be taken into account (Bird and Bildstein 2007).

It is possible to mark chicks at nest but climbing skills and gear are required. Furthermore, a detailed nest monitoring, from hatching date, is crucial because early arrival to the nest will prevent deployment of tagging equipment as it may affect body growth, whereas delayed arrival to the nest may entail premature fledging (and death).

After fledging, the most common methods are based on attracting vultures to a food source as bait to catch them on the ground (Fig. 1). There are several alternatives, most described in details by Bloom et al. (2007), and summarized in Table 1. According to the review detailed in Table 1, the most common method of capture was walk-in traps (10 studies), followed by leg-hold traps, projected-nets and nestling (4 studies each) and rehabilitated birds (2).

Tagging and Tracking Vultures

The possibilities to tag vultures are diverse and so are the costs and the potential output for management and research (Fig. 2). Darvic rings (numbered plastic rings) are good for mass tagging and allow long distance observation, typically up to 400 m on large vultures (even 700 m for color-ring codes). However because they might be broken or lost (Mihoub et al. 2013), they must be complemented by metal rings that would allow a re-identification of bird if recaptured. In addition, leg rings are not suitable for Cathartidae due to their habit to defecate on their legs, presumably for thermoregulation (Houston 1994). Wing tags (also called "patagial tags") enhance resighting of birds at long distance and in flight (Wallace et al. 1980; Reading et al. 2014). However, it must be reminded that the impact of wing tags on bird flight (by increasing drag) has never been properly investigated in large soaring raptors (Trefry et al. 2013). Radio Frequency Identification Tags (RFID) allow detection (presence-absence) by passive tags at focal sites (Bonter and Bridge 2011). RFID can be either passive and provide presence data at short distances (up

Fig. 1 Four different systems of traps for capturing vultures: (**a**) Walk-in trap in an aviary (active because an observer is hidden nearby to trigger the sliding door, France; Photo credit O.Duriez); (**b**) Cage trap (passive because vultures fall into the cage through openings in the roof, Israel, Photo credit I. Shaked); (**c**) leg-hold trap (here more specifically Phai traps as noose surrounding a bait; the rope linking the nooses is hidden under the snow or ground; France, Photo credit O. Duriez); (**d**) Californian condor chick equipped at the nest with GPS tag embedded in patagial tag (photo courtesy of Joseph Brandt, USFWS California Condor Recovery Program)

to 0.5 m), or active (powered by batteries) and detected at longer distance (up to 100 m). If base stations are deployed in advance around carrions, data collected can provide information of arrival time, duration of stay and departure time.

Several techniques can be used to track vultures and we review here only the methods useful to study vulture foraging ecology. For foraging studies, we advise to use interval <10 min between consecutive positions because vultures can cover several km in a few minutes (see below) and sometimes can spend <15 min to land, feed and take off (Harel et al. 2016a; Monsarrat et al. 2013). VHF radio-tracking should not be advised nowadays for foraging studies because of its low precision and relative inefficiency (as flying vultures move much faster than land-based observers) despite intense fieldwork requirement (observers need to follow each bird and manually find its position by triangulation). Satellite telemetry using Argos transmitters provide geographic positions anywhere in the world but are not recommended for detailed foraging studies, due to low precision (typically >1 km) and time interval between positions often >1 h. GPS tracking is the most widely used system nowadays which allows precise tracking (typically <100 m error) at short time interval. Solar powered devices are preferred for multi-annual studies

Table 1 Summary of methods used to trap vultures, with main advantages and disadvantages

Trapping methods	Use	Advantage	Disadvantage	Source
Walk-in cage trap (active)	Mass capturing	Easy to handle, safe for the vultures	Time consuming as triggered by observers	Bloom et al. (2007), Barber and Bildstein (2011), Avery et al. (2011), Dodge et al. (2014), Harel et al. (2016a), Monsarrat et al. (2013), Garcia-Ripolles et al. (2011), Phipps et al. (2013a), Bamford et al. (2007), Kane et al. (2016), Vasilakis et al. (2016)
"Fall-in" cage trap (passive)	Mass capturing	Easy to handle, safe for the vultures, passive trapping (vultures enter in "their free time")	May be hard to attract vultures, not efficient where food is available in large quantities	Iezekiel et al. (2003), Harel et al. (2016a)
Cannon/rocket-projected/pneumatic-nets	Mass capturing	Highly efficient, selective (triggered by observer)	Projectiles can rarely hurt or kill birds, stressful, big team to handle, handling explosives may be complicated	Bamford et al. (2009), Lambertucci et al. (2014), Gil et al. (2014), Carrete et al. (2013), Lopez-Lopez et al. (2014)
Leg-hold traps ("padded jaws")/nooses carpets/Phai traps	For single birds/small numbers	Simple, small team required	Non species selective, need constant presence	Bloom et al. (2007), Dodge et al. (2014), Urios et al. (2010), Spiegel et al. (2013a)
Pit trap	For single birds (bearded vultures, condors)	Can be performed by only one person	Time consuming	Reid et al. (2015)

Credit: the chapter authors

(the longest tracking duration in France and in Israel was 5 years; Duriez and Hatzofe, unpublished). GPS transmitters usually store data on-board but several companies have developed systems to remotely download data using a radio-link, GSM or satellite transmission.

Tracking devices should be placed near the center of gravity on the bird's back. The common attachment method includes a Teflon ribbon harness which is robust and non-abrasive for skin. Some elasticity can be added by inserting a silicon tube inside the ribbon. Back-pack and body harness are efficient methods of attachment, but there is a risk of entanglement of the bird if one string is broken (Bögel et al.

Fig. 2 Eurasian griffon vultures *Gyps fulvus* carrying a GPS tag, (top) attached by a back-pack harness in silicon threaded Teflon ribbon (photo by I. Shaked); and (bottom) attached by a leg-loop harness (photo by O. Duriez). The position of both tags is located close to the bird's center of gravity. For long-distance visual identification, these birds are also marked with wing tags (top) and with a darvic coded ring (bottom)

2000). Leg-loop harness (Rappole and Tipton 1991) is fast and easy to fit, secure for the birds (as it prevents entanglement when broken), but power charging can sometimes be made more difficult because solar panels can be covered by wings when vultures are perched (Duriez unpublished). In condors telemetry devices have been mounted on wing-tag (Wallace et al. 1980), but this method is questionable because it adds weight on a sensitive place of the wing and can potentially change lift or drag forces.

Additional sensors coupled with the GPS transmitter can enhance the information collected on the behavior of individuals. Accelerometers are useful to estimate body posture and movements, and ultimately feeding behavior and thus determining the location of carrions found and eaten by vultures (Spiegel et al. 2013b; Nathan et al. 2012). Magnetometer data can provide additional data on body position and help in discriminating between different behaviors (Williams et al. 2017). In the near future, we believe that cameras can be deployed also on vultures' back to collect data on social behavior around carrions, as done on seabirds (Tremblay et al. 2014), however this technique is currently limited by the technical difficulty to download large amount of image data remotely without recapturing the bird. Nowadays the use of cameras is limited to the ground, with camera traps that can record attendance of scavengers (obligate and facultative) around carrions and help population monitoring by reading rings (Mateo-Tomas et al. 2017; Wilson 2015; Moreno-Opo et al. 2015; Ogada et al. 2012).

How Carrion Affect Foraging Movement of Vultures?

A main research question concerning the study of vulture movements in space and time is "how vultures make use of unpredictable food like carrions, and how does this affect their foraging strategies depending on the environmental context?"

We reviewed 39 studies published between 2007 and 2018 about vulture foraging movements using telemetry (Table 2), excluding studies about migration, dispersal or demography. 34 studies (87%) used GPS tracking devices and the 5 others used combinations of Argos or VHF telemetry units. 31 studies (79%) concerned Old-World vultures (family Accipitridae) and 8 concerned New-World vultures (family Cathartidae). Studies on Old-World vultures concerned 9 species out of 16 described (56%) and studies on Cathartidae concerned 4 species out of 7 described (57%). The 10 species where no data on foraging movements has been published (thus absent from Table 2) are *Gypohierax angolensis, Gyps indicus, G. ruppellii, G. tenuirostris, Necrosyrtes monachus, Sarcogyps calvus, Trigonoceps occipitalis*, and *Cathartes burrovianus, C. melambrotus, Sarcoramphus papa.* From a biogeographical perspective, while all species from Palearctic and Nearctic have been studied (at least once), only half of the species from Afrotropical and Neotropical regions have been studied and only one species from the Indo-Malayan region (Fig. 3). A similar pattern can be found regarding the main habitat, where all species from temperate-mountainous habitat have been studied, half of species from arid-Savanna habitats have been studied and only one species from tropical forest (Fig. 3). There are studies from all IUCN Red-list groups, however only a third of the species listed as Critically Endangered have been studied, as well as 42% of Least-Concerned species (Fig. 3). Finally, most studies concerned species displaying social behaviors for nesting or feeding, and few studies concerned territorial species (the notable exceptions being the bearded, lappet-faced and Egyptian vultures).

Therefore, our review points a lack of studies of vultures from the tropics, where vulture communities are (or were) the most abundant and diverse. This sample is

Table 2 Summary of methods used to track vultures and main results of home range size, daily distance travelled and daily maximum displacement (radius of feeding events from colony or roost), in 39 studies dealing with vulture foraging ecology

Species	Country	Tracking	Fix interval	home range[a]	Bird status[b]	Season	Home range (km^2)[d]	daily distance travelled (km)[d]	Radius of displacement[c,d] (km)	Source
Cathartes aura	USA	GPS	1 h	95% FK	Imm and ad	Summer	Med 551 [R 26–2860]			Avery et al. (2011)
					Imm and ad	Winter	Med 68 [R 36–25056]			
			Variable	95% MBK	Ad	Breed	53 ± 34			Holland et al. (2017)
					Ad	Non breed	65 ± 63			
	USA	GPS	1 h	–	Br ad	Summer		25 ± 25		Dodge et al. (2014)
	Argentina			–				59 ± 60		
	USA				Non-br ad	Winter		36 ± 36		
	S America							43 ± 34		
Coragyps atratus	USA	GPS		95% FK	Imm and ad	Summer	Med 143 [R 2–3106]			Avery et al. (2011)
					Imm and ad	Winter	Med 11 [R 1–3985]			
			Variable	95% MBK	Ad	Breed	46 ± 40			Holland et al. (2017)
					Ad	Non breed	23 ± 23			

(continued)

Table 2 (continued)

Species	Country	Tracking	Fix interval	home range[a]	Bird status[b]	Season	Home range (km²)[d]	daily distance travelled (km)[d]	Radius of displacement[c,d] (km)	Source
Gymnogyps californicus	USA	GPS	1 h	99% FK	Ad	Annual	562 [CI 427–741]			Rivers et al. (2014)
					Imm		413 [CI 300–568]			
Vultur gryphus	Chile, Argentina	GPS	1 h	MCP	Ad	Annual	16,284 ± 14,370			Lambertucci et al. (2014), Alarcon et al. (2017)
Gypaetus barbatus	Spain	Argos	1 h	95% FK	Imm	Annual	11,765 ± 7999			Gil et al. (2014)
				90% FK	br Ad	Annual	57 ± 36	24 ± 22	8 ± 13 (M)	Margalida et al. (2017)
					Non-Br ad		12,063 ± 5513	39 ± 38	23 ± 27 (M)	
	S Africa	GPS	1 h/15 min	MCP	Imm	Annual	38,500	66		Urios et al. (2010)
			3 h	95% FK	Imm	Annual	11,466 ± 4540			Krüger and Amar (2017)
			3 h	90% FK	Imm	Annual	21,880 ± 8187			Krüger et al. (2014), Reid et al. (2015)
					Br ad	Annual	95 ± 19		5.5 (M)	
Neophron percnopterus	Spain	VHF	Irregular	MCP	Imm and non-br ad	Summer	2370 ± 1493			Carrete et al. (2013)
	West Africa	Argos	1 h	MCP	Imm and non-br ad	Winter	78,016 ± 72,909			

	Spain	GPS	2 h	95% FK	Br ad	Summer	253 [202–1670]			Lopez-Lopez et al. (2014)
	Sahel	GPS	1 h	95% FK	Imm	Winter	211,777 ± 189,896			Oppel et al. (2015)
Gyps africanus	Namibia	GPS	10 min	–	Ad	Annual		120 ± 10	55 ± 3 (M)	Spiegel et al. (2013a)
	S Africa	GPS	1 h	95% FK	Imm	Annual	334,923 ± 286,062	34 ± 11		Phipps et al. (2013a)
Gyps bengalensis	Pakistan	GPS	1 h	MCP	Non-br ad	Annual	24,155 ± 29,625	33 ± 14		Gilbert et al. (2007)
Gyps fulvus	Spain	GPS	1 h	95% FK	Non-br ad	Annual	4078 ± 8630	Med 18 ± 18		Garcia-Ripolles et al. (2011)
			10 min	95% FK	Br ad	Annual	23,924 ± 25,216			Arrondo et al. (2018)
					Non-br ad	Annual	33,864 ± 36,714			
	France	GPS	10 min	95% MBK	Br ad	Spring	1272 ± 752	92 ± 6		Monsarrat et al. (2013), Fluhr et al. (2017)
			1 h	95% MBK	Br ad	Winter	473 ± 237	29 ± 6		
			3 s	–	Br ad	Summer		141 ± 92	24 ± 12 (M)	Harel et al. (2016a)
	Greece	VHF	Irregular	95% FK	Br ad	Winter-spring	692 ± 299		29 ± 10 (M)	Xirouchakis and Andreou (2009)
	Israel	GPS	10 min	–	Br ad and imm	Annual		90 ± 30	30 ± 13 (M	Harel et al. (2016a, b)
			1 s	–		Annual		160 ± 110	32 ± 15 (M	

(continued)

Table 2 (continued)

Species	Country	Tracking	Fix interval	home range[a]	Bird status[b]	Season	Home range (km^2)[d]	daily distance travelled (km)[d]	Radius of displacement[c,d] (km)	Source
Gyps himalayensis	Asia		30 min	95% FK	Imm	Summer	61,130 ± 20,062			Sherub et al. (2017)
	Buthan					Winter	13,973 ± 6507			
Gyps coprotheres	Namibia	GPS	1 h	MCP	Ad	Annual	38,327			Bamford et al. (2007)
				MCP	Imm	Annual	482,276			
	S Africa	GPS	4 h	99% FK	Ad	Annual	80,752 ± 40,095	48 ± 26		Phipps et al. (2013b)
					imm	Annual	401,107 ± 244,042	37 ± 7		
	S Africa	GPS	15 min/4 h	95% FK	Ad	Dry	31,040 ± 45,598			Kane et al. (2016)
				95% FK	Ad	Wet	75,162 ± 130,832			
Aegypius monachus	Spain	Pellets		95% FK (colony)	Ad	Annual	1523		26 ± 36 (F)	Moreno-Opo et al. (2011)
			1 h	95% FK	Imm	Annual	3307 ± 2948			Castaño et al. (2015)
			3 h	95% FK	Imm and non-br ad	Annual	49,508 ± 34,236			Arrondo et al. (2018)
	Turkey	GPS	1 h	90% FK	Imm	Pre-migratory	356 ± 54	5 ± 4		Yamaç and Bilgin (2012)
	Caucasus-Middle east	GPS, Argos	1 h	95% FK	Imm	Summer	3847 ± 3120			Gavashelishvili et al. (2012)

				95% FK	Imm	Winter	3870 ± 2541			
	Greece	GPS	45 min to 2 h	95% FK	Imm and ad	Annual	1763 ± 839			Vasilakis et al. (2016)
	Mongolia	Argos	Irregular	MCP	Ad	Summer	[R 540–2653]			Batbayar et al. (2008)
Torgos tracheliotus	Namibia	GPS	10 min	–	Ad	Annual		122 ± 11	48 ± 9 (M)	Spiegel et al. (2013a)

Credit: the chapter authors

[a]*MCP* minimum convex polygon, *FK* fixed Kernel, *MBK* movement based Kernel

[b]*ad* adult, *imm* immature birds, *br* breeding, *non-br* non breeding

[c]Radius of displacement can refer to (M) maximal distance of displacement from central place, or (F) radius of feeding place from central place

[d]Unless specified, results are presented as Mean ± SD, otherwise, Med is for median; in brackets: R is for range, and CI is for 95% confidence intervals

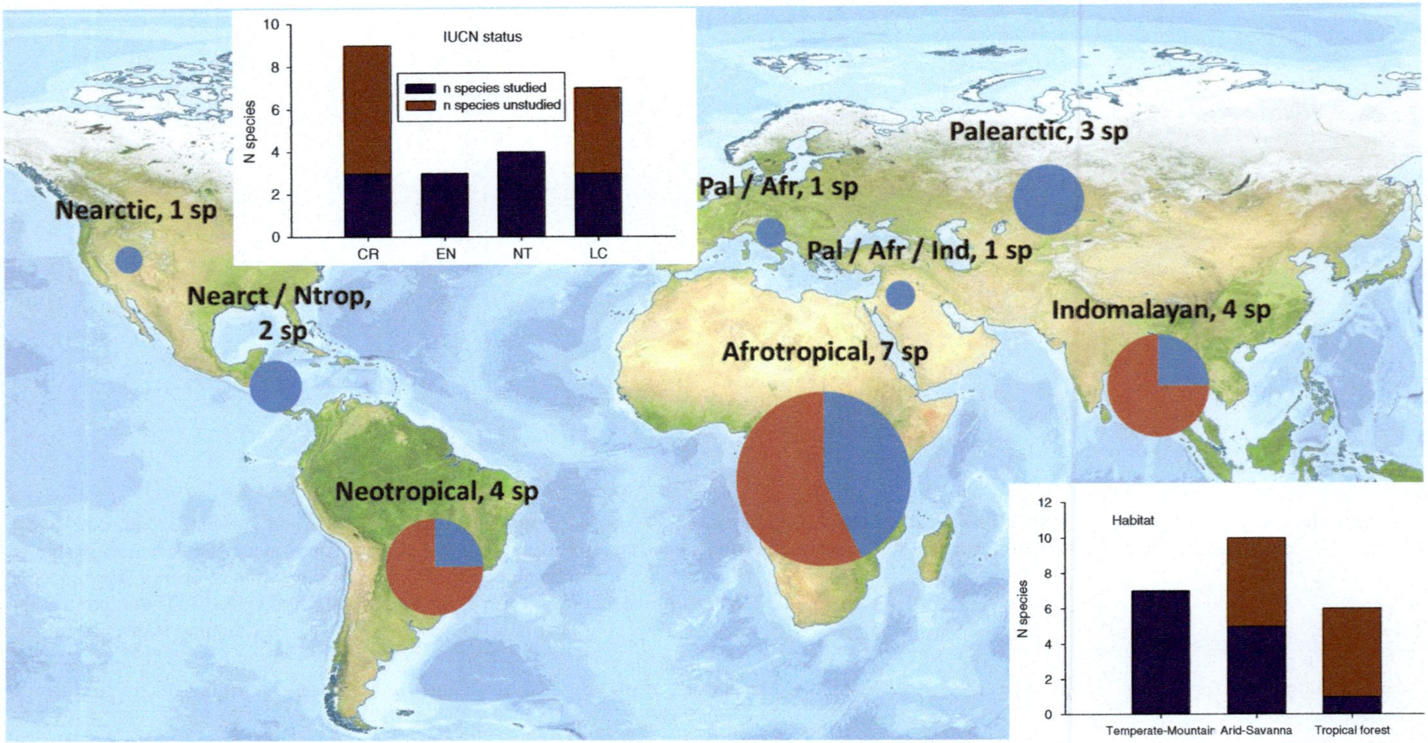

Fig. 3 Number of species in which detailed foraging studies have been performed (in blue) and species unstudied (in red), according to IUCN red list status (*CR* critically endangered, *EN* endangered, *NT* near threatened, *LC* least concerned), biogeographical region (distinguishing when species are found in different regions) and main habitat used (from studies detailed in Table 2). Credit: the chapter authors

therefore biased taxonomically, geographically and behaviorally and our knowledge of vulture movements mostly represents the behavior of social species in temperate mountains or arid areas (savannas or deserts).

With the aim to compare foraging movement behaviors from various species and derive general patterns, we reported in Table 2 the most widely computed parameters: the radius of displacement from central place, the daily distance travelled and home range size.

The radius of displacement from central place (roost or nest) has been reported in 9 studies only, however it is probably the most repeatable metric because it is less dependent of technology and sampling rate. Yet, caution is needed in interpretation because some studies measured the daily maximal distance from central place while other measured the mean radius of feeding event from the central place (e.g. even possible to compute from pellet analyses in which ear-tags of livestock can be found; (Moreno-Opo et al. 2011). As expected, the foraging radius is shorter in territorial species than in social species, among which it varies between c. 30 km in European vultures and c. 50 km in African species (Table 2).

The daily distance travelled per day has been computed in 13 studies. It varies between 5 km day^{-1} in *A. monachus* juvenile birds in Caucasus before migration (Yamaç and Bilgin 2012) and 160 km day^{-1} in *G. fulvus* in Israel (Harel et al. 2016a). However the value reported is highly dependent on the interval between fixes, as illustrated in Israel in *G. fulvus* at the same time of year, with values of 90 km day^{-1} when recording at interval of 10 min and 160 km day^{-1} when recording at interval of 1 s (Harel et al. 2016a). Such difference is due to the high flight speed of vultures (mean "cross-country" speed ranging 20–40 km h^{-1}, combining vertical soaring phases and horizontal gliding phases, but instantaneous ground speed >50 km h^{-1}; (Pennycuick 1972)): at interval of 1 h between fixes, two fixes could be spaced of a few km while the bird could have travelled in reality several tens of km in one direction and be back on its sinuous path. At interval shorter than 5 s, the soaring circles become visible and further increase the distance covered (Harel et al. 2016a; Treep et al. 2016). When considering only studies with interval between fixes 1–10 min, the mean daily distances varied between 90 km (*G. fulvus* in France and Israel) and 120 km (*G. africanus* and *T. tracheliotus* in Namibia) (Harel et al. 2016a; Spiegel et al. 2013a). When considering only studies with interval between fixes around 1 h, the mean daily distances varied between 5 and 66 km, with an average at 34.9 ± 21.6 km. This range of values between 30 and 60 km day^{-1} are remarkably similar for *C. aura* in the Americas, *G. africanus* in Africa and *G. fulvus* in Europe.

The home range is defined as the area in which an animal lives and moves on a periodic basis. The simplest method uses Minimum Convex Polygon (MCP) to delineate all recorded positions of an animal, although it is well known to greatly overestimate home range size. More recent methods use kernel density methods to estimate Utilization Distributions (UD) as the probability that an animal has to be in each cell of its home range (Worton 1989). When positions are supposed to be independent (i.e. collected at large interval of time) it is possible to use fixed-kernel method, while movement-based kernels are advised when data are collected at short time intervals (Benhamou and Cornélis 2010). Home range methods have been the most popular methods to describe vultures' space use since it was used in 46 sub-populations or seasons and in 31 studies. However inter specific and intra-specific comparisons are made difficult because of the heterogeneity of methods, and within a method, the researcher's decision to use a probability threshold of 99%, 95% or 90% to define his home range (Table 2). When considering only studies using the most popular 95% kernel estimators, individual home range sizes of vultures varied tremendously between 11 km^2 and 334,923 km^2, with a mean of 42,680 ± 105,244 km^2. The smallest home ranges (<100 km^2) belonged to the two New-World species *C. aura* and *C. atratus*, while the largest home ranges, >10,000 km^2, all belonged to species in Africa: *G. africanus*, *N. percnopterus* and *G. barbatus*. The season often played a role in home range size, but there was no clear general trend. In temperate and sedentary species, like *G. fulvus, G. coprotheres* or *C. atratus*, home ranges tended to be larger in summer than in winter, when flight is more constrained by adverse aerological conditions (Kane et al. 2016; Avery et al. 2011; Monsarrat et al. 2013). In migratory species like *N. percnopterus*, home ranges were smaller in breeding season in temperate summer than in tropical winter, where movements are not constrained by returning to a nest (Carrete et al. 2013; Lopez-Lopez et al. 2014). However the opposite pattern was found in the long-distance migrant *C. aura* (Dodge et al. 2014). The status or age of the birds also played a role, with breeding adults tending to have smaller home ranges than non-breeding adults or immature birds.

Foraging activity of avian scavengers may serve as an estimator for carrion distribution (Fig. 4). When searching for food, avian scavengers may follow specific

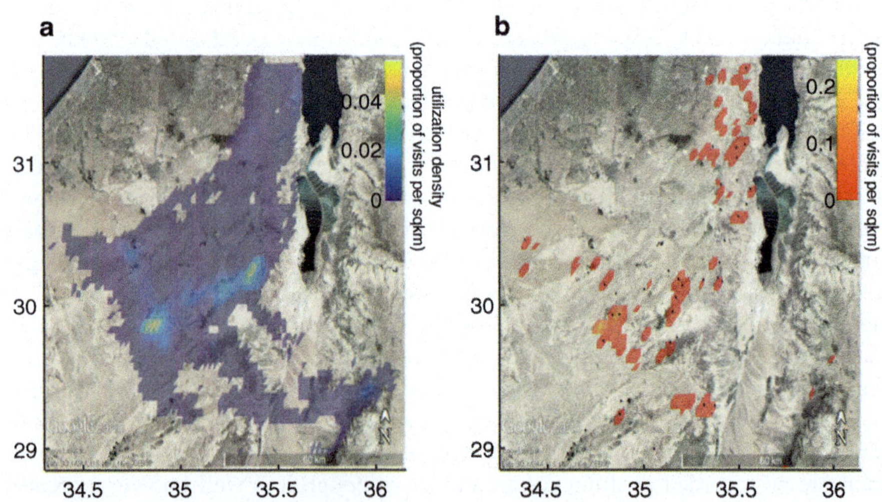

Fig. 4 Space-use patterns of griffon vultures *Gyps fulvus* in Israel as an estimator for carrion distribution patterns. (**a**) The spatial utilization distribution of foraging flights of GPS-tracked vultures may serve as a proxy for the area scanned by scavengers, while (**b**) feeding events (classified by accelerometry and movement data) represent the actual use of carrions available in the region. Colors in both panels represent the proportion of visits in a 2 × 2 km cell grid over the study period. Black dots in panel (**b**) represent active feeding stations. Credit: the chapter authors

land cover features, which are correlated with the presence of carrions (Houston 1974; Monsarrat et al. 2013) and may be limited by other factors, such as the roost site locations. However, within a suitable habitat, vultures may search in a random manner for food resources, or using their personal knowledge of the most likely places to find carrion (Kendall et al. 2014; Fluhr et al. 2017). To locate carrions, most vultures also use information retrieved from conspecifics (Houston 1974) or from other species (Kane et al. 2014; Spiegel et al. 2013a). Some gregarious species may use communal roosts as information centers to follow specifically the individuals that have been successful in foraging the previous day, probably using the state of the crop (extended) as proxy (Buckley 1996; Harel et al. 2017). Therefore, in order to establish the link between scavenger movements and the actual carrion distribution, the composition of foraging groups should be taken into account. Nevertheless, competition, diet preferences, human disturbance, topography and land cover may affect scavengers access to carrions and is expected to cause mismatches between the observed feeding behavior and the actual carrion distribution. Few modeling studies have compared the importance of self and social-information for vultures' foraging efficiency and population persistence (Jackson et al. 2008; Deygout et al. 2009, 2010; Cortes-Avizanda et al. 2014). They suggested that acute reduction in population densities of avian scavengers may impair their foraging efficiency. Such models could be developed to predict the actual effect that avian scavengers can have on carrion distribution, but should incorporate individual characteristics, such as fasting period (Spiegel et al. 2013b), dominance hierarchies within species (Bosè et al. 2012) and between species (Houston 1975; Moreno-Opo et al. 2016; Sebastián-González et al. 2016), and different scenarios of population or community densities (Dupont et al. 2011).

Conclusions and Future Perspectives

In conclusion it is still premature to derive general conclusions about vulture foraging movements in space, partly due to the different methods used for tracking and estimating movement metrics, but also due to large differences in ecology and habitat of vulture species around the world. The only general conclusion possible to say so far is that vultures are able to cover large distances of tens of km per day, in search of carrions, over very large areas, and thus can extend scavenging services well beyond reach of their colonies. But we still lack data on many species, particularly from the tropics. However, the methods to catch vultures are plentiful and the development of GPS technology and remote download of data offers avenues for promising future research about the spatial use of carrions by specialized scavengers.

More practically, to improve intra-specific and inter-specific comparisons, we encourage researchers to standardize their methods. The most important would be to homogenize the statistical methods to describe space use, and systematically report for each individual or each group the daily maximal displacement (which is the less biased metric), daily travelled distance and 95% kernel UD. To study fine details about foraging behavior, vulture biologists must aim at collecting tracking data at intervals <5 min (Fluhr et al. 2017). Regarding the choice of loggers, the GPS system with automatic download with GSM network is by far the most accurate and cheaper. The most recent GPS devices, powered by solar panels over several years, allow recording GPS positions at intervals <5 min, and including burst of high resolutions recording 1 position per second, where it becomes easy to distinguish fine behavioral details (Harel et al. 2016a, b; Sherub et al. 2016). Two additional features can be very useful to study vulture feeding behavior if integrated into the device. The accelerometer can record body posture and movements associated with feeding events, and eventually can be used as a proxy of energy expenditure (Nathan et al. 2012; Spiegel et al. 2013b; Duriez et al. 2014). The possibility to define a "GPS fence" around areas of interest (like feeding stations, or wind farms) can allow to use special settings to be applied (like high resolution GPs recording), to better understand vulture behavior in these areas.

Acknowledgements The authors want to thank all their colleagues who helped capturing and tagging vultures in the field, A. Camiña, I. Shaked and J. Brandt for providing photos of trapping and tagging techniques, and the editors of the book for giving the opportunity to write this chapter.

References

Alarcon PAE, Morales JM, Donazar JA, Sanchez-Zapata JA, Hiraldo F, Lambertucci SA (2017) Sexual-size dimorphism modulates the trade-off between exploiting food and wind resources in a large avian scavenger. Sci Rep 7:11461. https://doi.org/10.1038/s41598-017-11855-0

Arrondo E, Moleon M, Cortes-Avizanda A, Jimenez J, Beja P, Sanchez-Zapata JA, Donazar JA (2018) Invisible barriers: differential sanitary regulations constrain vulture movements across country borders. Biol Conserv 219:46–52. https://doi.org/10.1016/j.biocon.2017.12.039

Avery ML, Humphrey JS, Daughtery TS, Fischer JW, Milleson MP, Tillman EA, Bruce WE, Walter WD (2011) Vulture flight behavior and implications for aircraft safety. J Wildl Manag 75(7):1581–1587. https://doi.org/10.1002/jwmg.205

Bamford AJ, Diekmann M, Monadjem A, Mendelsohn J (2007) Ranging behaviour of cape vultures *Gyps coprotheres* from an endangered population in Namibia. Bird Conserv Int 17(04):331–339. https://doi.org/10.1017/S0959270907000846

Bamford AJ, Monadjem A, Diekmann M, Hardy ICW (2009) Development of non-explosive-based methods for mass capture of vultures. S Afr J Wildl Res 39(2):202–208. https://doi.org/10.3957/056.039.0201

Barber DR, Bildstein KL (2011) A lightweight portable, walk-in trap for catching vultures. Vulture News 60:22–25

Batbayar N, Reading R, Kenny D, Natsagdorj T, Kee PW (2008) Migration and movement patterns of cinereous vultures in Mongolia. Falco 32:5–7

Benhamou S, Cornélis D (2010) Incorporating movement behavior and barriers to improve Kernel home range space use estimates. J Wildl Manag 74(6):1353–1360. https://doi.org/10.2193/2009-441

Bird DM, Bildstein KL (2007) Raptor research and management techniques. Hancock House Publishers, Surrey

Bloom P, Clark W, Kidd J (2007) Capture techniques. In: Bird DM, Bildstein KL (eds) Raptor research and management techniques. Hancock House Publishers, Surrey, pp 242–248
Bögel R, Prinzinger R, Karl E, Walzer C (2000) A multisensor telemetry system for studying flight biology and energetics of free-flying griffon vultures - *Gyps fulvus*. A case study. In: Chancellor RD, Meyburg B-U (eds) Raptors at risk. WWGBP/Hancock House, Johannesburg, pp 879–883
Bonter DN, Bridge ES (2011) Applications of radio frequency identification (RFID) in ornithological research: a review. J Field Ornithol 82(1):1–10. https://doi.org/10.1111/j.1557-9263.2010.00302.x
Bosè M, Duriez O, Sarrazin F (2012) Intra-specific competition in foraging griffon vultures: 1. The dynamics of feeding in groups. Bird Study 59:182–192. https://doi.org/10.1080/00063657.2012.658639
Buckley NJ (1996) Food finding and the influence of information, local enhancement, and communal roosting on foraging success of North American vultures. Auk 113(2): 473–488
Carrete M, Bortolotti GR, Sánchez-Zapata JA et al (2013) Stressful conditions experienced by endangered Egyptian vultures on African wintering areas. Anim Conserv 16(3):353–358. https://doi.org/10.1111/acv.12001
Castaño JP, Sanchez JF, Diaz-Portero MA, Robles M (2015) Dispersal and survival of juvenile black vultures Aegypius monachus in central Spain. Ardeola 62(2):351–361
Cortes-Avizanda A, Jovani R, Donazar JA, Grimm V (2014) Bird sky networks: how do avian scavengers use social information to find carrion? Ecology 95(7):1799–1808. https://doi.org/10.1890/13-0574.1
DeVault TL, Beasley JC, Olson ZH et al (2016) Ecosystem services provided by avian scavengers. In: Sekercioglu CH (ed) Why birds matter. University of Chicago Press, Chicago, p 36
Deygout C, Gault A, Sarrazin F, Bessa-Gomes C (2009) Modeling the impact of feeding stations on vulture scavenging service efficiency. Ecol Model 220(15):1826–1835
Deygout C, Gault A, Duriez O, Sarrazin F, Bessa-Gomes C (2010) Impact of food predictability on social facilitation by foraging scavengers. Behav Ecol 21(6):1131–1139. https://doi.org/10.1093/beheco/arq120
Dodge S, Bohrer G, Bildstein K, Davidson SC et al (2014) Environmental drivers of variability in the movement ecology of turkey vultures (*Cathartes aura*) in North and South America. Philos Trans R Soc London Ser B 369:1643. https://doi.org/10.1098/rstb.2013.0195
Dupont H, Mihoub JB, Becu N, Sarrazin F (2011) Modelling interactions between scavenger behaviour and farming practices: impacts on scavenger population and ecosystem service efficiency. Ecol Model 222(4):982–992
Duriez O, Kato A, Tromp C et al (2014) How cheap is soaring flight in raptors? A preliminary investigation in freely-flying vultures. PLoS One 9(1):e84887
Fluhr J, Benhamou S, Riotte-Lambert L, Duriez O (2017) Assessing the risk for an obligate scavenger to be dependent on predictable feeding sources. Biol Conserv 215:92–98
Garcia-Ripolles C, Lopez-Lopez P, Urios V (2011) Ranging behaviour of non-breeding Eurasian Griffon Vultures *Gyps fulvus*: a GPS-telemetry study. Acta Ornithol 46:127–134
Gavashelishvili A, McGrady M, Ghasabian M, Bildstein KL (2012) Movements and habitat use by immature cinereous vultures (*Aegypius monachus*) from the caucasus. Bird Study 59:449–462. https://doi.org/10.1080/00063657.2012.728194
Gil JA, Baguena G, Sanchez-Castilla E et al (2014) Home ranges and movements of non-breeding bearded vultures tracked by satellite telemetry in the Pyrenees. Ardeola 61:379–387
Gilbert M, Watson RT, Ahmed S et al (2007) Vulture restaurants and their role in reducing diclofenac exposure in Asian vultures. Bird Conserv Int 17(1):63–77. https://doi.org/10.1017/s0959270906000621
Harel R, Duriez O, Spiegel O et al (2016a) Decision-making by a soaring bird: time, energy and risk considerations at different spatio-temporal scales. Philos Trans R Soc London Ser B 371(1704):20150397. https://doi.org/10.1098/rstb.2015.0397

Harel R, Horvitz N, Nathan R (2016b) Adult vultures outperform juveniles in challenging thermal soaring conditions. Sci Rep 6:27865

Harel R, Spiegel O, Getz WM, Nathan R (2017) Social foraging and individual consistency in following behaviour: testing the information centre hypothesis in free-ranging vultures. Proc R Soc Lond B 284:1852. https://doi.org/10.1098/rspb.2016.2654

Holland AE, Byrne ME, Bryan AL, DeVault TL, Rhodes OE, Beasley JC (2017) Fine-scale assessment of home ranges and activity patterns for resident black vultures (Coragyps atratus) and turkey vultures (Cathartes aura). PLoS One 12(7):e0179819

Houston DC (1974) Food searching in griffon vultures. East Afr Wild J 12:63–77

Houston DC (1975) Ecological isolation of African scavenging birds. Ardea 63:55–64

Houston DC (1994) Family cathartidae (new world vultures). In: del Hoyo J, Elliott A, Sargatal J (eds) Handbook of the birds of the world, New world vultures to guineafowls, vol 2. Lynx Edición, Barcelona, pp 24–41

Iezekiel S, Woodley B, Hatzofe O (2003) Cage traps for *Gyps fulvus*. Vulture News 49:14–16

Jackson AL, Ruxton GD, Houston DC (2008) The effect of social facilitation on foraging success in vultures: a modelling study. Biol Lett 4(3):311–313

Kane A, Jackson AL, Ogada DL, Monadjem A, McNally L (2014) Vultures acquire information on carcass location from scavenging eagles. Proc R Soc Lond B 281:1793. https://doi.org/10.1098/rspb.2014.1072

Kane A, Wolter K, Neser W, Kotze A, Naidoo V, Monadjem A (2016) Home range and habitat selection of cape vultures *Gyps coprotheres* in relation to supplementary feeding. Bird Study 63:387–394. https://doi.org/10.1080/00063657.2016.1214105

Kendall CJ, Virani MZ, Hopcraft JGC, Bildstein KL, Rubenstein DI (2014) African vultures don't follow migratory herds: scavenger habitat use is not mediated by prey abundance. PLoS One 9(1):e83470

Krüger S, Amar A (2017) Insights into post-fledging dispersal of bearded vultures *Gypaetus barbatus* in Southern Africa from GPS satellite telemetry. Bird Study 64(2):125–131. https://doi.org/10.1080/00063657.2017.1295019

Krüger S, Reid T, Amar A (2014) Differential range use between age classes of Southern African bearded vultures *Gypaetus barbatus*. PLoS One 9(12):e114920

Lambertucci SA, Alarcon PAE, Hiraldo F, Sanchez-Zapata JA, Blanco G, Donazar JA (2014) Apex scavenger movements call for transboundary conservation policies. Biol Conserv 170:145–150. https://doi.org/10.1016/j.biocon.2013.12.041

Lopez-Lopez P, Garcia-Ripolles C, Urios V (2014) Food predictability determines space use of endangered vultures: implications for management of supplementary feeding. Ecol Appl 24(5):938–949. https://doi.org/10.1890/13-2000.1

Margalida A, Perez-Garcia JM, Moreno-Opo R (2017) European policies on livestock carcasses management did not modify the foraging behavior of a threatened vulture. Ecol Indic 80:66–73. https://doi.org/10.1016/j.ecolind.2017.04.048

Mateo-Tomas P, Olea PP, Moleon M, Selva N, Sanchez-Zapata JA (2017) Both rare and common species support ecosystem services in scavenger communities. Global Ecol Biogeo 26(12):1459–1470

Mihoub J-B, Prince K, Duriez O, Lécuyer P, Eliotout B, Sarrazin F (2013) Comparing release method effects on post-release survival of the European black vulture *Aegypius monachus* reintroduced population in France. Oryx 48:106–115

Monsarrat S, Benhamou S, Sarrazin F, Bessa-Gomes C, Bouten W, Duriez O (2013) How predictability of feeding patches affects home range and foraging habitat selection in avian social scavengers? PLoS One 8(1):e53077

Moreno-Opo R, Arredondo A, Guil F (2011) Foraging range and diet of Cinereous vulture *Aegypius monachus* using livestock resources in Central Spain. Ardeola 57:111–119

Moreno-Opo R, Trujillano A, Arredondo A, Gonzalez LM, Margalida A (2015) Manipulating size, amount and appearance of food inputs to optimize supplementary feeding programs for European vultures. Biol Conserv 181:27–35. https://doi.org/10.1016/j.biocon.2014.10.022

Moreno-Opo RN, Trujillano A, Margalida A (2016) Behavioral coexistence and feeding efficiency drive niche partitioning in European avian scavengers. Behav Ecol 27:1041–1052. https://doi.org/10.1093/beheco/arw010

Nathan R, Spiegel O, Fortmann-Roe S, Harel R, Wikelski M, Getz WM (2012) Using tri-axial acceleration data to identify behavioral modes of free-ranging animals: general concepts and tools illustrated for Griffon vultures. J Exp Biol 215:986–996

Ogada DL, Torchin ME, Kinnaird MF, Ezenwa VO (2012) Effects of vulture declines on facultative scavengers and potential implications for mammalian disease transmission. Conserv Biol 26(3):453–460. https://doi.org/10.1111/j.1523-1739.2012.01827.x

Oppel S, Dobrev V, Arkumarev V, Saravia V, Bounas A, Kret E, Velevski M, Stoychev S, Nikolov SC (2015) High juvenile mortality during migration in a declining population of a long-distance migratory raptor. Ibis 157:545–557. https://doi.org/10.1111/ibi.12258

Pennycuick CJ (1972) Soaring behaviour and performance of some East African birds observed from a motor glider. Ibis 114:178–218

Phipps WL, Willis SG, Wolter K, Naidoo V (2013a) Foraging ranges of immature African white-backed vultures (*Gyps africanus*) and their use of protected areas in Southern Africa. PLoS One 8(1):e52813

Phipps WL, Wolter K, Michael MD, MacTavish LM, Yarnell RW (2013b) Do power lines and protected areas present a catch-22 situation for cape vultures (*Gyps coprotheres*)? PLoS One 8(10):e76794

Rappole JH, Tipton AR (1991) New harness design for attachment of radio transmitters to small passerines. J Field Ornithol 62:335–337

Reading PP, Maude G, Hancock P, Kenny D, Garbett R (2014) Comparing different types of patagial tags for use on vultures. Vulture News 67:33–42

Reid T, Krüger S, Whitfield DP, Amar A (2015) Using spatial analyses of bearded vulture movements in Southern Africa to inform wind turbine placement. J Appl Ecol 52:881–892. https://doi.org/10.1111/1365-2664.12468

Rivers JW, Johnson JM, Haig SM, Schwarz CJ, Burnett LJ, Brandt J, George D, Grantham J (2014) An analysis of monthly home range size in the critically endangered California Condor *Gymnogyps californianus*. Bird Conserv Int 24:492–504. https://doi.org/10.1017/S0959270913000592

Ruxton GD, Houston DC (2004) Obligate vertebrate scavengers must be large soaring fliers. J Theor Biol 228(3):431–436

Sebastián-González E, Moleón M, Gibert JP, Botella F, Mateo-Tomás P, Olea PP, Guimarães PR, Sánchez-Zapata JA (2016) Nested species-rich networks of scavenging vertebrates support high levels of interspecific competition. Ecology 97(1):95–105. https://doi.org/10.1890/15-0212.1

Sherub S, Bohrer G, Wikelski M, Weinzierl R (2016) Behavioural adaptations to flight into thin air. Biol Lett 12:10. https://doi.org/10.1098/rsbl.2016.0432

Sherub S, Fiedler W, Duriez O, Wikelski M (2017) Bio-logging - new technologies to study conservation physiology on the move: a case study on annual survival of Himalayan vultures. J Comp Physiol A 203(6):531–542. https://doi.org/10.1007/s00359-017-1180-x

Spiegel O, Getz WM, Nathan R (2013a) Factors influencing search efficiency: why do scarce Lappet-faced vultures outperform ubiquitous white-backed vultures. Am Nat 181:5. https://doi.org/10.1086/670009

Spiegel O, Harel R, Getz W, Nathan R (2013b) Mixed strategies of griffon vultures' (*Gyps fulvus*) response to food deprivation lead to a hump-shaped movement pattern. Mov Ecol 1(1):5

Treep J, Bohrer G, Shamoun-Baranes J, Duriez O, Prata de Moraes Frasson R, Bouten W (2016) Using high resolution GPS tracking data of bird flight for meteorological observations. Bull Am Meteorol Soc 97(6):951–961. https://doi.org/10.1175/bams-d-14-00234.1

Trefry S, Diamond A, Jesson L (2013) Wing marker woes: a case study and meta-analysis of the impacts of wing and patagial tags. J Ornithol 154(1):1–11. https://doi.org/10.1007/s10336-012-0862-y

Tremblay Y, Thiébault A, Mullers R, Pistorius P (2014) Bird-borne video-cameras show that seabird movement patterns relate to previously unrevealed proximate environment, not prey. PLoS One 9(2):e88424

Urios V, Lopez-Lopez P, Limiñana R, Godino A (2010) Ranging behaviour of a juvenile bearded vulture (*Gypaetus barbatus meridionalis*) in South Africa revealed by GPS satellite telemetry. Ornis Fenn 87:114–118

Vasilakis DP, Whitfield DP, Schindler S, Poirazidis KS, Kati V (2016) Reconciling endangered species conservation with wind farm development: cinereous vultures (*Aegypius monachus*) in South-Eastern Europe. Biol Conserv 196:10–17. https://doi.org/10.1016/j.biocon.2016.01.014

Wallace MP, Parker GP, Temple SA (1980) An evaluation of patagial markers for cathartid vultures. J Field Ornithol 51:309–314

Williams HJ, Holton MH, Shepard EL, Largey N, Norman JA, Ryan P, Duriez O, Scantlebury M, Quintana F, Magowan E, Wilson RP (2017) Identification of animal movement patterns using tri-axial magnetometry. Mov Ecol 5:6. https://doi.org/10.1186/s40462-017-0097-x

Wilson B (2015) An introduction to camera trapping of wing-tagged vultures in Southern Africa. Vulture News 69:3–22

Worton BJ (1989) Kernel methods for estimating the utilization distribution in home-range studies. Ecology 70(1):164–168

Xirouchakis SM, Andreou G (2009) Foraging behaviour and flight characteristics of Eurasian griffons *Gyps fulvus* in the island of Crete, Greece. Wildl Biol 15(1):37–52. https://doi.org/10.2981/07-090

Yamaç E, Bilgin CC (2012) Post-fledging movements of cinereous vultures *Aegypius monachus* in Turkey revealed by GPS telemetry. Ardea 100(2):149–156. https://doi.org/10.5253/078.100.0206

Synthesis and Future Perspectives on Carrion Ecology and Management

Pedro P. Olea, Patricia Mateo-Tomás, and José A. Sánchez-Zapata

Contents

Main Conclusions

Carrion, dead animal matter, is a high-quality ephemeral resource present in all ecosystems, from arctic to tropical biomes in terrestrial and aquatic (freshwater and marine) environments (Fig. 1). Hundreds to thousands of invertebrates and vertebrates species are estimated to be facultative scavengers, i.e. consuming carrion opportunistically, yet a minority of them (<1%) have adapted to exclusively feed on this trophic resource (i.e. obligate scavengers). Carrion underpins therefore a rich food web comprised of microbial, arthropod and vertebrate taxa; a necrobiome that is responsible for the decomposition of carrion (Benbow et al. 2013). Carrion decomposition is a key

Pedro P. Olea and Patricia Mateo-Tomás contributed equally to this chapter.

P. P. Olea (✉)
Centro de Investigación en Biodiversidad y Cambio Global (CIBC-UAM), Universidad Autónoma de Madrid, Madrid, Spain

Departamento de Ecología, Universidad Autónoma de Madrid, Madrid, Spain
e-mail: pedrop.olea@uam.es

P. Mateo-Tomás
Centre for Functional Ecology (CFE), Department of Life Sciences, University of Coimbra, Coimbra, Portugal

Research Unit of Biodiversity, (UMIB, UO/CSIC/PA), Mieres, Spain

J. A. Sánchez-Zapata
Departamento de Biología Aplicada, Universidad Miguel Hernández, Elche, Spain
e-mail: toni@umh.es

P. P. Olea et al. (eds.), *Carrion Ecology and Management*, Wildlife Research Monographs 2, https://doi.org/10.1007/978-3-030-16501-7_12

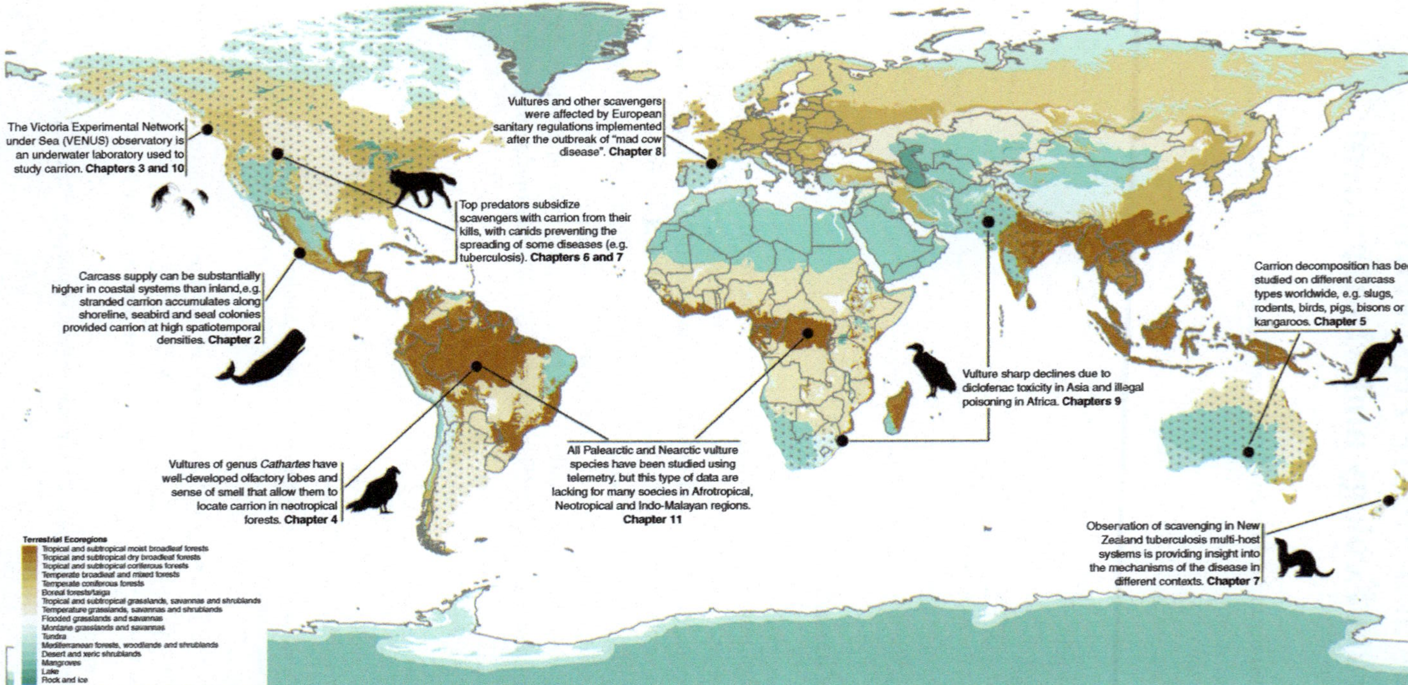

Fig. 1 Scavenging is a widespread strategy present in all ecosystems and this book contains examples on carrion ecology and management from all around the world. Detailed studies on carrion ecology are described in this book for at least the dotted countries. Credit: Patricia Mateo-Tomás

ecological process for ecosystem functioning, related to the recycling of energy and nutrients, and scavenging—i.e. the consumption of carrion by scavengers – plays a crucial role in this process (chapters "Carrion Decomposition" and "Ecological Functions of Vertebrate Scavenging"; Barton et al. 2013). Carcass characteristics (e.g. size) and the biotic (e.g. habitat) and abiotic (e.g. temperature, moisture) environment act as moderators of the process of carrion decomposition, which differs between terrestrial systems and aquatic environments (Beasley et al. 2012; chapters "Carrion Decomposition" and "Ecological Functions of Vertebrate Scavenging")

Scavenging plays an important role in structuring food webs (Wilson and Wolkovich 2011; chapter "Ecological Functions of Vertebrate Scavenging"). Scavengers feeding at carcasses can form structured assemblages that are driven by ecological succession processes in invertebrates and the creation of nested networks in vertebrates (chapters "Invertebrate Scavenging Communities" and "Vertebrate Scavenging Communities"). These feeding relationships involving animal carcasses increase complexity and stability of food webs (chapter "Ecological Functions of Vertebrate Scavenging").

Carcasses represent hotspots of biological diversity. The inter- and intraspecific interactions (e.g. facilitation, competition) that occur at carrion among species and between kingdoms (microbe-animal) drive key ecosystem functions such as biodiversity maintenance and nutrient recycling. Coexistence of species sharing the same carrion underpins biodiversity through facilitation processes and resource partition among species (e.g. species utilizing different parts within a carcass or carcasses of different sizes). The relationships between diversity of scavengers and ecosystem functioning (carrion decomposition, nutrients recycling) provides in turn important benefits for human societies, i.e. ecosystem services, such as carcass disposal and disease control (chapter "Ecological Functions of Vertebrate Scavenging").

Humans have shifted from carrion consumers to carrion suppliers, and thus the relationship of humans and wild scavengers has changed from competition or predation to mutualism and commensalism (Gangoso et al. 2013, chapter "Human-Mediated Carrion: Effects on Ecological Processes"). Humans are increasingly becoming a fundamental piece in carrion ecology due to the growing presence of human-mediated carrion in natural ecosystems worldwide (Oró et al. 2013). The ubiquity of dead livestock, hunting remains or fisheries discards on Earth ecosystems not only alters the spatiotemporal distribution and quality of carrion, but also have consequences for biodiversity maintenance, ecosystem functioning and human well-being (chapter "Human-Mediated Carrion: Effects on Ecological Processes"). Human-mediated carrion has also become a reservoir for toxic residues such as pharmaceutical drugs, lead and pesticides, and thus scavenging this contaminated carrion is likely to propagate the toxics across food webs (chapter "What Makes Carrion Unsafe for Scavengers? Considerations for Appropriate Regulatory Policies and Sound Management Practices"; Figs. 1 and 2). The sustainable management of human-mediated carrion emerges therefore as a pressing issue. To be effective, this management should be science-based, i.e. grounded on the best available knowledge on carrion ecology. Nevertheless, effective carrion management demands also additional knowledge to fill major gaps that still exist regarding not only the conser-

Fig. 2 Illegal poisoning threatens biodiversity worldwide, and especially obligate scavengers such as vultures and condors; 34 Andean condors *Vultur gryphus* were killed with a single Carbofuran-baited sheep carcass set out by ranchers to combat mammalian predators in Argentina in January 2018 (Alarcón and Lambertucci 2018). Credit: Eva López García

vation of biodiversity and ecosystems, but also a better understanding on key ecological functions and services such as carrion decomposition, nutrient recycling or disease control. This knowledge will help to face major drivers of global change that are expected to exert a noticeable impact on carrion ecology, such as land use intensification and degradation, climate change or invasive species.

Under the current scenario of global environmental change, a sharply growing human population demands more food and also generate more waste, including enormous amounts of carrion into ecosystems. We are now starting to assess how these subsidized ecosystems absorb the inputs of human-mediated carrion and the ecological consequences from populations and species to communities and ecosystems (chapters "Human-Mediated Carrion: Effects on Ecological Processes" and "What Makes Carrion Unsafe for Scavengers? Considerations for Appropriate Regulatory Policies and Sound Management Practices"; Oro et al. 2013). However, this issue remains still poorly understood.

What's Next?

Besides providing an updated summary of current knowledge on the topic, the authors of this book identify some "hot topics" that research in carrion ecology should try to answer in the next decades.

Despite the observed increase in scientific research on carrion and scavenging during the last decades, the relative importance of this field within the environmental and biological sciences remains constant (Fig. 1.4) and far below the figures reached by other topics in ecology and conservation (e.g. predation; see chapter "Introduction to the Topic of Carrion Ecology and Management"). Therefore most authors in this book highlight several major knowledge gaps still existing in carrion ecology, identifying some key questions to be answered by researchers in the discipline in the near future.

Data on scavenger assemblages and scavenging dynamics are still lacking in many ecosystems around the world. Authors in this book provide specific recommendations in relation to the research of scavenging communities particularly in island, arctic, tropical, and aquatic ecosystems, as well as in the marine-terrestrial interface (see chapters "Carrion Availability in Space and Time" and "Ecological Functions of Vertebrate Scavenging"). The role of carrion in supporting ecological processes across levels of organization (from individuals to ecosystems) remains still understudied. Further attention is demanded too for the intra- and inter-specific interactions occurring at carrion, highlighting interesting topics such as the microorganism-invertebrate dynamics (see chapter "Carrion Decomposition").

The great potential of new technologies for broadening knowledge on carrion ecology and supporting the effective management of carrion and carrion-eaters is also recognized by several authors in this book (e.g. chapters "Invertebrate Scavenging Communities" and "Carrion Decomposition"). In fact, the methodological section provides some examples of how different technologies (e.g. GPS tracking, underwater laboratories) can be used to support research on the topic. Underwater cabled laboratories such as Ocean Networks Canada's VENUS Observatory has demonstrated high usefulness for studying carrion ecology in challenging environments such as marine waters (Fig. 1).

A better quantification of the spatiotemporal availability of carrion is a major topic to deal with, as it would lead to a better understanding of scavenging dynamics.

The more explicit use of experimental approaches is also remarked to better assess ecological theories regarding the impact of different drivers of global change on carrion ecology (see above). Disease dynamics at carrion are an emerging issue with important questions to be addressed (e.g. from the use of carrion by scavengers and the subsequent exposure to different diseases to the role of human related factors in such disease dynamics; see chapter "The Role of Scavenging in Disease Dynamics"). It is needed to know the role that scavengers play in the control and dissemination of disease (Fig. 1).

A major claim of several authors in this book asks for developing and strengthening interdisciplinary collaboration among professionals from the different scientific disciplines involved in the study of carrion and its consumption. Further insight into carrion ecology is expected to result from combining molecular and ecological perspectives (see chapter "Carrion Decomposition"). While the collaboration of microbiologists, entomologists, vertebrate zoologists and ecologists is necessary to better understand carrion decomposition, the study of disease dynamics at carcasses would benefit from close collaboration between wildlife ecologists and public health professionals.

Scavenging in the Anthropocene

Scavengers and scavenging processes face major changes in the increasingly humanized ecosystems that characterize the Anthropocene. Provided that temperature is recognized as a modulator of carcass persistence and scavenging dynamics, this factor (along with habitat type, and species and carrion size) should be considered in the design of studies of carrion ecology (chapter "Ecological Functions of Vertebrate Scavenging"). Knowing how temperature determines the fate of carcasses will be also useful to predict how climate change will likely affect both carrion availability, through altering mortality patterns (e.g. disease, extreme weather), and the rates of carrion decomposition in ecosystems.

Assessing the potential effect of global change on both carrion availability and its consumption across the globe is of paramount importance in carrion ecology (see chapter "What Makes Carrion Unsafe for Scavengers? Considerations for Appropriate Regulatory Policies and Sound Management Practices"). The increasing presence of humans in natural ecosystems is severely impacting carrion availability and consumption at all ecological levels (i.e. from individuals and populations to communities and ecosystems; see chapter "Human-Mediated Carrion: Effects on Ecological Processes"). In this regard, scavengers, particularly obligate scavengers, face major carrion-related threats such as contamination of carrion by multiple toxic compounds and abrupt changes of food sources driven by regulatory measures (Fig. 2). The effect of toxic residues in carcasses on micro-scavengers (arthropods) should be further acknowledged and explored in order to develop mitigation measures accordingly (chapter "What Makes Carrion Unsafe for Scavengers? Considerations for Appropriate Regulatory Policies and Sound Management Practices").

The conservation of species-rich guilds of scavengers is a key issue for ecosystem functioning. Reinforcing the science-management interface in carrion ecology is therefore needed to tackle scavenging conservation, especially considering the critical situation of some scavenging species such as vultures worldwide (i.e. 16 out of 22 species threatened according to IUCN; chapter "Vertebrate Scavenging Communities") or top predators. The emerging recognition of the ecosystem services provided by scavengers might help to reverse their decline. More efficient policies and management based in science is therefore needed to better conserve scavengers and scavenging processes in ecosystems in a rapidly changing world.

References

Alarcón PAE, Lambertucci SA (2018) Pesticides thwart condor conservation. Science 360:612

Barton PS, Cunningham SA, Lindenmayer DB, Manning AD (2013) The role of carrion in maintaining biodiversity and ecological processes in terrestrial ecosystems. Oecologia 171:761–772

Beasley JC, Olson ZH, Devault TL (2012) Carrion cycling in food webs: comparisons among terrestrial and marine ecosystems. Oikos 121:1021–1026

Benbow ME, Lewis AJ, Tomberlin JK, Pechal JL (2013) Seasonal necrophagous insect community assembly during vertebrate carrion decomposition. J Med Entomol 50:440–450

Gangoso L, Agudo R, Anadón JD, de la Riva M, Suleyman AS, Porter R, Donázar JA (2013) Reinventing mutualism between humans and wild fauna: insights from vultures as ecosystem services providers. Conserv Lett 6:172–179

Oro D, Genovart M, Tavecchia G, Fowler MS, Martínez-Abraín A (2013) Ecological and evolutionary implications of food subsidies from humans. Ecol Lett 16:1501–1514

Wilson EE, Wolkovich EM (2011) Scavenging: how carnivores and carrion structure communities. Trends Ecol Evol 26:129–135